Alicia Amherst

A History of Gardening in England

Alicia Amherst

A History of Gardening in England

ISBN/EAN: 9783337080389

Printed in Europe, USA, Canada, Australia, Japan

Cover: Foto ©ninafisch / pixelio.de

More available books at **www.hansebooks.com**

OF

GARDENING IN ENGLAND

BY

THE HON. ALICIA AMHERST

"*They set great store by their gardeins*"
SIR THOMAS MORE

LONDON

BERNARD QUARITCH, 15 PICCADILLY, W.

1895

London
O. Norman & Son, Printers Flora Street
Covent Garden

Dedicated

TO

MY FATHER AND MOTHER.

PREFACE

KNOWING that I was fond both of practical gardening and the study of old garden literature, Mr. Percy Newberry suggested to me in the Spring of 1891, that I should edit some articles he had written on the History of Gardening in England down to the reign of Elizabeth, which had appeared in the *Gardener's Chronicle* in 1889 ; and that I should carry on the History from that point. I became so much interested in the subject, and had collected so much new material, that I decided to enlarge on the original plan, and not only to continue the history, but to traverse again all the earlier part, drawing my information afresh from the original authorities. I wish, therefore, to acknowledge my indebtedness to Mr. Newberry, who so kindly put his articles and notes at my disposal in the first instance.

This work does not pretend to be a history of the Gardens of England, which would indeed be a delightful task to carry out, therefore many well-known gardens have not been mentioned in the following pages, only a few examples having been cited to serve as illustrations of each successive fashion; and to enumerate others would only have been to multiply instances. It is hoped rather that this work, inadequate though it is in comparison with the vastness of the subject, may in some measure serve as a handbook by which to classify gardens, and fix the dates to which they belong. In many cases it must always be difficult to assign an exact date to a garden, as although frequently a garden adjoining the house has existed from very early times, the changes, though few, have been so gradual that it is almost impossible to determine for certain the time at which they assumed their present condition. I have to thank the many friends who have very kindly afforded me information respecting their gardens, and provided me with plans or photographs, or who have given me ready access to the MSS. in their possession in public or private collections.

I also wish gratefully to acknowledge the kindness of Mr. J. G. Baker, F.R.S., in looking over the following pages whilst still in proof sheets. The correction of the proofs had been rendered an easy task for me by the kind co-operation of my friend, Miss Margaret MacArthur. My thanks are also due to Professor Skeat and Mr. James Britten for their help in the identification of some of the plants mentioned in the fifteenth century MSS., and to Mr. R. E. G. Kirk who assisted me in decyphering some of the earlier Latin ones, also to Mr. Michael Kerney for revising my bibliography of printed books on gardening to the end of the seventeenth century. I regret that the continuation from the year 1699 has not received as much time and attention as I wished to bestow upon it, as I have had to complete it rather hurriedly on account of my having been absent abroad for several months.

ALICIA M. T. AMHERST.

DIDLINGTON HALL, NORFOLK.
September, 1895.

CONTENTS

CHAPTER		PAGE
I.	MONASTIC GARDENING	1
II.	THIRTEENTH CENTURY	30
III.	FOURTEENTH AND FIFTEENTH CENTURIES	42
IV.	EARLY GARDEN LITERATURE	62
V.	EARLY TUDOR GARDENS	74
VI.	ELIZABETHAN FLOWER GARDEN	105
VII.	KITCHEN GARDENING UNDER ELIZABETH AND JAMES I.	131
VIII.	ELIZABETHAN GARDEN LITERATURE	160
IX.	SEVENTEENTH CENTURY	174
X.	GARDENING UNDER WILLIAM AND MARY	206
XI.	DAWN OF LANDSCAPE GARDENING	227
XII.	LANDSCAPE GARDENING	256
XIII.	NINETEENTH CENTURY	280

APPENDIX

SURVEYS OF WIMBLEDON AND THEOBALDS, 1649	307
LIST OF PRINTED BOOKS ON GARDENING CHRONOLOGICALLY ARRANGED DOWN TO THE YEAR 1837	323
NAMES OF THE AUTHORS OF WORKS ON GARDENING	381
LETTER FROM JOHN EVELYN TO SAMUEL PEPYS, 1686	387
INDEX	389

ILLUSTRATIONS

		PAGE
1.	Plan of Bicester Priory	13
2.	Vine Pruners. From an Anglo-Saxon MS., eleventh century	21
3.	Ashridge. From a photograph	27
4.	The Eagle Pond, Newstead Abbey. From a photograph	29
5.	Garden in a Town. From French MS., late fifteenth century	34
6.	Grafting. From the Arte of Planting and Grafting, by Leonard Mascall, 1592	50
7.	Turfed Seat in a garden wall. From Roman de la Rose, Flemish MS., late fifteenth century	51
8.	Arbour, from the same MS.	53
9.	Fountain, from an English MS., "Speculum," c. 1450	55
10.	Garden, from Flemish MS. of the Roman de la Rose, late fifteenth century	60
11.	The Feate of Gardening, by Ion Gardener, MS. c. 1440, Trinity College, Cambridge	67
12.	Railed Flower Bed. From French MS. of the Roman de la Rose, c. 1450	75
13.	The Mount, Rockingham. From a drawing by Howard Carter	76
14.	Old Yew Walk and Mount, Rockingham. From a photograph	77
15.	Garden with a gallery. From "The Second Booke of Flowers, Fruicts, Beastes, Birds and Flies," 1650	79
16.	Garden House, Loseley. From a drawing by Howard Carter	82
17.	Knot from the "Gardener's Labyrinth"	84
18.	Picture at Hampton Court, showing the railed beds and beasts	89
19.	Apricot Trees on old garden wall, Littlecote. From a drawing by Howard Carter	93
20.	From "Gardener's Labyrinth"	95
21.	Tools used in Grafting	102
22.	Pleached Alley at Drayton. From a photograph	113
23.	Boscobel in 1894. From a photograph	115
24.	Boscobel in 1660. From an old print	117
25.	Maze. From the Art of Gardening	119
26.	Heslington. From a photograph	120
27.	Example of Topiary Work in Cottage Garden, Haddon. From a drawing by Lady William Cecil	121
28.	Charter of the Gardeners' Company	133
29.	Castle Bromwich. From a photograph	141
30.	Orange Court at Burghley House. From a Picture at Burghley	153
31.	L'Obel. From an Engraving in the Tyssen Library, Hackney	163

ILLUSTRATIONS.

		PAGE
32.	Gerard. From the title-page of his "Herbal," 1597	165
33.	Parkinson. From the title-page for his "Paradisus," 1629	168
34.	Bulwick. From a photograph	181
35.	Hunstanton. From a drawing by Hon. Margaret Amherst	183
36.	Orangerie and Canal, Euston. From a sketch by Edmond Prideaux, c. 1716	194
37.	Orangerie at Chiswick. From an engraving by Rocque, 1736	195
38.	Levens. From a Picture by Geo. S. Elgood	198
39.	Sundial, Euston, with the Arlington Arms, about 1671. Drawing from a photograph by Miss Ethel FitzRoy	200
40.	Parterre, from London and Wise	210
41.	Parterre, from London and Wise	211
42.	Plan of Canons Ashby. Drawn by Sir Henry Dryden	213
43.	Netherton. From a Sketch by Edmond Prideaux	220
44.	Apothecaries' Garden, Chelsea, in 1894. From a photograph	224
45.	Ingestre, the Seat of Lord Viscount Chetwynd. From Plot's "Staffordshire," 1686	228
46.	Cashiobury, the Seat of the Rt. Honble. the Earle of Essex, in Hartfordshire. From an engraving by Kip	229
47.	Brome Hall, Suffolk, one of the Seats of the Right Hon. Charles Lord Cornwallis. From an engraving by Kip	230
48.	Bramham. From a drawing by Vernet Carter	231
49.	Hall Barn. From a drawing by Vernet Carter	233
50.	Belton, in Lincolnshire. From an engraving by Badeslad	236
51.	Palladian Bridge at Stow. From a photograph	240
52.	Title-page of Catalogue of the Society of Gardeners, 1730	247
53.	Castle Ashby. From a photograph	265
54.	Burghley. The Temple designed by Brown. From a photograph	267
55.	Roche Abbey. Drawn from a photograph by Hon. Margaret Amherst	269
56.	Woodford, No. 1. From a drawing by H. Repton	272
57.	Woodford, No. 2. From the same drawing by H. Repton, showing the suggested improvements	273
58.	Narford, No. 1. From a sketch by Edmond Prideaux about 1716	276
59.	Narford, No. 2. 1894. From a photograph	277
60.	Arley, a garden laid out fifty years ago in the old formal style. From a photograph	281
61.	Harewood. From a photograph	296
62.	Shrublands. From a photograph	298
63.	Rock Garden, Batsford. From a photograph	300
64.	Narcissus in the Scilly Isles. From a photograph	302
65.	Lilies in a wild garden. From a photograph	304

A HISTORY
OF
GARDENING IN ENGLAND

CHAPTER I.

" Forsitan, et pingues hortos quæ cura colendi
Ornaret, canerem, ... "
VIRGIL, *Geor.*, iv. 118.

THE history of the Gardens of England follows step by step the history of the people. In times of peace and plenty they increased and flourished, and during years of war and disturbance they suffered. The various races that have predominated, and rulers that have governed this country influenced them in a marked degree. Therefore, as we trace their history, we must not lose sight of the people whose national characteristics or whose foreign alliances left a stamp upon the gardens they made.

Nothing worthy of the name of a garden existed in Britain before the Roman Conquest. The Britons, we know, revered the oak, and held the mistletoe sacred, and stained their bodies with woad,* but of any efforts they may have made for the cultivation of these or any other plants we know nothing. The history of Horticulture in this country cannot fairly be said to begin before the coming of the Romans. In this, as in other sciences, the Romans were so far advanced that it was centuries before they were surpassed, or even equalled by any other nation.

* Mr. Baker points out that woad is not wild in Britain.

They cultivated most of the vegetables with which we are still familiar. At Rome, said Pliny the Elder, "The garden constituted of itself the poor man's field, and it was from the garden that the lower classes procured their daily food." The rich indulged in luxury and extravagance in the garden, and vegetables and fruits were raised at great cost for their use, which were not enjoyed by the community at large. But most of the vegetables which are still in general use were common to all classes, and many of these plants were brought by the Romans to this country. Some of them took so kindly to this soil, and were so firmly established, that they survived the downfall of the Roman civilization. A curious example of this is one species of stinging-nettle, which tradition says was introduced by the Romans as an esteemed pot-herb.

Tacitus, writing in the first century, says that the climate of Britain was suitable for the cultivation of all vegetables and fruits, except the olive and the vine. Before long, even the vine was grown, apparently with some success. It is generally believed that the Emperor Probus, about the year 280 A.D., encouraged the planting of vineyards in Britain. Pliny tells us that the cherry was brought in before the middle of the first century. Perhaps this was some improved variety, as this fruit is indigenous in this country.

We cannot suppose that the Roman gardens in Britain were as fine as those on the Continent. Gardens on such an elaborate scale as that at Pliny's Villa, or at the Imperial Villas near Rome, with their terraces, fountains, and statues, could scarcely have been made in this country. But the remains of Roman houses and villas which have been found in various places in England, so closely resemble those found in other parts of the Empire, that doubtless the gardens belonging to them were laid out as nearly as possible on the same lines as those of Italy and Gaul. The South of England could afford many a sheltered spot, where figs and mulberries, box and rosemary, would grow as well as at " Villa Laurentina," seventeen miles from Rome. A " terrace fragrant with the scent of violets," trailing vines and ivy ; or enclosures of quaintly-cut trees in the forms of animals or letters filled with roses, would not there seem out of place. If the

Roman gardens in Britain were like this—and why should it be doubted when we see the remains of villas, mosaic pavements, baths, roads, and bridges left by that nation?—it was fully a thousand years before anything as beautiful was again seen in our Island.

The fall of the Roman Empire, and the invasions of barbarians, struck a death-blow to gardening as well as to all other peaceful arts. During the stormy years which succeeded the Roman rule in Britain, nearly all knowledge of horticulture must have died out. Only such plants as were thoroughly naturalized and acclimatized would be strong enough to continue to grow when not properly cultivated.

The few Saxon names of plants which can be traced to the Latin seem to identify these hardy survivors, or at any rate show that the Anglo-Saxons were well acquainted with many of the Roman plant-names. The following list, given by Mr. Earle in *English Plant Names*, clearly shows their Latin origin :—

Latin.	Anglo-Saxon.	English.
Amigdala	Magdula treow	Almond
Beta	Bete	Beet
Buxus	Box	Box
Cannabis	Hænep	Hemp
Caulis	Caul	Kale
Coliandrum	Celendre	Coriander
Chærophyllum	Cerfille	Chervil
Castanea	Cisten beam*	Chestnut
Cornus	Corn treow	Cornel
Crotalum	Hratele	Yellow rattle
Cuminum	Cymen	Cummin
Cerasus	Ciris beam*	Cherry
Febrifugia	Feferfuge	Feverfew
Ficus	Fic beam*	Fig
Feniculum	Finul	Fennel
Gladiolum	Glædene	Gladden
Lactuca	Lactuce	Lettuce
Laurus	Laur beam*	Laurel
Linum	Lin sæd	Linseed
Lilium	Lilige	Lily
Lubestica	Lufestice	Lovage
Malva	Mealwe	Mallow
Morus	Mor beam*	Mulberry

(* Beam—the living tree, as Ger. "Baum.")

Latin.	Anglo-Saxon.	English.
Mentha	Minte	Mint
Napus	Naep	T*ur*-nip
Papaver	Popig	Poppy
Persica	Persoc treow	Peach
Petroselinum	Petersilie	Parsley
Pirus	Pirige	Pear
Porrum	Por leac	Leek
Prunus	Plum treow	Plum
Radix	Raedic	Radish
Rosa	Ro-se	Rose
Ruta	Rude	Rue
Sinapi	Senap	Mustard
Unio	Yul leac	Onion
Ulmus	Ulm treow	Elm
Vinea	Win treow	Vine

It may be that some plants, such as the cherry, cabbage, lettuce, leek, onion, radish, rose, and parsley, continued in this country; although many species which were in cultivation in Britain, in Roman times, had to be re-introduced into England at a later date, having been entirely lost during the years of Teutonic invasion. On the Continent, the same state of things followed the dissolution of the Roman Empire, and horticulture only revived with the spread of Christianity and the establishment of monasteries after a lapse of centuries.

In this country the revival was due to the same cause, and in the early years of England's history undoubtedly the monks were better skilled in horticulture than any other class of the community. The lines in which their lives were cast tended to maintain this superiority. They were left quiet, and, to a great extent, undisturbed by wars; and when other property was destroyed and plundered, that of the monks was respected. Many of them were men of skill and intelligence, and they were able to learn, not only from books, but from their intercourse with the Continent, both what plants to grow and how to grow them.

The earliest records of gardens on the Continent (after Roman times) date from the ninth century. In the list of Manors of the Abbey of Saint Germain des Prés, Saint Armand and Saint Remy, in the time of Karl the Great, mention is made of various gardens.* At other places, as at Corbie, in Picardy, and at St. Gall, near the lake of Constance, there remains more than a mere mention of the existence of a garden. At Corbie the garden was very large; either divided into four, or else four distinct gardens, and ploughs, which had to be contributed annually by certain tenants, were used to keep it in order; while other tenants had to send men from April to October, to assist the monks in weeding and planting.† At St. Gall, the "hortus" is a rectangular enclosure, with central path leading from the gardener's house and shed for tools and seeds situated at one end, with nine long and narrow beds of equal size on either side. The "herbularis," or physic garden, is smaller, with a border of plants all round the wall, and four beds on either side of the central walk; and the plants contained in each of these beds are carefully noted.‡

In England we have not such an exact description of any garden, and it is only by carefully examining the records of the various monasteries that the existence of gardens or orchards in the eleventh and twelfth centuries, and a few of even earlier date, can be proved.

A garden was a most essential adjunct to a monastery, as vegetables formed such a large proportion of the daily food of the inmates. Therefore, as soon as monasteries were founded, gardens must have been made around them, and these were probably almost the only gardens, worthy of the name, in the kingdom at that time. Still, the number of plants they contained was very limited, and probably many of those grown on the Continent had not found their way into this country. The monks may have received plants from abroad, as some connexion with religious houses on the Continent was kept up; and in bringing back treasures for their monasteries or churches the garden would not be forgotten. But plants were chiefly brought for medicine, and we may infer that they were imported in a dry state, as our word "drug" is simply part of the Anglo-Saxon verb "drigan," to dry.

* *Polyptyque de l'Abbé Irminon.* Ed. by M. B. Guérard. Paris, 1844.
† *Ibid.*
‡ *Archæological Institute Journal.* Vol. V.

Soon after monasteries had been established in this country, missionary monks set forth to convert their Teutonic kinsfolk on the Continent. It has been suggested by Mr. Earle that some of the German names of plants which resemble old English, are not cognates, but were derived from words used by the Saxon missionaries, who first brought with them the knowledge of the virtues of those plants.*

The old word for garden was "wyrtȝerd," a plant yard, or "wyrttun," a plant enclosure. Also the form "ortȝerd" or "orceard," which is the same as our word orchard, though the meaning is now confined to an enclosure planted with fruit trees. "Wyrt" or "wurt" was used for any sort of vegetable or herb, and is the same as the modern word "wort," suffixed to so many names of plants, as "St. John's Wort," or "herb John." Sometimes a special plant filled most of the enclosure, thus the kitchen garden was occasionally called the "leac tun," or leek enclosure. We still speak of an appleyard, the old "appultun," or "appulȝerd," but we say a cherry orchard, while the old word was equally simply "cherryȝerd."†
A part of the monastery garden was sometimes called the "cloysterȝerd," and the garden laid down in grass where flowers were not grown was the "grasȝerd."

At this early period, and for many centuries later, gardens were planted chiefly for their practical use, and vegetables and herbs were grown for physic or ordinary diet. Flowering plants were but rarely admitted solely on account of their beauty. But it does not necessarily follow that bright and pretty flowers found no place within the garden walls. Roses, lilies, violets, peonies, poppies, and such like, all had medicinal uses, and therefore would not be excluded.

The beauty of flowers appeals to nearly every one, and even in the most disorderly periods of our early history they may have exercised some softening influence. A pretty story is told of William Rufus, which shows that monarch, as it were for a

---

* The German for Plantago s "Wegbreit," the A. S. "Waegbrede." The old German for Camomile was "meghede," the A. S. "magede."
† Gardener's Accounts, Norwich Priory.

moment, in a more gentle light than perhaps any other incident during his turbulent reign. Eadgyth, or Matilda, afterwards the wife of Henry I., was being educated at the convent of Romsey, where her Aunt Christina was Abbess. When the child was twelve years old, the Red King wished to see her, and one day the Abbess was distressed to hear him and his knights demanding admission at the convent gate. The good lady, fearing some evil purpose towards the child, made her wear a nun's veil; then she opened to the king, who entered, "as if to look at the roses and other flowering herbs." While the rough king thus inspected her flowers, the Abbess made the nuns pass through the garden. Eadgyth appearing veiled among the rest the king suffered her to go by, and quietly took his leave.* The story was told by the Abbess to Anselm, who narrated it to Eadmer, in whose history this most picturesque scene is recorded.

While the Abbess Christina was adorning her cloister gardens with roses and flowering herbs, other monasteries were being beautified in like manner. The first Abbot of Ely, Brithnodus, was famed for his skill in planting and grafting, and improved the Abbey by making orchards and gardens around it.†

It seems as if there were gardens at Ely earlier than his time (twelfth century), as the following quaint story implies the existence of some sort of garden in the neighbourhood of Ely. It is related among various miracles wrought at the tomb of St. Etheldreda ‡ how the hand of a girl was cured. She was servant to a certain priest, and "was gathering herbs in the garden on the Lord's Day, when the wood which she held in her hand, and with which she desired to pluck the herbs unlawfully, so firmly adhered (to her hand) that no man could pluck it out for the space of five years, by the merits of St.

---

* Migne, *Patrologiæ cursus completus*, tom. 159-160, sec. xii. "Eadmer," p. 427. Also D'Achery, *Spicilegium* (Paris, 1723), Vol. II., p. 893. Freeman, *Wm. Rufus*, Vol. II., p. 32.

"Rex siquidem propter inspiciendas rosas et alias florentes herbas, claustrum nostrum ingressus."

† Gale, *Historiæ Britannicæ*, 1691. "Hist. Eliensis," Liber II., chap. ii.

‡ Dugdale, *Monasticon*, Vol. I., p. 473 (new ed.).

Etheldreda was cured." The Saint died in 679, and, although of no historical value, surely such a curious legend is worth relating.

Few records of a very early date have come down to us, but monastic life did not quickly change, and probably the gardens of the fourteenth century differed little from those of the twelfth. To gain a fuller knowledge of these gardens, we must pass over two centuries to the time when written accounts begin. As we get into the fourteenth century there is more material on which to work. The outlines of the management of these gardens is clear, although the details can only be filled in by imagination.

Each department within the monastery was directed in a regular and orderly way, and was presided over by an officer, with set duties to perform: who had to keep the accounts of his office, and was responsible for its management. There was a Gardener, or Hortulanus or Gardinarius, or Garden Warder, just as much as there was an Almoner, Sacristan, Precentor, or any other officer.

In some instances the accounts of the Hortulanus have been preserved, and further references to gardening matters are scattered throughout various chartularies. Two very perfect series are those of Norwich Priory and Abingdon Abbey,* and they are doubtless fair examples of the Gardener's accounts in the majority of monasteries. There are four accounts at Abingdon, the earliest for the year 1369-70. The Norwich series is far more numerous, there being some thirty rolls, the earliest 1340, the last 1529; the first years of the fifteenth century being well represented.

These accounts show the receipts and expenses of the office, the cost of repairs, the money received from the few products sold, but they throw no light on the processes of cultivation, nor do they particularize the plants which were grown.

Like the other officers, or obedientiars, the Hortulanus had

---

* Those at Norwich are only in MS. Those at Abingdon are printed by Camden Soc., *Accounts of the Obedientiars of Abingdon Abbey*, R. E. G. Kirk, 1892.

his "famulus" to assist in the work, and was also allowed to employ labourers, and money was forthcoming for their payment from the rent of some small piece of land, or some tenements which belonged to the office. At Ramsey Abbey* there were two "famuli" in the garden, and their payment (circ. 1170 A.D.) was "to each of them fourteen loaves," and two acres of land.† But in spite of various small rents and money received from the surplus garden produce, or grain grown on the lands belonging to the garden office, the accounts do not always show a balance on the right side, and the receipts not infrequently failed to cover the expenses.

In early times the monks seem to have worked better, or at any rate managed more carefully, for the garden paid its expenses; but at Norwich as the years went on, the office got more and more into debt. In 1429 "the expenses exceed the receipts, £8. 2s. 8½d.;" in 1431 there is a deficit of £13. 16s. 8¾d. Then a new plan began, and the garden was let to a certain William Draper, who paid 40s. for the farm of it ; ‡ this state of things continued to the end of the period covered by these accounts. The following are transcriptions of some of the rolls, the greater part are translated from the Latin but the words in quotation marks are spelt as they occur in the originals.

The earliest roll, A.D. 1340, is here given complete.

Account of brother Peter di Donewich of the garden in the 14th year of Dan William di Claxton Prior.

Receipts—
Remainder of preceeding account, 73s. 8d.
Of rent of assize that is to say from Adam Gilbirt now holding one shop in Nedle rowe, 18d.—of "fagot" branches and roots, 28s. 2¼d.—of rods [of "osiers," 13s. 4d.—of timber "Stamholt and wrong," 9s. 8d. —of hay, 36s. 10d.—of beans, 15d.—of herbs, 13d.—of garlick, 11d.—of

---

* *Cartularium Monasterii de Rameseia,* Wm. Hart. List of Monastic officers.

† At Durham monastery the payment was to " Robert Kyrvour, ortulanus, per annum 5s.," together with a few other small payments amounting to about another 5s. (Surtees Society.)

‡ Examples of the entries:—
1471. Receipts " From the great garden demised to John Plomer for the term of 20 years this being the sixth, 25s."
1487. " From Robert Castyr for the farm of the great garden, demised to him for the term of 10 years this being the second, 26s. 8d."

apples & pears, 13s. 4½d.—of "Sandice" (*Sandal wood?*), 5s. 6d.—of eggs, 14s.—of "hempsede," 1d.—of wax, 9s. 7d.—of "forage," 2s.—of "lapp," 3s.
Sum of receipts, £8. 19s. 6d.
Of 1 cow, 3 bullocks, 27s.—of calves, 8s. 2d.—of milk, 65s. 9d.—of the farm of 1 cow demised to farm, 2s.—sum, 102s. 11d.
Sum of whole receipts, £17. 16s. 1½d.

Expenses—
In the wages of servants, 13s.—in their stipends, 10s.—in the wages of the "garcionem," 14s. 11d.—in stipends, 2s.—also given to the same of favour, 20s.—given to a certain "pageto" by the year, 2s. 3d.

Pensions and Contributions—
In the O of the gardener, 20s. 8d.—in oblations of servants and certain men of the Court, by favour, 2s. 11d.—in alms, 2s. 2½d.—to the Scholars of Oxford, 2s.—to the Sub-Prior for the cloister, 2s.—to the cellarer for the cutting of herbs, 2s.—to the Almoner for the tithe of the garden, 12s.—in one tenth to the Lord the King, 1½d.—to the Cardinals, ½d.—in oblations and "flaunis," 13d. (= *flaun = custards*, or *pancakes, at rogations*)—to the reapers of the Lord Prior, 6d.—to John de Leverington, 6d.—to the Carpenter of John de Berney, 6d.—to the "boscar" (= *woodman*) of the Lord Bishop, 6d.—in gloves, 7s —sum, 60s. ½d.

Mowings and other things—
In the mowing of the meadow for both crops and of the court and paths, 3s. 5d.—in peas for pottage of the convent and servants, 3s. 3d.—in mustard seed, 3s. 3d.—in beans, 2s. 2d.—also in beans and butter in the convent, 15d.—in cherries, 8½d.—in milk, 16d.—in forage, 12s. 11d. —sum, 28s. 3½d.

Weeding and Hiring—
In weeding and "aids," 30s. 2d.—in the stipend of Ralph Brenetour and others working upon the bank and cleaning the ditch in the meadow for 12 days, 8s.—taking by the day, 8d.—in their drinks and other expenses, 12d.—in the stipend for one carpenter carving timber and mending other divers things, 2s.—in one tiler roofing with tiles and doing other things, 3d.—sum, 41s. 5d.—in "Pikerell" and roach for stock, 2s. 5d.—in lard, tallow and candle, 8d.—in iron spades and fixing "bills," 3d.—in mending an "axes" and in one new "Hachet," 7d.—in "skalerons" (*? escallions = scalions—small onions*), 1d.—In dung, 3s. 3d.—in keys, 4½d.—in "Juncis" (= *rushes*) for the Infirmary, 10d.—in 2 "tribul" (= *sieves or rakes*), 2 spades, 2 dung forks, new ironed, 12d.—in 1 scythe, 1d.—in "moles," 1d.—in a wooden measure, 3½d.—in 2 "clayes" for the bank, 10d.—in the "dentation" of a scythe handle, 1d.—in cord (or string), 1d.—in one earthen pot, 1d.—4350 tiles with carriage, 10s. 2d.—in pasture allowed from the Lord Bishop, that is to say "le hundhill," 2s. 2d.—in three "limours cenonette" (*? whitewashers*), 11d.—given to the "raton" (= *rat-catcher*). 1d.—in parchment, 1d.—Sum, 24s. 5d.—in boots of the gardener with repairs, 2s. 6d.—in wine sent to the Lord

Prior and given to the brethren with divers expenses made in blood lettings and at St. Leonard's at times, 11s. 10½d.—in divers spices and almonds, 2s. 7½d.—in foreign expenses, 8d.—in "Wardecorgard," 2s. 6d.—Sum, 20s. 2d.—Sum of all the expenses, £10. 18s. 1¾d., so the receipt exceeds the expenses, £6. 17s. 11½d.

A.D. 1402. Account of brother Thomas Rughton of the office of gardener. 22nd year of Prior Alexander, Michaelmas to Hilary.
Receipts—
  Excess of the account of the year preceding, 43s. 10d.—for pears and nuts ("avelanis"), 4s. 4d.—for apples, 16d.—for herbage, 15d.—for dry trees, 18s. 3d.—"pro faggots and Astel" (= *shavings*), 11s. 3d.—for willows, 9d.—for plants of herbs, 2s. 3d.—for cock and hens, 18d.—for onions, 3s. 10d.—for osiers, 3s. 4d.—for leeks, 6d.—for "tasel" (= *teasel*), 5s. 10d.—for trees sold to the Master of the Cellar, 35s. 4d.—for "lawyr of crabthorn" and other things sold to the Master of the Cellar, 35s. 4d.—"pro lawyr of wythis," 10d.—Sum, £4. 18s. 7d.—Sum total of the receipts, £7. 2s. 5d.
Expenses—
  First for rushes and carriage of the same, 5s.—for garlick and onions, 2s.—for mustard seed, 8s. 6½d.—for beans, 3d.—for planting beans, 12d.—for parchment, 3d. In stipends of the servants to one of them, 10s. 6d.—to the other of them, 10s.—for their "tunics," 8s—to the labourers in the garden, 2s. 4d.—to other labourers about "tasel," 3s. 5d.—to the scholars of Oxford, 18d.—in the presence of the Lord Prior for small things and other recreations, 2s. 6d.—for milk for the convent, 2s. 2d.—to the cellarer for knives, 2s.—in oblations at Christmas, 3s.—for the boots of the gardener, 15d.—for spades, shovels and other utensils, 13½d.—in gifts, 6d.
  Sum of all expenses, £3. 8s. 1d.
  And so the receipts exceed the expenses, £3. 14s. 4d.

A.D. 1403. Extract from the account of brother Thomas de Corpsty, Michaelmas, 5th year of Henry IV., to the same feast in the 23rd year of Prior Alexander.
Receipts—
  "Pro albell" (= *abele*, *white poplar*), 8s. 8d.—for timber, 6s. 8d.—"pro crabdractis and ok" (? *crab draughts, cartloads of crab-trees and oak*), 3s. 9d.—"pro tasles," 6s. 8d.—"pro" star (= *sedge*) and reed, 16d.—"pro lillys" (= *lilies*), ½d.—for the small garden, 8d.—for the meadow demised, 37s. 8d.—sum of Receipts, £10. 3s. 9d.
Expenses—
  Arrears, 59s. 4d.—seed of onion bought, 12d.—for nails and keys, 6d.—tenths of the Lord the King, 1½d.—gloves, 7d.—"Pro tribul" (= *rake*), spades, &c., 3s. 9½d.—for O O of the gardener, 26s. 8d.—stipends of the servants: To one of them, 16s., to the other of them 15s. 2d.—for their tunics, 8s. 10d.—on the day of the account, 12d.

Sum of the expenses, £8. 8s. 0d.—with the arrears, £11. 7s. 10d.
The expenses exceed the receipts, 24s. 1d.

A.D. 1427 (complete). Account of brother William Metygham from S. Michael to S. Gregory, 6th Henry VI., in the 1st year of Prior William.

Receipts—
For herbs, "lekys" and "Porrettes," 4s.—for faggots ("fasciculis"), Astill and "ozyerys" (= *osiers*), 8s. 2d.—for the meadows from the cellarer, 20s.—for the garden between the gates, 12d.—sum of receipts, 33s. 2d.

Expenses—
Arrears of preceding year, 68s.—for mustard seed, 7s. 4d.—to the Almoner, 12s.—for milk in Advent and Quinquagesima, 4s. 3d.—for planting garlic and beans and for weeding, 2s.—to workmen hired at times, 15d.—in medicines of the gardener, 2s., in the presence of the Lord Prior, and at St. Leonards and elsewhere, 3s. 2d.—in gifts to the servants at Christmas, 18d.—in the repair of the houses, utensils, "schelvis," and boards bought, 5s. 4d.—in the boots of the gardener, 12d.—to Thomas the servant for stipend, 12s.—to John the servant, 9s.—for their tunics, 10s.

Sum of expenses, 70s. 10d.
Sum of all expenses with arrears, £6. 18s. 10d.
So the expenses exceed the receipts, £5. 5s. 8d.

A.D. 1484. Account of brother John Metham, from Michaelmas 1st, Richard III., and Michaelmas 2nd, Richard III. Prior John Bonwell.
Remainder of account of preceding year.

Receipts, 3s. 5¼d. First from the Lord Prior for the parcel of the garden annexed by the separation of the great "fosse" (*ditch*) [to, "le ortȝerd" of the same, 16d. For beans sold, for the straw of the same for "eldyng" (=*fuel*), 6s. 10d.—Onions sold, 16d.—Sum of receipts with remainder, £4. 7s.

Expenses—
Tithes, and no more, because certain tenements are built on the soil of the garden, in "Holmstrete,"—scholars, brother John Helgey and brother William Gedney.—Robert Cook for pottage made of peas and spices for the convent, 6d.—for "frixures" (=*fritters*).—for labour of labourers in extracting the "mosse" from the cloister green, 6d. —for cleaning the great ditch that goes round the garden with the small ditch which is next the "scaccarium" (= *exchequer*) of the gardener, 18d.—(several payments to labourers mentioned by name.)—for "gryffing," 4d.—for digging and other things, 10½d.— pay to Thomas Mylys and Henry Cobyller, of the Parish of St. John of Matermarket, for thrice mowing the garden and "bina" (= *twice*) mowing the cloister, 3s.—For one "wyndowstal" for the orto cěrsőr (= *cherry garden*)—for "flagello" (=*flail*), 1d.—for labourers for in-gathering mustard seed with the threshing of the same, 7d.

Sum of expenses, £4. 7s. 7½d. Receipts exceed the expenses, 2s 10¼d.

Some items occur without variation every year, such as the payments to the servants; and their tunics, boots and gloves. The gloves are not uncommon entries; they appear among the accounts of Bicester,* Bury, Holy Island, and other places. They were probably thick gloves for weeding.

The O of the gardener is also of regular occurrence, as it was expenses at a yearly feast, and it is thought the O refers to the

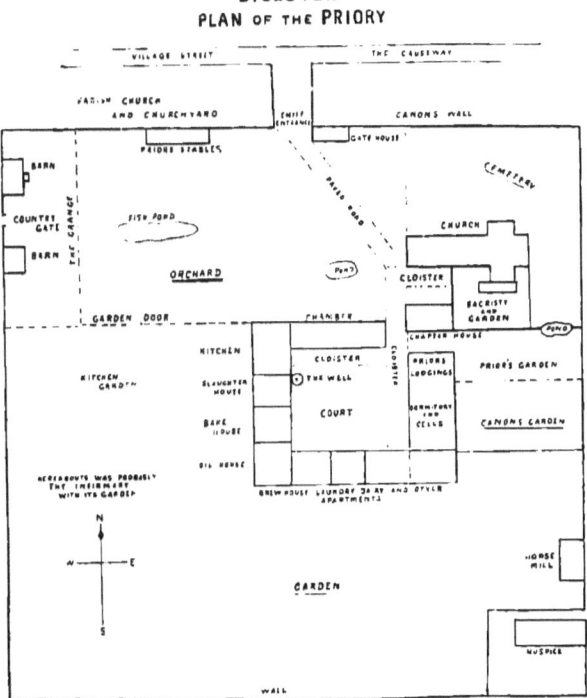

BICESTER
PLAN OF THE PRIORY

Psalm sung on the occasion by the Hortulanus, commencing " O Radix Jesse." In the Abingdon Accounts it is entered, "To O Radix, 6s. 10d.," and another time (A.D. 1388) still more at length, " In expensis factis pro mittent exeunia ad O Radix XVId."

* Blomefield, *History of Bicester*.

It will be noticed also that in these and other accounts the tithe is deducted. The year in which it first was enacted that tithe should be paid "of fruit trees and every seed and herb of the garden," was A.D. 1305, the decree insisting on the payment, being issued by the Council at Merton, in Surrey.*

The chief variations are in the tools bought, and the repairs. "For a saw," "knives for herbs," "mending a hatchet," "repairs of the garden wall," "lock and keys for the gates," &c.; and sometimes fruit, apples, cherries, beans, onions, or such like, had to be bought when the garden supply fell short. But this "great garden" under the care of the Hortulanus was not by any means the only garden. Many other office holders had gardens too.

The plan † on page 13 is compiled from the remains and the records of Bicester Priory: the various gardens and their relative positions are all mentioned in the chartulary, and the quantity of distinct gardens is not in excess of the usual number. As a rule the Prior had an enclosure of his own. At Melsa there was both "the garden which is called the Prior's," and "the garden of the Abbot's chamber."‡ At the Abbey of Haghmon, in Shropshire, the Prior was allotted "for his recreations a certain chamber under the dormitory, . . . with the garden of old called 'Longenores gardine,' annexed to the chamber before-mentioned, together with the dovecote in the same." §

At Norwich, payments occur to the gardener from the Lord Prior for a "parcel of the garden," or small piece reserved for his special use. The "little garden," or "garden within the gates," at Norwich, was let to the cellarer. The Sacristan, the Treasurer, the Precentor, and the "Custos operum," all had separate gardens at Abingdon, and paid rent for them to the gardinarius. At Winchester, the payment to the gardener, "Roberto Basynge, custodi gardini conventus," occurs in the

---

* Wilkin's *Concilia*, Vol. II., p. 278; "Mertonense," 1305, "et de fructibus arborum et seminibus omnibus et herbis hortorum."

† Reproduced from the *History of the Deanery of Bicester*, by the kind permission of the author, J. C. Blomefield.

‡ Abbot Burton's *Chronicle of Melsa*, Vol. III., p. 242.

§ Dugdale, *Monasticon* (new ed.), Vol. VI., p. 112.

Receiver's account (A.D. 1334) as well as charges for mowing the Almoner's garden, and besides these the 'custos operum" defrayed the expenses of a garden called "Le Joye." The Infirmarian's garden was usually an important one, as in it he grew healing herbs for the sick of the monastery, and for convenience this plot was, as a rule, placed near the infirmary or hospital.

In all countries, heathen and Christian, and in all ages, flowers have played an important part in ceremonies, such as funeral rites and marriage feasts. England in the Middle Ages was no exception; and the use of flowers in the services of the Church, in crowning the priests, wreathing candles, or adorning shrines, was very general.

The gardens within the monastery walls for providing these flowers were under the care of the Sacristan. At Abingdon, he paid the gardinarius four bushels of corn for the rent of his garden.* At Norwich, the Sacristan seems to have had more than one garden, as a very cursory glance at the MS. accounts of that office shows the names of both "St. Mary's" and the "green garden."† There was a "gardina Sacristæ" at Winchester as early as the ninth century,‡ and to this day a piece of ground on the east side of the north transept of the cathedral bears the name of "Paradise," and marks the site of the Sacrist's garden. The fifteenth century doorway, which was the entrance to the enclosure, is still standing.

Such a garden as this is referred to when the Abbot of Ramsey, between 1114-1130, had to come to some agreement about certain pieces of land in London which adjoined the property of the Priory of the Holy Trinity; and the Prior consented§ "to give

---

* Abingdon Accounts. R. E. G. Kirk:
1388-9, et de iiij hussellis frumenti de Sacrista pro orto suo, nichil hic in denarijs quia recipuntur in sua specie ut patet extra.

† Sacrist Account, MS. Norwich :
1431. " In weeding in the garden of St. Mary, 2s."
1428. " For weeding in the ' green garden.' "
1489. " Received for the trunk of a pear-tree blown down by the wind, 11d." Gardener's account, 1472. " For farm of the garden of the Sacrist, 2s."

‡ Wharton, *Anglia Sacra.* Part I., p. 209.

§ *Cartularium Monasterii de Ramesia.* Vol. I., p. 133.

up his claim which he had upon the chapel of the Abbot, and the garden which is before the chapel." These "gardini Sacristæ" were not only found within monastic precincts, but were attached to many churches and chapels. The Hortulanus of Abingdon let out a garden "next to St. Nicholas' Church," to the Rector, for a term of years.* There is an interesting record of the chapel garden in the Manor of Wookey, in Somersetshire, which belonged to the Bishops of Bath and Wells, in the account of the Reeve of that place for the year 1461-2.† Three men were employed for four and a half days at two pence a day, "digging and cleaning the chapel garden."

Henry VI. left such a garden to the church of Eton College. The clause in his will runs thus : " The space between the wall of the church and the wall of the cloister shall conteyne 38 feet, which is left for to sett in certaine trees and flowers, behovable and convenient for the service of the same church," and it was to be surrounded by "a good high wall with towers convenient thereto." ‡ Many other such examples of gardens connected with churches could be enumerated.

At all great functions, both during the processions or while performing the services, the priests were crowned with flowers. This was specially the custom at St. Paul's,§ in London : and when on June 30th, 1405, Bishop Roger de Walden was installed there, he and the Canons of the Cathedral walked in solemn procession, wearing garlands of red roses. ||

The use of these " coronæ sacerdotales," or wreaths worn by the priests on feast days, continued for many centuries,¶ and their prevalence up to the time of the Reformation is apparent from various churchwardens' accounts. These entries, however, are not frequent, as the gardens attached to the churches were evidently, as a rule, able to supply sufficient flowers for ordinary

---

* 1413. Accounts, by Kirk.
† *History of the Parish and Manor of Wookey*, by T. S. Holmes.
‡ Nichols' *Wills of the Kings and Queens of England*. Ed. 1780, p. 298.
§ Polydore Vergil. *De rerum Inventoribus*. Lib. II.
Historia di Episcopis et Decanis Londiniensibus, by H. Wharton, 1695 (p. 150).
¶ " Ceremonial use of Flowers," *Nineteenth Century*, 1880.

use, and it was only for great occasions, or on special feast days, when larger quantities were required, that they had to be bought.

For instance, at St. Mary Hill, where some entries are found in the accounts, there was a garden near the church.*

A.D. 1483-1497. St. Mary Hill. Churchwarden account. "For birch at Midsomer, 8d.—Box and palme on Palmesonday, 1s.—Polis on Estir evyne, 10d.—Garlondes on Corpus Christi day, 10d.—A dozen and a half rose garlondes on St. Barnebe's day, 8½d. for rose garlondis and wodrove garlondis on Seynt Barnebe's day, 11d.—for two doss. di boese garlondes for prests and clerkes on St. Barnebe daye.

1510. For palme flowrys and cake on Palme Sunday, 10d.
Also at St. Martin Outwich, London, 1524.

Item—For rose garlands on Corpus Christi day, 6d.—Item—For byrche at Midsomer, 2d.—Item—For rose garlands, brede, wyne, & ale on ij Sent Marten's days, 15½d.—Item—For holy and ivy at Chrystmas, 2½d.

1525. Paid for palme on Palme Sunday, 2½d. Paid for brome ageynst Ester, 1d. Payd for rosse garlonds on Corpus Christi daye, 6d.

When such decorating of churches was considered unlawful after the Reformation, these gardens would naturally fall into disuse, even where the lands they covered were not at once appropriated for other purposes.

In 1618, James I. set forth a declaration permitting certain " lawfull recreations . . . after divine service,† and allowed that women should have leave to carry rushes to the church for the decoring of it according to their old custome." These rushes may have been simply for the floor, and not for the altar or walls, as, for example, we find in 1580, churchwardens at Wing, in Buckinghamshire, spent 1d. for "one burden of roshes to strewe the church howse agaynst the comyssyoners sate there." ‡

Coles, writing as late as 1656, says : " It is not very long since the custome of seting up garlands in churches hath been left off with us: and in some places setting up of holly, ivy, rosemary, bayes, yew, &c., in churches at Christmas. is still in use." § This, however, is looking too far ahead, and at the time we are

* Nichols, *Illustrations of the Manners and Expenses in England . . . deduced from Accounts of Churchwardens, &c.* 1797.
† Fuller, *Church History.* London, 1655. Book X., p. 74.
‡ *Archæologia.* Vol. XXXVI., p. 238.
§ *The Art of Simpling*, by W. Coles. 1656.

considering, the monks within the quiet cloister, week by week and year by year, supplied the best flowers their skill and knowledge could produce, to adorn their churches and chapels.

But to return to the consideration of the department of the gardinarius. He had more than the garden under his care, for his jurisdiction extended over both the orchard and vineyard.

The orchard, or "pomerium," supplied not only apples and pears for eating and cooking, but also apples for cider. Large quantities of cider were made each year, except when in an unusually bad season the apple crop failed. This was the case in 1352, when the Almoner at Winchester made the following note in his accounts, "Et de ciserat nihil quia non fuerunt poma hoc anno." 1412 was another bad apple year, and no cider was made at Abingdon, and the not unfrequent purchase of apples and pears for the use of some of the monasteries, shows they did not always grow sufficient for their consumption, although in some years there was enough and to spare.* The Wardon pear, which was such a favourite for many centuries, originated at the Cistercian monastery of that name in Bedfordshire, and they bore three Wardon pears for the arms of the house.† It was a kind of cooking pear, and every early cookery-book contains receipes for "Wardon pies," or pasties. They are usually mentioned quite as a distinct fruit, as "apples, pears, Wardons, and quinces," because they were the best known variety.

Some of the orchards must have been of considerable size. In the time of King John the grant of land to Lanthony Priory included twelve acres of orchard. An oft-quoted example to prove the early existence of orchards is a Bull of Pope Alexander III., dated 1175, confiscating the property of the monks of Winchenley, in Gloucestershire, with the "town of Swiring and all its orchards."

The cherry was, from the date of its introduction by the

* Gardener's Accounts, Abingdon, 1388, "Et de xiii s. iiii d. di cicera vendita per estimacione et de xxxii s. vi d. ob de fructibus venditis, viz. : pomis wardon et nucibus."

† Dugdale, *Monast.*, Vol. V., p. 371, says they were also called Abbot's pears, but assigns no authority.

Romans, a popular fruit in this country. The "ciris beam," or cherry-tree, continued to be grown in early Saxon times. In the twelfth century it was one of the fruit trees praised by Necham, Abbot of Cirencester, in his poem, "De laudibus divinæ Sapientiæ," and this fruit was not forgotten in any monastic garden.

At Norwich, besides the "Pomerium," the appleyard or orchard, there was a "cherryȝerd," or, as it is called in another place, "orto cersor," or cherry-garden, and in spite of this we find cherries had to be bought "for the convent" from time to time, so great was the demand for this fruit. Perhaps it was the too frequent use of it that suggested to Necham the advisability of warning his readers that "cherries, mulberries, and grapes should be eaten fasting, and not after a meal." *

The third department, of the "Garden Warder," must now be considered. It has been already pointed out that vines were grown by the Romans in Britain, and, with the exception of the gap immediately following Roman rule, their history is continuous. Tradition points to a place called Vine, in Hampshire, as having taken its name from the vines planted there during the time of the Emperor Probus. Vines, the "Winestreow," are noticed as boundaries or landmarks at several places in Saxon charters of the tenth century, and these might have been survivals of Roman vineyards.†

Bede, writing early in the eighth century, says that Britain "excels for grain and trees . . . it also produces vines in some places." ‡ In the laws of Alfred,§ which were chiefly compilations of existing ones, it was notified that anyone who "damaged the vineyard or field of another, should give compensation." In the tenth century King Edwy confirmed the grant of a vineyard at Pathenesburgh, in Somerset, to the Abbey of

---

* Necham, *De Naturis Rerum*.
† Kemble's *Codex Diplomaticus*, Vol. V.
　MCXLVI. Eadmund, 943. Lechamstide.
　MCLXXVII. Ealdred, 949. Boxoram.
　MCXCVIII. Eadwig, 950. Welligforda, &c.
‡ Bede, *Hist. Eccle. gentis Anglorum*. Ed. 1848, p. 108.
§ *LL. Saxon* Wilki: p. 31. *LL. Aelf*: 26.

Glastonbury. The grapes were gathered in October, and that month was called "Winter filling moneth," or "Wyn moneth," another proof of the extent to which vines were cultivated. The pruning of the vine took place in February. The picture of vine pruners taken from an Anglo-Saxon MS. in the British Museum, illustrates that month in the calendar.

Necham devotes a chapter of his *De Naturis Rerum*, to the vine, but he chiefly moralizes, and does not treat his subject in its practical sense. He tells us that in gathering grapes, having reached the final row, the workers in the vineyard break into a song of rejoicing, but, unfortunately, he does not satisfy our curiosity by handing down the words of their chant.

In Domesday Book, the "vinitor," or vine-dresser, is only once mentioned, but some idea of the size of the vineyards may be gathered from the survey, as about thirty-eight in many different counties are described.* They are usually measured by "arpendi," the arpends being equal to about an acre, or less. The largest was at Bitesham, in Berkshire, on the land of Henry de Ferrieres, and covered twelve arpends. Some vineyards were old, others but newly-planted, as at Westminster four arpends are described as "vinea novella," and at Ware another vineyard as "nuprime plantatæ." Some of the vineyards bore grapes, while others did not, and these are distinguished as "vinæ portantes," or "vinæ nonportantes." The quantity of wine yielded by a vineyard of six arpends in Essex was as much as twenty "modii," or about forty gallons, if the season was favourable.

If England could boast of so many vineyards before the Norman Conquest, it was only natural that the influx of foreigners from a grape-growing country should infuse fresh ardour into vine-culture, and monasteries, with Abbots or Priors from the Continent, lost no time in improving the old and making new vineyards on their lands. The name "vineyard" was often retained long after the monks who planted it had passed away.

---

* In Kent, Hampshire, Wiltshire, Dorset, Gloucester, Berkshire, Hertford, Essex, Norfolk, Suffolk, &c.
A General Introduction to Domesday Book, by Sir Henry Ellis, 1833.

VINE PRUNERS, FROM AN ANGLO-SAXON MS. ELEVENTH CENTURY. B.M. COTTON. TIBERIUS, f. 5.

Thus "Vineyard," near Gloucester, described in *Camden's Britannia* as the seat of the Bridgemans, "on a hillet" to the west of the town, was once the vineyard belonging to the Abbots of Gloucester.* Gloucestershire was famous for its vines, which, wrote William of Malmesbury in the twelfth century, are "more plentiful in crops, and more pleasant in flavour than any in England :" for the wines do not "offend the mouth with sharpness, since they do not yield to the French in sweetness." † Again, we find in towns a "Vine Street," as in London, Grantham, Peterborough, and many others. Perhaps, at the latter place, the name marks the site of the vineyards planted by Abbot Marten, early in the twelfth century.

‡ At Hereford, sloping to the South-west, is the spot known as the "Vinefields," where the terraces, laid out for the vines, can still be distinguished. The accounts of the Diocese of Hereford, when the See was vacant by the death of Louis de Chorlton, in 1369, and the lands were in the hands of the King (Edward III.) until the next appointment, show the existence of a vineyard within the Manor of Ledesbury : while in a similar account for the year 1536-7,§ although the costs of the garden are entered, there is no mention of a vineyard ; and at another Manor on the same roll (Prestbury), the "herbage of the pasture called Vyneyarde" was sold, thus proving the former existence of vines on the spot, and showing how gradually they died out. But with our climate, what strikes one as more wonderful than their passing away, is that they were, at one time, so numerous throughout England. Even as far north as Cheshire, in the twelfth century, although there does not appear to have been any actual vineyard, the vine was not unknown, for Reginald of Durham notices, at Lextune in that county, a little church built of timber with vines climbing over it.‖

It is difficult to realize the appearance of Ely in the eleventh

* Gough's *Camden*. Vol. I., p. 392. Ed. 1806.
† De Gesta, *Pontif.* Book IV.
‡ Ministers' Accounts, B. 1138, No. 4. Bishops' Temporalities, Hereford Diocese. Record Office.
§ Exchequer Q. R., Hereford Diocese, No. 133 (R.O.).
‖ Reginaldi, *Mon ; Dunelm ; Lib. ; de Admirandis Beati Cuthberti*. Surtees Soc., 1835.

century in the days "when Cnut the King came sailing by," as it rose from out the dreary and undrained fen land, the sunny slopes around the cloisters, so thickly planted with vineyards, tended by those monks who sang so merrily, that the Normans gave it the name of the "Isle des Vignes."

Another old rhyme thus celebrates these vines :—

> "Quatuor sunt Eliæ: Lanterna Capella Mariæ,
> Et Molendinum, nec non claus Vinea vinum."

"Englished" thus by Austin, in 1653 :—

> "Foure things of Elie towne, much spoken are.
> The Leaden Lanthorn, Marie's chappell rare
> The mighty Milhill in the Minster field,
> And fruitful vineyards which sweet wine do yeeld."[*]

Ely long continued to be famous for its grapes. From time to time, when the manors were in the king's hands, during some interregnum caused by the death of the Bishop; the papers relative to the administration of the lands give evidence of the vineyards as well as of the orchards and gardens belonging to the See, from which a profit was derived.[†] The chief entries refer to the "herbage of the garden," "apples," "pears" and nuts sold, also hemp and reeds. The farm of the "rosery" often occurs, but the word is disappointing; and it stands for "roseria," "rosar," or bed for reeds or rushes, at places in the Fens.[‡]

In the "Bailiwick of Cambridge, except the island," and at Somersham Manor, there were vineyards which yielded grapes, but the principal one was at Ely itself. In 1298 as much as twenty-seven gallons of verjuice, "viridi succo," from the grapes, were sold; and the next year, twenty-one gallons.

---

[*] Ralph Austin, *A Treatise on Fruit Trees*. 1653.

[†] Exchequer Q. R. Bishops' Temporalities, ⁂; and Ministers' Accounts, ¹¹⁵². (Record Office.)

[‡] "Litilport 40s. of yearly rent of the 'Roseria' at the Annunciation," A.D. 1302.

The following are examples of the entries of most of the Manors:—

1286. Downham. 9s. of apples and nuts sold there.
1286. Littlelburi. 7s. 2d. of apples and pears sold there during the same time.
1286. Derham. 15s. of apples sold there.
1298. Feltevelle. 55s. 9d. of herbage and fruit of the garden and pasture sold.

The entry runs thus:—

"And of 10gs. 8d. of pasture and herbage sold in the vineyard and elsewhere in divers places in the summer. And of 25s. 3d. of fruit in two gardens and the vineyard, "besides the grapes, with 21 gallons of verjuice sold. And of £10 for 9½ butts of wine sold, of the remainder of the preceding year."

From another passage in 1302 it appears that cherries were the other fruit, besides the grapes, which grew in the vineyard,* and also we find in the same year the charges for the livery of the vine-dresser and the labourer under him, which was paid for in corn.†

The Bishops of Ely also had a vineyard attached to the garden, "Ely Place," of their house in Holborn, the site of which the present "Vine Street" commemorates. The earliest records of these gardens date from the reign of Edward III., and they are preserved at Ely. They are most interesting from the names of streets and houses in London mentioned in them, some with gardens attached,‡ for which rent was paid to the Bishop. But it is only in a few of the earliest ones that we find any details of the garden or vineyard, for from the year 1379-80 to 1480-81, they were let at the yearly sum of 60s. The rent of the garden alone was 20s. The accounts until the year 1419 are preserved at Ely; the continuation from 1423 to 1483 are in the Record Office.§ Among the latter in the time of Bishop John Morton, 20 to 21 Edward IV., we find the garden is at last again in the Bishop's hands; the entry states that there is no rent, "quod occupatur ad vsum Domini proprium hoc anno."

The following is the earliest of the rolls at Ely:—

Account of Adam Vynour, gardener ("ortolani") of the Lord Bishop of Ely, in his Manor of Holbourne, and collector of the rents, belonging to the said manor, from Michaelmas in the 46th to 7th June in the 47 year of Edward III.

* "Of 20d. from cherries in the vineyard sold."

† 20 March to 18 July—30th Edward I. "Wheat and barley—In the livery of one 'vinitor' during the same time, 2 qrs. 1 bus., he taking 1 quarter for 8 weeks. In the livery of his 'garcionis' during the same time 6¼ bushels 1 peck taking 1 quarter for 20 weeks."

‡ 1312. "In lez railes in gardino apud Faryndonesin."

§ Ministers' Accounts, Bishops' Temporalities, 14³⁷.

(1372-3). (Then follow rents of assize, and payments for the farm of shops. 77s. 6d.)

Issues of the Garden—And of 16s. for onions and garlick sold.—And of 9s. 2d. for herbs, "lekes," parsley, and herbage sold. And of 48s. 6d. for pasture in "le grasjerd" sold, and of 5s. 4d. for beans in the husks sold, sum, 79s., also £6. 6s. 8d. from Sir Thomas Wylton, sum total of receipts, £14. 3s. 2d.

Expenses—Rents repaid to various churches, &c.

Costs of the Vineyard and Curtilage and in divers labourers and women for digging the vines and curtilage, and also for cleansing and pulling up weeds in the curtilage, as appears by the parcels sewn to this account, 69s. 1½d., and in thorns bought, viz. 4 cartloads of thorns for making the hedges round the great garden, 6s. 8d., and in the stipends of 2 men making 6 score and 1 perches of hedges round the same garden, 35s. 3¼d., by the perch, 3½d. 111s. 1d.

Costs of repairs, &c. :—

Wages of the Bailiff—In the wages of the accountant for 35 weeks and 6 days, 62s. 9d., taking by the day, 3d. In the wages of 1 boy digging in the vineyard, and in the curtilage from the last day of December until 17th day of April, in the feast of Easter, for 106 days, 17s. 8d., taking by the day, 2d. In the stipend of the same boy for the same time, 5s. And in the stipend of the accountant for the half-year, 13s. 4d.—Sum, £14. 18s. 9d.

Small Expenses—Paid to the Rector of the Church of St. Andrew, in Holbourne for the tithe of the pasture of the great garden, 4s. 10d. Sum of all the expenses, £15. 12s. 6¼d.

Afterwards there is allowed to the same [accountant] 21s. 6¾d., which he paid to Sir Walter de Aldebury, Prebendary of the Prebend of Holbourne, for the rent of the vineyard of the Prior of Ely for 6 years and for one quarter of a year last past, viz. 3s. 5½d. by the year, viz. for the whole time during which the Lord Bishop held the said vineyard of the Prior at farm, and there is allowed to the same 9s. 4d. for his stipend from the day of the death of the Lord until the feast of Michael for 16 weeks taking by the week 7d. for the custody of the said vineyards and the pasture aforesaid.—And so the sum of both surpluses is 60s. 3¼d., which he received of Sir Roger Beauchamp.—And so he departed content.

(On the dorse) *Verjuice*—The same answers for 30 gallons of verjuice of the issues of the vineyard—sum, 30 gallons—thereof in tithe 3 gallons —And for one peck of parsley seed ("seminis petrosilli"), and for one quart hyssop ("ysop") seed—And for 1 quart savoury ("savori") seed, and for 1 quart leek ("lekes") seed.

Dead Stock—There remain there two iron spades ("vange ferree"), 1 rake ("tribul"), 4 hoes ("howes"), and 1 lamp (lucerna), 1 "shave," 1 axe ("bolex"), 1 box for candles, 1 box for spices, the latter broken.

The Bishop of Ely's Holbourne vineyard did not stand alone in that locality. Hard by was another belonging to the Earl of Lincoln, from which about fifty gallons of verjuice were sold in one year (1295-6).* A little further on, in Smithfield, a vineyard was planted by Geoffrey, Earl of Essex, on the land belonging to the "Canons of Trinity Church, London," which was restored to that body in 1137.†

It would be tedious to enumerate all the vineyards belonging to monastic houses which are known to have existed, and of which there is merely the name or some slight record surviving, as at Canterbury, Beaulieu, Ramsey, Abingdon, Spalding, Bury St. Edmunds, and many others.‡ Enough has been told to show how important an item the vineyard was in the gardener's department. His cares, however, did not quite end there; as the moat and the ponds were also under his charge.§ At Norwich the gardener's office bore the expense of cleaning the ditches which divided the various gardens, the Prior's from the chief garden, and so on.|| At Abingdon we find also he defrayed the cost of cleaning out the moat, and both there and at Ramsey the gardener purchased nets and baskets for catching the fish in the moat and ponds.¶

To get at the details of the management of monastic gardens, we have to go so constantly to the accounts of the office, and to look solely at the business side of the question, that one is apt to forget the other aspect, namely, the pleasure they afforded. But, alas! there are few gardens in existence which can give any idea of what these were really like. A thick hedge or a fish pond is generally the only survival. The wall enclosing a corner of the garden at Ashridge is part of the old cloister, and near it there is also a thick yew hedge surrounding another small piece of garden. These, if not actually the same as in the

* Duchy of Lancaster account. Bundle 1. No. 1.
† Syllabus of Rymer's Fœdera. Vol. I., p. 3.
‡ The total cost of the vintage one year at Abingdon was 4s. 4d. In 1388-9 the profits from the vineyard were—"from wine, 13s. 4d., from grapes, 20s. 0½d., from verjuice, 2s., from vines, 4d."—*Accounts of Abingdon Abbey*, by R. E. G. Kirk.
§ See plan on page 13.
|| 1483-4. "For cleaning the great ditch that goes round the garden with the small ditch which is next the 'scaccarium' (= *exchequer*) of the gardener, 18d." (There is an entry, 1516, "for making a window of glass in the 'scaccarium,' 2od.")
¶ Abingdon, 1450-1: Et in welez emptis pro piscibus capiendis in fossato conventus 4s. 10d. et in factura unius tronke pro piscibus custodiendis 3d.

ASHRIDGE.

days when the place was a monastery, are on the same lines, and have been kept as gardens ever since the days when the monks enjoyed the solitude of the cloister. The times we have been considering were periods of constant strife, when the cloister was the only place in which quiet and retirement could be found, and to those who sought refuge within its walls, how dear must those peaceful hours in their gardens have been. Perhaps some inmate of Sopwell (a cell of St. Albans) was too fond of early morning or late evening strolls in the garden, for Abbot Michael (about 1338) made the rule that in winter "the garden-door be not opened (for walking) before the hour of prime, or first hour of devotion:—and in summer that the garden and the parlour doors be not opened until the hour of none (? nine) in the morning:—and to be always shut when the corfue rings."*

Even the warlike Hospitaller Orders, the Templars and Knights of St. John, contributed something towards the improvement of Horticulture. In their wanderings in the East during the Crusades, they may have remembered some garden in England, and brought back plants for it, as, for example, the splendid Oriental plane at Ribston, the planting of which tradition attributes to the Templars. The surveys of the manors all over the kingdom belonging to these Orders show the large number of gardens of which they were possessed. At the Chancery of the Order of St. John of Jerusalem in England, in Clerkenwell, there was a garden in the time of Prior Philip de Thame (in 1338) which was still existing in the reign of Henry VII.,† and the Hospitallers had also a house with gardens attached at Hampton, on the site of the present gardens of Hampton Court.‡ In many ways through those troublous times the monastic orders kept alive the science of Horticulture, and spread the knowledge of it to those around them. Thus by practising, as well as by preaching, they showed by their useful lives that "to labour was to pray."

* Rev. Peter Newcome, *History of St. Albans*, p. 468.
† Close Roll, Henry VII., A.D. 1486.
‡ The Knights of St. John of Jerusalem and the Templars, also the Cistercians, were exempted from payment of the tithe of the gardens.— FULLER, *Church History*.

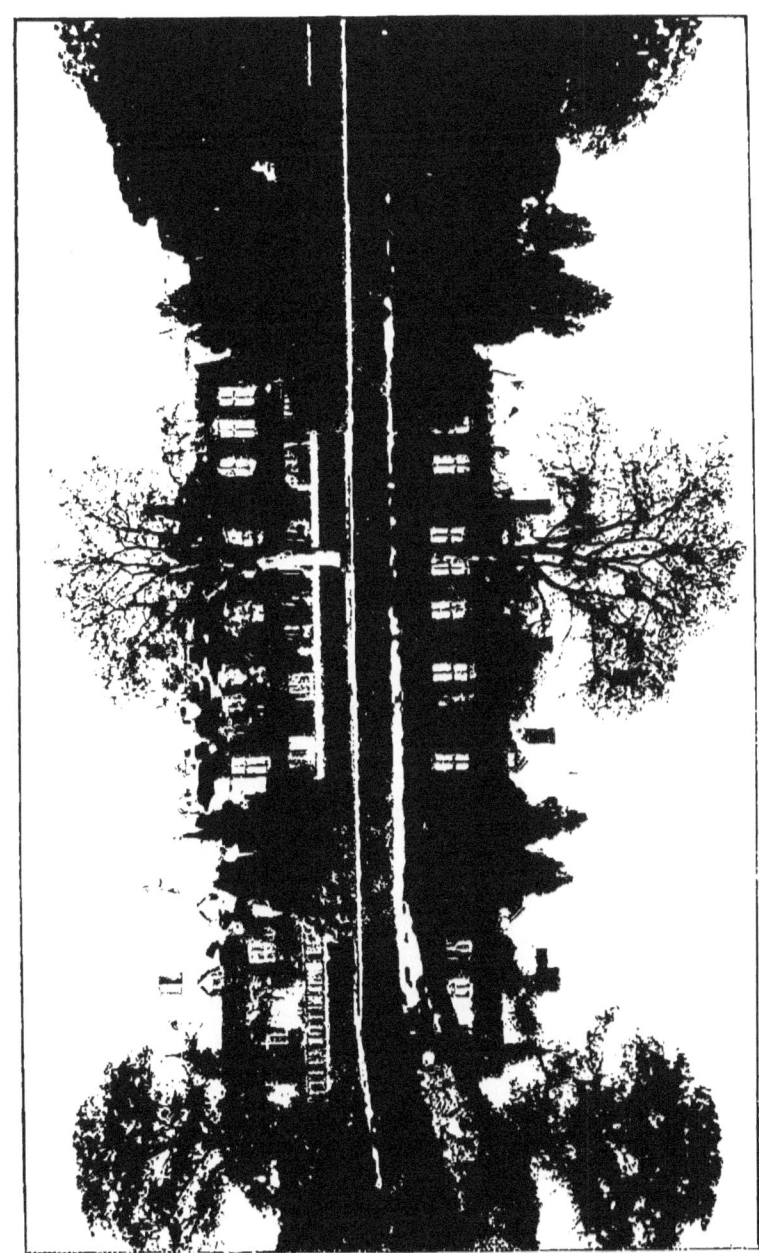

CHAPTER II.

> " *The* rose rayle*th* hire rode
> *The* lenes on *the* lyhte wode
> Waxen al wi*th* wille
> *The* mone mande*th* hire bleo
> *The* lilie is lossom to seo
> *The* fenyl and the fille."
>
> *Springtime*, MS., c. 1300.

DURING the years which succeeded the Norman Conquest, the country was constantly plunged in wars abroad and troubles at home. There could be little thought of the quiet pleasures of a garden while William I. and his sons ruled the conquered English with a rod of iron ; while Stephen was fighting for the crown against " the Empress Maud" ; while men's minds were occupied by Crusades to the Holy Land ; or while the Constitution of England was being slowly built up, and her liberties gradually secured by bloodshed and ceaseless struggles.

It was necessary, in these troublous times, for security of life and property, to live in as inaccessible a position as possible. Castles were built on the tops of hills, or protection was sought by placing the dwelling behind some river or marsh, when no high ground or escarpments of steep rocks afforded a suitable defence. This was the opposite course from that pursued by the monks, who, as a rule, chose a fertile valley in which to place their cloister, and plant their orchards, gardens and vineyards. There was no room for much garden within the glacis of a feudal castle, and as it was not safe for any of the inmates to venture beyond, it was scarcely worth while making any garden or orchard outside, merely to see it plundered by some turbulent neighbour.

But, in spite of all these disadvantages, some attempt at cultivation of fruit was not unfrequently made.

At Carlisle there must have been gardens round the town, and outside the castle walls, if the old rhyming Chronicle of the Wars in 1173 and 1174, between Henry II. and William the Lion, of Scotland, is to be believed. The supposed author, Jordan Fantosme, describes the siege of the Castle of Carlisle. The translation of one verse runs thus* :—

> "They did not lose within, I assure you I do not lie,
> As much as amounted to a silver denier.
> But they lost their fields, with all their corn
> [And] their gardens [were] ravaged by those bad people,
> And he who could not do any more injury took it into his head
> To bark the apple trees;—it was bad vengeance."†

Scattered throughout the Pipe rolls and Exchequer rolls and Liberate rolls, there are to be found a few entries which indicate some of the royal gardens in the twelfth and thirteenth centuries. In 1158-9 occur payments to the king's orchardman, "Henricus Arborarii," in London, and to the vine-dressers at Windsor and elsewhere.‡ In 1259, Henry III. made extensive alterations at the Palace of Westminster, and among payments to workmen and carpenters and others,§ occur several to labourers for "levelling the area of the garden with a roller."

In the reign of Edward I. further entries occur for keeping the garden, and for dressing the vines in the vineyard at Westminster, and of payment of the daily wage of 2¼d. to "Roger le Herberur," "formerly servant to the Lord the King Henry, the king's father." In 1276-7, we find the king paying as much as £97. 17s. 7½d. to Master Robert de Beverley, keeper of the king's woods, "for divers necessary things . . . to make

---

* Surtees Society, 1840, p. 77.
† A curious confirmation of the gardens at Carlisle even earlier, 1131, is in the Pipe Roll, 31st Henry I. (printed ed., p. 141).
Receipt from Crown lands—"William Fitz Baldwin renders account of 30s. of old farm of the king's garden of Carlisle. He delivered the same] into the Treasury—And he is quit. And the same William owes 30s. of the farm of the same garden of this year past."
‡ Pipe Roll Society. Vol. I., 1884.
§ Devon Issue Rolls of the Exchequer, 1837.

mews at Charing, and likewise to make the king's kitchen-garden there." Henry III.'s chief garden was at Woodstock, but he was not the originator of it, as there had been a garden there in the time of the second Henry. In it was the labyrinth which concealed the "Bower," made famous by the tragic fate of the "Fair Rosamond." A halo of romance and mystery hangs round this hiding place, but in reality labyrinths were by no means uncommon. There is evidence of the existence of labyrinths in very early times, and they, presumably, suggested the maze of more modern date. The first labyrinths were winding paths cut in the ground, and the survival of these is still traceable in several places in England. Of these, Saffron Walden, with its encircling ditch, is the most striking example. Camden describes one existing in his time in Dorsetshire, which went by the name of Troy Town or Julian's Bower.*

In 1250, Henry III. improved the gardens at Woodstock for his queen. Among certain works which he commanded the Bailiff of Woodstock to perform, were the following :—"To make round about the garden of our Queen two walls, good and high, so that no one may be able to enter, with a becoming and honourable herbary near our fish pond, in which the same Queen may be able to amuse herself;—and with a certain gate from the herbary which is next the chapel of Edward our son, into the aforesaid garden." † Again, on August 19th, 1252, the order was given to turf the "great herbarium." ‡ The word herbarium may simply mean a place where herbs were grown, but in this case it seems as if it were used for "herber," the old English word for arbour, which only means a shelter, or "harbour."

The same year, among other works at Clarendon the queen's "herbarium" was to be "remade and amended." §

---

* *Camden's Britannia*, by Gough, 1806. Vol. I., p. 73.

† Liberate Roll, 34 Hen. III., m. 6—Dated at Wodestok, 20 June, "cum herbario decenti et honesto prope vivarium nostrum, in quo ipsa Regina possit spaciari."

‡ Ibid., 36 Hen. III., m. 4.

§ Ibid., 36 Hen. III., July 9th, m. 6.

This looks as if it was what is usually understood by an arbour, a covered-in place. There are many descriptions of such arbours in the fourteenth century, and it was the custom to turf them. The herbarium may, however, have been a small private garden, planted with herbs, with high thick hedges. The garden at Clarendon was enclosed by a paling,* while those of Windsor† and Kennington‡ were enclosed by a ditch. In 1260 more alterations were carried out in the garden outside Windsor Castle; the gardener's house was moved, and a further wall built. During many successive reigns this garden at Windsor was kept up, and from time to time improved, and the orchard or vineyard was extended. Entries of the wages paid to the gardener and the vine dresser occur in many of the household accounts preserved in the Record Office. The gardener received 100s. a year, the labourers 2½d. a day. It is curious to note that the produce of these gardens was sold, and it seems to have been the exception when all the fruit was consumed by the king's household. In 1332 there is the following entry among the receipts—"6s. 6d. received for the fruits and herbage of the king's garden outside the Castle,"§ and other like entries occur. In "the account of Walter Hungerford, Knight, Steward of the Household of King Henry V. and Constable of the Castle of Wyndsore" ‖ (1419-22), "for any issues arising from fruits of the garden and vines of the king there in the two second years (sic) in the time of this account, he does not answer, for that the fruits of the said garden were delivered to the Household of the Lord the King there, and the grapes of the vines there were eaten by the Ladies and others of the King's Household then being there, so that the same Constable had not and could not have any profit thereof, as he says upon his oath."

---

* Liberate Roll, 37 Hen. III., m. 13.
† Ibid., 37 Hen. III., m. 17.
‡ Ibid.
§ Ministers' Accounts, Bundle 753, No. 9.
‖ Ibid. Bundle 755, No. 10.

Besides the royal gardens at Westminster, Charing, and the Tower, there were others around London. We get a glimpse of the smaller gardens belonging to the citizens, from a description of the town by FitzStephen in his life of Thomas à Becket, whose contemporary he was. The passage

GARDEN IN A TOWN, FROM FRENCH MS. LATE FIFTEENTH CENTURY.
LE RUSTICAN DES PROFITS RURAUX. P. CROISSENT. B.M. ADD. 19,720.

(translated) runs thus:—" On all sides outside the houses of the citizens who dwell in the suburbs, there are adjoining gardens planted with trees both spacious and pleasing to the sight." The only other large garden near London, not belonging to a religious house, of which there is any record, is that of Henry de Lacy, Earl of Lincoln, in Holbourne. There is an account of all the manors held by the Earl in the year 1295-6.* At all the places, lists occur of the produce sold, such as hemp, corn, beans, pulse, &c., but Holbourne appears to be the only garden of sufficient size to allow of the sale of any of its produce. At "Grante sete Manor," 7s. 4d. was paid for cutting and cultivating the vines, but at most of the other large manors, such as Thoresby and Pontefract, there is no mention of a garden at all. The Holbourne accounts are most interesting, and show the wages paid to the gardener and labourers, the number of gallons of verjuice made from the vines, and the large quantities of pears and apples sold; while other varieties, probably more choice than those grown in the garden, were purchased and sent to the Earl, and slips of apples and pears were bought to replenish the garden.

Accounts of the Possessions of Henry de Lacy, Earl of Lincoln, 23 and 24 Edward I., Holburne; William de Donyneton, Serjeant, renders his account at Holburne on the day of Saint Clement the Pope in the 25th year of the reign of King Edward before the same ʳSir William de Nony⸱ and for the same time Michaelmas 23 to Michaelmas 24 Edw. I.⸱.

Of £9 for pears, apples, and great nuts of the garden sold, the tithe being deducted. Of 2s. 3d. for cherries of the garden sold, the tithe being deducted. Of 8s. 9½d. for herbs and "Jeritis" of the garden sold, the tithe being deducted. Of 6s. for beans of the garden sold, the tithe being deducted. Of 20½d. for verjuice "in fobis," the tithe being deducted. Of 12s. 3d. for 40 gallons of verjuice of issue, the tithe being deducted. Of 3s. 2d. for roses sold, the tithe being deducted. Of 4s. 6d. for herbage of the garden, the tithe being deducted. Of 2s. 3d. for hemp of the garden, the tithe being deducted. Of 4s. 1½d. for onions and garlick sold, the tithe being deducted. Of 2s. 6d. for little plants (plancettis or plantettis?) of the vines sold. (There are also receipts for deer sold.)

Expenses—52s. 2d. in the wages and robe of the gardener by the year. And 60s. 8d. in the wages of the Serjeant by the year. And 10s. in the robe of the same. And 43s. 8d. paid to the Warden of the Gaol of Flet, for the yearly farm due to him. And 39s. 8½d. in

---

* This very fine large roll, which consists of several sheets nearly 3 ft. long, and about 15 ins. wide, is preserved at the Record Office, Duchy of Lancaster Ministers' Accounts. Bundle 1, No. 1.

3 *

the stipends of divers ⌈men⌉ working in the garden, as well for the vines as for herbs, leeks, and other curtilages ⌈and⌉ for carrying and spreading dung. And 5s. 7d. in two bushels of beans ⌈and⌉ seed of hemp, onions, and garlic bought for planting. And 22d. in the stipend⌈s⌉ of ⌈men⌉ helping to make verjuice with salt bought for the same. And 3s. 2½d. "in 2 insitis de Rule, 2 de Martin, 5 de Caloel, et 3 de pesse pucele," bought for planting. And 2s. 6d. in mending the paling of the garden. And 44s. 4½d. in one "kay" newly made for the support of the paling from the stable to the north head of the greater ditch in the garden. And 8s. 0½d. in small fish, small frogs, and eels bought for the sustenance of the pikes *(luporum aquaticorum)*. 27s. in 100 "caloels," 100 "pesse puceles," 200 "Rules," 300 "Martyns," and 300 "quoynz," bought and sent to the Earl at Ambr' (Amesbury, Wilts?) with the carriage of the same . . . . . . 17s. 0½d. in 1500 onions and ⌈ 1½ load of garlic bought and sent to Camford: and 11s. in the carriage of the same.

Many of the pears mentioned in this and other accounts appear to be of French origin. The "caloel" occurs in other places as "cailloel" for "caillou," a pebble, so called, let us hope, from its shape and not from its hardness. The "pesse," or "passe pucelle" is also evidently French. The "S. Rule" pear was probably named after St. Regolo, or Rule, who was Bishop of Arles, and first Bishop of Senlis. Rochelle, in France, was celebrated for its pears, and one year the Sheriffs of London imported some from thence to present to Henry III. Further information regarding these kinds of pears, and the prices paid for them, is to be gained from some other most interesting documents preserved in the Record Office. These papers are bills for the fruit bought for Henry III. and Edward I. at different times. The earliest is probably for the year 1223; the beginning of the document is missing, but it is dated in the seventh year of some king unnamed. From the internal evidence afforded by the names of places and dates, it appears that Henry III. is the king. He was still a minor, and his movements during the seventh year of his reign are uncertain, but the itineraries of all the other possible kings in their seventh year are known, and do not correspond with the dates in this document. The first entry is for April 20th, at "Pois," when six hundred apples, costing 12s., one hundred pears of "S. Rule," for 10s., and five hundred nuts for 2s., were brought from Paris. Henry was journeying towards England, and at each place, "Arenes," "Abeville," "Gart," and "Bolone," he was supplied with large quantities of fruit from Paris daily. On April 27th he was at Dover, and the apples, pears and nuts were still supplied daily until he reached London.* From a similar document for the year 1292-3, of which the following are extracts, the names of several other sorts of pears can be gathered.†

> Memorandum that John the Yeoman of Nicholas the Fruiterer on Tuesday next before the feast of All [Saints?] led a certain horse-load of fruit from Cambhus, where the ship . . . . to the Castle of Berwyk. First 900 "Calluewell" pears, price of the hundred 4s. [and with] the same load 500 "pas pucell," price of the hundred 2s. In panēr (paniers?) and cords 8d. In the hire of the horse and expenses of the same and of one man for four days 3s. 6d. Also on Wednesday next before the feast of Saint Edmund the king from the town of Berwyk to [the [Castle] 700 Regul' pears, price of the hundred 3s.—also 300 costard apples, price of the hundred 12d. . . . . . In porterage ½d.
> Sent to the Lord the King at Bernwell, on Monday next after Palm Sunday, 800 and a-half of Regul' pears, price of the hundred 10d. also 900 apples, price of the hundred 3d. Also 1200 "Chasteyns" [price] of the hundred 2d. In paniers and cords 6d. In the hire of one horse and expenses of the same, and of one groom going and returning 2s. 1d. sum 13s. 11d. proved. . . . . .
> Sent to the Lord the King by Stephen Mewe on Friday after the Lord's Epiphany, 1700 Regul' pears, price of the hundred 10d. Also 1400 and a-half of "Martin" pears, price of the hundred 8d. Also 700, price of the hundred 3d. . . .
> Sent to the Lord the King in the North parts, 4500 "dieyes" (or dreyes?) pears, price of the hundred 3d. also 1200 "sorell" pears. . . .
> Sent to the Lord the King at York . . . 6000 " gold knopes" pears, price of the hundred 2d. also 5000 "Chyrfoll" pears. . . .

The fruit was supplied to Edward I. at Newcastle, York, Pontefract, Berwick, and various places in the North. This date was the commencement of the wars with Scotland, at the time of Bruce and Baliol, when Edward held his parliament at

---

* "Item le VJme iour de May a la Tour de Londres pour ii c. de poumes ii s. esterlins et pour i c. de poires ii s. esterlins et pour iii c. de nois vi d. esterlins."—*Exchequer Q. R. Miscellanea,* 4/3.

† Extract from Exchequer. Treasury of the Receipt Miscellanea, 4/3. 20-21 Edward I.

Newcastle, and then at Berwick. It is curious to think that such great events should be the means of revealing the names of the best known pears of the period. We still find most of the S. Rule or " Regul pears," as they are written in this account, and they are bought in quantities, as in the earlier bills, the cost being usually 3s. per hundred, but sometimes only 10d. for the same amount. The pears which come next after the " Regul," in the frequency of the entries and quantities, are the "Calluewell" or " Calwell," and the " pas pucell " or " pase pucell," and we also find " Martins "; all these four sorts being also found in the Earl of Lincoln's accounts, the prices varying from 4s. to 8d. per hundred. Besides these. there occur " Dieyes " (or dreyes), " sorell," " chyrfoll," and " gold knopes " pears—also apples, quinces. called " coynes," chestnuts, " chasteynes," and " great nuts." The only kind of apple specially noticed is the Costard. The name of this variety, which was the most popular of apples for many centuries, has been preserved in the word " costermonger," originally a seller of this fruit. At Oxford, in 1296, the Costard apple was sold for 1s. per hundred, and the price of twenty-nine Costard apple-trees, in 1325, was 3s.*
It is spoken of by early writers as a distinct fruit, in the same way as Wardons and pears. Grosseteste mentions them as " apples and Costards." † Another popular variety of apple was the Pearmain. At an early date we find it being used for cider. In the sixth year of King John a certain Robert de Evermere held the lordship of Runham in the Hundred of East Flegg, in Norfolk, by petty serjeanty, by the payment of two hundred Pearmains and four hogsheads (modios) of wine, made of Pearmains, into the Exchequer, on the feast of St. Michael yearly.‡ These were still being paid annually in the ninth year of Edward II. One other kind of pear, the "Janettar," is noted in one of the Wardrobe accounts in the thirty-sixth year of the reign of Henry III., being bought

* Thorold Rogers, *Hist. of Agricultural Prices.*
† Sloane MS. 686. " Tretyse off Housbandry that Mayster Groshede made."
‡ Blomefield, *Hist. of Norfolk.* Vol. V., p. 1378. Ed. 1775.

with "sorells" and "cailloels" from "John the Fruiterer of London."*

Besides these fruits, which appear to have been common,† there were a few choicer sorts, such as cherries, mulberries, medlars, and even peaches. If proof were needed that this latter fruit was to be had in England, we have it in the fact that King John, at Newark, in the midst of his despair and disappointment, hastened his end by a surfeit of peaches and ale.‡

The various accounts which have been quoted, although tedious, there being so much sameness in them, are nearly the only trustworthy sources of information about the fruits and gardens of this period. To supply such large quantities of fruit, there must have been extensive orchards. For although a few may have come from abroad, it is impossible to imagine that the fruiterer to the king procured the thousands of apples and pears required for his royal master, from France. By the early part of the fourteenth century, many fine and old-established gardens and orchards must have existed in this country, and were being cultivated, not by the religious orders only, but under many secular owners of land. Gardens were being made around the various colleges at Oxford and Cambridge then coming into existence. Trinity Hall, Cambridge, had a good garden, with vines and a "herbaria," within a short time of its foundation, and Peterhouse, a few years earlier. The gardens round London have already been noticed; something further about them might be gained by searching old leases. The following sample gives some idea of the number of gardens in one part of London. It is a lease, dated 1373,§ for "A garden situate in Tower Ward, near the city wall, which John Seoh lately held: being between the garden which Geoffery Puppe holds on the North side, and the garden which William Lambourne holds on the South." There is no better proof of this great increase, than

* Exchequer Q. R. Ancient Miscell. Wardrobe and Household Account, 4/5. R. O.
† Pipe Roll (printed 1884. Vol. I.), 5th Henry II., 3s. for chestnuts (castaneâr) sent to the Queen at Salisbury.
‡ *Chronicle of Roger of Wendover.*
§ Letter Book. H. F. XIII. 40 Ed. III.

a discussion which took place between the gardeners in and near London and the Lord Mayor with regard to the locality in which they were allowed to sell the produce of their gardens.

It appears that for many years previous to 1345 the gardeners of the earls, barons, bishops, and citizens of London were accustomed to sell their "pulse, cherries, vegetables, and other wares to their trade pertaining," on a piece of ground "opposite to the church of S. Austin near the gate of S. Paul's churchyard." By 1345, however, this fruit and vegetable market had grown to such an extent, and had become so crowded as to hinder "persons passing both on foot and on horseback," and the "scurrility, clamour, and nuisance of the gardeners and their servants" had become so obnoxious "to the people dwelling in the houses of reputable persons there," and "such a nuisance to the priests who are singing matins and mass in the church of S. Austin, and to others, both clerks and laymen, in prayers and orisons there serving God," that the mayor and aldermen were petitioned to interfere, and to remove the market to some more suitable place. The result of this petition was a meeting of the mayor and aldermen, and an order "given to the said gardeners and their servants, that they should no longer expose their wares aforesaid, for sale in that place, on peril which awaits the same." But the gardeners were not to be so easily defeated. They, in their turn, petitioned the mayor to reverse his sentence, and their petition runs thus:—"Unto the Mayor of London, shew and pray the gardeners of the earls, barons, and bishops, and of the citizens of the same city, may it please you, sire, seeing that you are the chief guardian of the said city, and of the ancient usages therein established, to suffer and to maintain that the said gardeners may stand in peace in the same place where they have been wont in times of old, in front of the church of S. Austin, at the side of the gate of S. Paul's churchyard, in London, there to sell the garden produce of their said masters, and make their profits as heretofore they have been wont to do, seeing that they have heretofore been in the said place unmolested, and that as they assert they cannot serve the commonalty, nor yet their masters, as they were wont to do. As to which they pray for redress." But the mayor would not give way at first, though it appears that he afterwards held "a conference between his aldermen," at which it was agreed that "all the gardeners of the city, as well aliens as freemen, who sell their pulse, cherries, vegetables, and other wares aforesaid in the city, should have as their place the space between the south gate of the churchyard of S. Austin's, and the garden wall of the Friar's Preachers at Baynard's Castle, in the same city, that so they should sell their wares aforesaid in the place by the said mayor and aldermen thus appointed for them, and nowhere else." *

* Letter Book F, fol. cxi, of the Guildhall, and Riley's *Memorials of London Life*.

CHAPTER III.

> "And in the gardin at the sonne uprist
> She walketh up and down wher as hire list
> She gathereth floures, party whyte and reede
> To make a sotil gerland for hire heede."
> CHAUCER, *Knight's Tale.*

GREAT changes were taking place in England during the latter half of the fourteenth, and beginning of the following century. Trades and industries increased, and in like manner horticulture revived. During the years which had passed since the Norman Conquest, the conquerors and conquered had become welded into one nation, and this had not been effected peacefully. But we now come to a period when the battles were being fought on foreign soil, while the nation was enjoying comparative peace at home. In the country itself, the poorer sections of the community were gradually asserting their rights against the lords of the soil. There was a class growing up, of farmers who farmed lands, merely paying some yearly tribute in service, or in kind, to their overlord. Round these small farms and manors, gardens and orchards were planted, and thus it can be seen how such movements would affect the progress of gardening.

From incidental references in writings of the time it appears that the poorer classes chiefly lived on vegetables, as the following quotations from Langland serve to show :

> "Alle the pore peple pesecoddes fetten *
> Benes and baken apples thei brou3te in her lappes
> Chibolles and cheruelles and ripe chiries manye." †

Again, he says, the poor folk

> "With grene poret and pesen to poysonn hunger thei thought." ‡

---

* Fetch. † *Piers Ploughman.* ‡ *Ibid.*

Also " Two loves of benes and bran
            Y baked for my children." *

In picturing the utter destitution of the patient Griseldis, Chaucer lays stress on the fact that she was dependent on vegetables for food, and being without a garden, had resort to the wayside herbs :—

> " Whan she homward cam she wolde bringe
> Wortes or othere herbes tyme ofte
> The which she shredde and seeth for her livinge." †

At the beginning of this period there was great distress, as the country was swept by a scourge worse than war, the fearful plague known as the Black Death. As if to add to the horrors of the time, and the sufferings of the survivors, there were bad seasons, and many crops failed. Even what harvest there was, could not be gathered, labourers were so scarce. Doubtless many orchards and gardens suffered much from the neglect of those years. But in spite of this, they were increasing, and by the end of the fourteenth century, every small manor and farm could boast of a garden. For "that londe bereth fruyt & corn good ynoughe, that londe is well at ease as longe as men lyue in peas." ‡ This was certainly true, for while men lived in comparative peace, there was a revival in gardening and husbandry. This progress was again checked by the Wars of the Roses; and the next step in advance did not come till the restoration of peace in Tudor times.

In the Middle Ages, what we should now call the kitchen garden, was in most cases the only one attached to a house. The idea of a garden, solely for beauty and pleasure, was quite a secondary consideration. In early cookery-books, various recipes for serving up vegetables are given, though only a few of these dishes are vegetables cooked alone. But the wealthy, who could afford to get all the ingredients of these many recipes, had so much meat, and such an immense variety of game, cranes, herons, curlews, and other birds, besides those still in use, that

---

* *Piers Ploughman*.
† *Clerk's Tale*.
‡ Trevisa, description of Britain in his translation of Higden's *Polychronicon*, cir. 1387 (printed by Caxton, 1482).

they did not care for vegetables served separately, in any quantities, except on fast days. Gardens had chiefly to supply herbs for stuffing and flavouring, and these were freely used. For example, the first recipe in one book* is for cooking a "hare in Wortes." It begins, "Take colys, and stripe hem faire from the stalkes, take Betus and Borage, auens, Violette Malvis, parsele, betayn, pacience the white of the lekes and the croppe of the netle; parboile, presse out the water, hew hem small. And do thereto mele," and so on. Onions, leeks, and garlick were very largely used. Such mixtures as meat or fish cooked with pears or apples, spices and sugar, and to which, leeks ground small, porrettes minced, whole onions or garlick sauce is added, are by no means uncommon. The Sompnour, among Chaucer's *Canterbury Pilgrims*, is a type of the class among whom this taste prevailed,

"Wel lovede he garleek oynouns, and ek leekes."

All strongly flavoured herbs were popular in cooking, and every garden contained a good assortment. Fennel was one in very general use, and both the green leaves and also the seeds were eaten. As much as eight and a-half pounds of fennel seed were bought for the king's household for one month's supply.† And with the poor folk, it was chiefly used to relieve the pangs of hunger on fasting days. In *Piers Ploughman*, a priest asks a poor woman,

"Hast thou ought in thy purs?" quod he,
"Any hote spices?"
"I have peper and piones," ‡ quod she, "and a pounde garlike,
A ferthyngworth of fenel seed, for fastyng dayes."

In an old medical MS.,§ it is said of this plant,

"Fenel is erbe precyows,
* * *
Good in his sed so is his rote,
And to many thyngys bote.‖
* * *

---

* *Harl. MS.* 4016, c. 1450. Printed Early Eng. Text Soc. Ed. by T. Austin.
† Wardrobe Acc., Edward I., 1281. ‡ Peonies.
§ Fourteenth century MS. preserved in the Royal Library, Stockholm. Extracts *Archæologia*, Vol. XXX.
‖ Good.

>   Fenel in potage *and* in mete
>   Is good to done whāne yᵘ schalt etc,
>   All grene loke it be corwyn * small
>   In what mete yᵘ vsyn schall."

Mint was often used with fennel in sauces. Chaucer mentions them growing together:—

>   "Then went I forth on my right hond
>   Downe by a litel path I fond
>   Of Mintes full and Fennell greene."—*Romaunt of the Rose.*

Parsley is, perhaps, still more common than either of these. In the earliest English gardening treatise,† a section of the short poem is devoted to parsley, and the instructions for its culture are quite correct:—

>   " Percell kynde ys for to be
>   To be sow yn *the* monthe of mars so mote y the
>   He will grow long and thykke
>   And euer as he growyth *thu* schalt hym kytte
>   *Th*u may hym kytte by reson'
>   *Th*ryes yn one seson '
>   Wurtys to make and sewes ‡ also.
>   Let hym neuer to hye go
>   To lete hym grow to hye hit is grete foly.
>   *   *   *
>   Thay *th*at the sede schal bere the
>   Kytte hym nou3t but lete hym be."

The same practical poet, John Gardener, also gives directions for the planting of onions, garlick, and leeks. They were to be sown on St. Valentine's day, as they are " herbys vn-meke," or what we call " hardy." The onion plants which were required for seed, were to be sown in April or March, and when the heads began to grow tall, they were supported by ash-sticks :—

>   " Forkys y made of asche-tre
>   That none of hem downe nou3t fall
>   *   *   *
>   When they rype they wyl schow
>   And by the bollys thu schalt hem know
>   The sede wt͡ith͡yn wul schewe blake
>   Then thu schalt hem vp take
>   They wul be rype at the full
>   At lammasse of Peter Apostull."

---

* Cut-up.
† MS. Trinity College, Cambridge. Printed in *Archæologia*, 1894.
‡ Sauces.

Saffron was used in cookery in astonishing quantities, and the price paid for it was very high, from ten to twenty shillings per pound. It was chiefly grown in the Eastern Counties. Walsingham, in Norfolk, was famous for its saffron in early times, and the plant gave its name to the town in Essex, Saffron Walden. The beds of saffron required considerable care. John Gardener says the "Beddys" must be "y-made wel wyth dyng, For sothe yf thay schal bere." The bulbs, he goes on to say, must be set with "a dybbyl," and planted three inches deep.

> "Thay wold be sette yn the moneth of September
> Three days by-fore seynt mary day natyuyte."

Among the other herbs of the garden, cabbages, or kale, held a foremost place. They are spoken of as "caboges," "cabochis," "caul," or "kole-plantes," and sometimes "wurtes," or "wortes," stands for cabbages.* John Gardener speaks of "wortys" in that sense:—

> "How he schall hys sedys sowe
> Of euery moneth he most knowe
> Bothe of wortys and of leke
> Ownyns and of garleke
> Percely clarey and eke sage
> And all other herbage."

He devotes a paragraph of twenty-five lines to the culture of these "wortys." He says they could be had all times of the year by a careful succession of sowings.

> "Euery moneth hath his name
> To set and sow w'ouȝt eny blame
> May for somer ys al the best
> July for cruyst† ys the nexst
> Novembr' for wynter mote the thyrde be
> Mars for lent so mote y the‡
>   *   *   *
> And so fro moneth to moneth
> Thu schalt bryng 'thy wurtys forthe."

In fifteenth century cookery books we find recipes for cabbages, both in "potage" or dressed with marrow, gruel, and saffron. In the lists of great banquets which have been

---

* "Brassica . . . wortes aut cole aut colewortes."— Turner's *Libellus*, 1538.
† = *harvest*.   ‡ = *so may I thrive*.

preserved, such dressed vegetables rarely, if ever, occur. At the third course of a banquet on the occasion of Henry the Fourth's marriage, "pescodde" and "strawberry" were among the dishes, but this is almost a solitary instance among bills of fare of that date.* Cabbages were, from the earliest times, grown in this country, but it may be some improved variety which is referred to in the following passage†:—" Sir Anthony Ashley, of Wimborne St. Giles, Dorset, first planted cabbages in this country, and a cabbage at his feet appears on his monument." The tomb is to be seen in the church to this day, dated 1627.

There was both a good variety and a fair supply of fruit in the fourteenth and fifteenth centuries. Several new kinds of apple and pear are mentioned by the poets of the day, and must have been well known. Lydgate speaks of the Pomewater,‡ Ricardon, Blaundrelle, and Queening apples. Gower of another kind, the Bitter-sweet :—

> " For all such time of love is lore
> And like unto the bitter-swete
> For though it think a man fyrst swete
> He shall well felen at laste
> That it is sower." §— *Confessio Amantis.*

In the "Miller's Tale," Chaucer incidentally alludes to the old custom of storing apples,—

> " Hire mouthe was swete as . . . . . .
> . . . hord of apples, laid in hay or hethe."

He gives us the name of a pear, evidently newly introduced, in the same description,—

> " She was wel more blisful on to see
> Than is the newe perjonete tree."

Wardons were still the most popular of cooking varieties. In recipes for dressing pears, the wardon is usually intended ; as,‖

---

* *Two Fifteenth Century Cookery Books.* By T. Austin. E. E. Text Soc.
† Isaac D'Israeli, *Curiosities of Literature.*
‡ Shakespeare—*Love's Labour's Lost.*—" Ripe as a Pomewater, who hangeth now like a jewel in the ear of cœlo."
§ *Romeo and Juliet.*—"Thy wit is a very bitter sweeting, it is a most sharp sauce."
‖ Harl. MS. 4016. E. E. Text Soc.

"Peris in Syrippe. Take Wardons, and cast hem in a fair potte," or, "Peris in Compost. Take pere Wardons and pare hem." At Henry the Fourth's wedding-feast, these pears in syrup occur twice, and are included in the same course as venison, quails, sturgeon, fieldfare, &c. At the coronation feast of the same king, we find quinces in "comfyte," and also "Pomedorreing" or golden apple, supposed in this case, to stand for oranges, as this rare fruit might be obtained for such a great occasion. Oranges probably were occasionally brought to this country at an even earlier date. It is said that in the eighteenth year of Edward I., the Queen bought, out of the cargo of a large Spanish ship, one frail of figs, one of raisins, one bale of dates, two hundred and thirty pomegranates, fifteen citrons and seven oranges.*

Cherries were cultivated very extensively. The season of gathering them is spoken of by Langland as "cherry-time." This cherry-harvest, coming at the height of summer, was a time of merry-making, and to it Gower compares the brief length of human life, which

> ". . . endureth but a throw
> Right as it were a cherry feast."—*Confessio Amantis.*

And Lydgate also uses the cherry-fair as a simile:—

> "This world is but a cherry fair."

Cherries and strawberries were hawked in the streets of London, and the cry of "ripe strawberries" was familiar even in Lydgate's time.

> "Then vnto London I dyd me hye
> Of all the land it beareth the pryse
> 'Hot pescodes' one began to crye
> 'Strabery rype' and 'cherryes in the ryse' †
> One bade me come nere and by some spyse
> Peper and safforne they gan me bede
> But for lack of mony I myght not spede."—*London Lyckpeny.*

Peaches are mentioned by Lydgate among "the fruits which more common be," but only inferior varieties were in cultivation.

---

* *Manners and Household Expenses.* Ed. Beriah Botfield. Roxburghe Club, 1841.

† = *branch, twig.*

Medlars were also grown, and were kept before being eaten, as is still the fashion. In the *Prologue to the Reeve's Tale*, Chaucer refers to this custom, speaking of the old age of the Reeve:—

> "But if I fare as doth an open-ers *
> That ilke fruit is ever lenger the wers
> Til it be roten in mullock † or in stre."†

In the description of the garden and arbour in *The Flower and the Leaf*, a medlar tree in full bloom, that "to the herber side was joyning," is thus picturesquely described:—

> "And as I stood and cast aside mine eie
> I was ware of the fairest medlar tree,
> That ever yet in all my life I sie.
> As full of blossomes as it might be.
> Therein a goldfinch leaping pretile
> Fro bough to bough; and as him list, he eet
> Here and there of buds and floures sweet."

Plums are not often mentioned, either by the poets, or in old accounts, but we know that both damsons and bullaces were grown in this country, though probably in no great quantities. In the *Romaunt of the Rose* Chaucer classes them among homely fruits:—

> "And many hoomely trees there were
> That peches, coynes,‡ and apples bere
> Medlers, ploumes, peres, chesteynis,
> Cheryse, of which many one fayne is,
> Notes, aleys and bolas
> That for to seen it was solas." (l. 1373.)

Gardeners of this date paid great attention to grafting. The art of grafting a pear on a hawthorn was known at a very early period. John Gardener directs the stocks for grafts of both apples and pears to be planted in January, the apple on an apple stock, and the pear "a-poñ a haw-thorne." The grafting, he says, should take place any time between September and April.

> "Wyth a saw thou schalt the tre kytte
> And *with* a knyfe smowth make hytte
> Klene a-tweyne the stok of the tre
> Where-yn that thy graffe schall be
> Make thy Kyttyng' of thy graffe
> By-twyne the newe & the olde staffe."

---

* = *medlars.*   † = *rubbish and straw.*   ‡ = *quinces.*

Clay had to be laid on the stock, "to kepe the rayne owte," and moss bound over the clay with "a wyth of haseltree rynde." Most of the early writers on gardening and husbandry devote a large share of their treatises to grafting, and various experiments to change the colour or flavour of the fruits were made. Robert Salle is quoted as an authority on grafting in the fifteenth century.* He says:—"Yf thou wilt make thyn apples reede, take the *graffe* of an appel tree and *graffe* hit on a stok of elme or aldyr and hit shall ber' reede apples." "Make an hole w^t a wymbyll' in a tree and what colour *thu* wilt distempre hit wi*th* wa*ter* and put hit in at the hole and the fruit shal be of the same colo*ur*." †

It was considered the most essential part of a husbandman's

GRAFTING. FROM THE ARTE OF PLANTING AND GRAFTING,
BY LEONARD MASCALL. EDITION 1592.

education, that he should be well-skilled in grafting, as the following lines, though of later date, so well describe:—" It is necessarye, profytable, and also a pleasure, to a housbande, to have peares, wardens, and apples of dyuerse sortes. And also cheryes, filberdes, bulleys, dampsons, plummes, walnuttes, and suche other. And therefore it is convenyent to lerne howe thou shalte graffe." ‡

Gardens of this date were usually square enclosures, bounded either by walls of stone, brick, or daub, or by thick hedges.

---

* Sloane MS., 122.
† The same recipes are also given in the Porkington Treatise printed for the Warton Club, 1855, ed. by Halliwell.
‡ *Book of Husbandry*. By Fitz Herbert, 1544. Ed. Skeat, 1882.

There were generally two entrances to them; one, a door opening from the house, the other giving access from the garden into the orchard or meadow. If high hedges and walls were retained in later times on account of their beauty or shelter, it was certainly with a view to security that they were originally adopted.

TURFED SEAT IN A GARDEN WALL. FROM ROMAN DE LA ROSE, FLEMISH MS. LATE FIFTEENTH CENTURY. B.M. HARL., 4425.

> "I saw a garden right anoon
> Full long and broad and everidele
> Enclosed was and walled wele
> With hie walles embatailed." *

* Chaucer, *Romaunt of the Rose*, l. 130.

Within the enclosure all was trim and neat. All round against the wall a bank of earth was thrown up, the front of which was faced with brick or stone, and the mould planted with sweet-smelling herbs. At intervals there were recesses with seats or benches covered with turf, "theck yset and soft as any velvet." Low mounds of earth were also made here and there, in the garden, "on which one might sit and rest," and these "benches" were also "turved with newe turves grene." The little paths throughout the garden were covered with sand or gravel, and kept free from weeds. Lydgate mentions a garden, in which "all the alleys were made playne with sand." *

No garden was considered complete without its arbour, its "privy playing place." They were either set in a nook in the wall, or in a part of the garden sheltered by a thick hedge. The arbour, or "herber," was made of trees thickly intertwined with climbing plants, to screen them from the eyes of the intruder. One is thus described in *The Flower and the Leaf*:—

"And at the last a path of little brede
I found, that greatly had not used be,
For it forgrowen was with grasse and weede,
That well-unneth † a wighte might it se:

Thought I, this path some whidar goth, parde,
And so I followed, till it me brought
To right a pleasaunt herber well y wrought."

That benched was and with turfes new
Freshly turved, whereof the grene gras,
So small, so thicke, so short, so fresh of hew,

That most like unto green wool wot I it was:
The hegge also that yede in compas ‡
And closed in all the greene herbere
With sicamour § was set and eglatere.

And shapen was this herber roofe and all
As a pretty parlour: and also
The hegge as thicke as a castle wall,

---

* *The Chorle and the Bird.* † = scarcely, hardly.
‡ = went round it. § = honeysuckle.

FOURTEENTH AND FIFTEENTH CENTURIES. 53

Then who that list without to stond or go
Though he would all day prien to and fro
He should not see if there were any wight within or no."

We see this same idea of seclusion, as the essential feature of

ARBOUR, FROM THE SAME MS. HARL., 4425.

an arbour, in the fifteenth-century poem, "La Belle Dame Sans Merci."*

"And sett me down by-hynde a traile
Fulle of levis, to see, a grete mervaile,

---

* E. E. Text Society, Vol. IV.

> With grene wythyes y bounden wonderlye
> The leeves wore so thicke with-out faile
> That thorough-oute myghte no mann me espye."

The flowers around an arbour are described in a fourteenth-century poem, entitled "The Pearl":—

> "I entered in that erber grene
> In augoste in a high seysoun
>   *   *   *
> Schadowed this worte; ful schyre * and schene
> Gilofre,† gyngure‡ & groomylyon §
> & pyonys powdered ay betwene."

Each garden contained some kind of cistern for water, and in many cases a fountain elaborately ornamented was placed in the centre, or in some conspicuous position. The illustration shows the ordinary fountain of a good garden of the day, introduced to represent Rebecca's well, and many good representations of such fountains are to be found in fifteenth century MSS. ǁ

The varieties of flowers planted in these gardens were not very numerous, but those few kinds grew in great profusion:—

> "Ther sprang the violete al newe,
> And fresshe pervinke riche of hewe,
> And floures yelowe, whyte and rede:
> Swich plentee grew ther never in mede,
> Ful gay was al the ground, and queynt
> And poudred, as men had it peynt.
> With many a fresh and sondry flour,
> That casten up a ful good savour." ¶

The periwinkle, or parwinke, was a general favourite. It was a plant well suited to cover and brighten the ground in the shady corners of the garden, and thus gained the appropriate name of "Joy of the ground."

> "Parwynke is an erbe grene of colour
> In tyme of May he beryth blo flour
>   *   *   *

---

* = bright.   † = clove-pinks.
‡ = tansy.   § = gromwell.
  See B. M. 14. E. 2. f. 77. &c.
¶ Chaucer, Romaunt of the Rose. l. 1431.

FOUNTAIN, FROM AN ENGLISH MS. "SPECULUM." C. 1450. B.M. 2838.

prowesse," the periwinkle being then used to typify excellence, in the same way as the pink in Elizabethan times, "The very pink of courtesy."

Among yellow flowers in the same garden, the marigold or gold, as it is called in old writers, would be conspicuous:

* = nearly. † Medical MS., Stockholm. *Archaeologia*, Vol. XXX.

"Golde is bitter in savour,
Fayr *and* 3clu is his flower
Ye golde flour is good to scene." *

Jealousy is described by Chaucer as decked with these flowers. "Jealousy that werede of yelwe guldes a garland."

Violets, as we learn from the former authority, were "herbs well cowth."† They were grown not only for their sweet fragrance, but also as salad herbs, and "Flowers of violets" were eaten raw, with onions and lettuce. Among the ingredients for a kind of broth they are mentioned with fennel and savoury.‡ They were also used to garnish dishes. In an old recipe for a pudding called "mon amy," the cook is directed to "plant it with flowers of violets, and serve it forth."§ In another MS. a recipe for a dish called "vyolette" is given. "Take flowrys of vyolet boyle hem, presse hem bray hem smal." This is to be mixed with milk, "floure of rys," sugar or honey, and "coloured" with violets. Not only were violets cooked, but hawthorn, primroses, and even roses, shared the same fate, and were treated in the same way. One recipe, called "rede rose," is simply, "Take the same sauc a-lye it with the yolkys of eyroun and forther-more as vyolet." The rose hips were also used, and in a dainty dish called "sauc saracen," "hippes" were the chief ingredient. It cannot have enhanced the beauty or poetry of such flowers, to feel that they were commonly cooked and eaten.

After this shock to sentiment, we are glad to find the rose still valued for its loveliness and perfume. Although a rosery

---

* Medical MS., Stockholm. *Archæologia*, Vol. XXX.
† = *known*.
‡ *Form of Cury*.
§ The following is the recipe of this excellent dish:—"Take thick creme of cowe-mylke, and boyle hit over the fire and then take hit up and set hit on the side:—and then take swete cowe cruddes and presse out the qway (*whey*), and bray hom in a mortar and cast hom into the same creme and boyle altogether—and put thereto sugre and saffron, and May butter—and take yolkes of eyren streyned, and betten, and in the settynge donne of the pot bete in the yolkes thereto, & stere it wel, & make the potage stondynge: and dresse five or seaven leches (*slices of bread*) in a dish, and plant with floures of violet and serve hit forthe."

of to-day would astonish the possessors of gardens in the Middle
Ages, and the varied forms and colours would bewilder them,
yet in some of our finest-looking roses they would miss, what
to them was the essential characteristic of a rose, its sweet
scent! Nothing more readily than the subtle fragrance of a rose
can conjure in our minds a dream of summer, and many a one
since the days of Chaucer has experienced what the poet felt
when, approaching a rose-garden, he exclaimed:

> "The savour of the roses swote
> Me smote right to the herte rote,"

or when crowns of roses and lilies perfume the air,

> "The swete smel, that in myn herte I find
> Hath changed me al in another kind."

There were both red and white double roses, as well as the
single, and the common dog-rose and sweetbriar. They were
planted along the walls, or singly, here and there in the garden,
or clambering over the arbour. The double-red (a variety of
Rosa Gallica) was the most prized, and as if this red rose was
the most lovely thing that could be imagined, it is thus brought
into an "Ave Maria" of the early fifteenth century:—

> "Heil be thou, Marie, that art flour of alle
> As roose in eerbir so reed!" *

Chaucer praises the buds of the double rose, which are more
lasting than the quickly-falling petals of the single kinds:

> "I love wel sweitie roses rede;
> For brode roses, and open also,
> Ben passed in a day or two;
> But knoppes † wilen fresshe be
> Two dayes atte leest or three."

When the red or white rose became the badge of two contending
parties, it doubtless depended on the side taken by the owner of
the garden which colour prevailed therein. The "fresh redde
rose newe, against the sommer sunne," ‡ or the "white rose of
England" that is frishe and wol not fade. Both the rote & the

---

* Early Eng. Text Soc.   † = *buds*.
‡ *Assembly of Fowles*. By Chaucer.

stalke that is of great honoure."* Roses were the commonest of all flowers, for weaving into wreaths and garlands:—

> "And on hire hed ful semely for to see
> A rose gerlond fresh and wel smelling." †

> "And also on his head was sette
> Of roses redde a chapelette." ‡

The periwinkle, with trailing leaves, was suitable for wreaths, and many other flowers were used. Emely in her garden gathered "floures, party whyte and reede, to make a sotil gerland for hire heede." § But these pretty chaplets of flowers were not only worn by beautiful maidens; we find even the far from prepossessing sompnour, among the Canterbury pilgrims, had "a garland set upon his heed." The annual rendering of a red rose was a common kind of " quit rent," also a flower or seed of the clove pink, or gilliflower, ‖ was frequently the payment. The lily ranked next to the rose in importance, in a garden,¶ and vied with the rose for a share in the poet's song. The white lily (*Lilium candidum*) served to typify all that was good, and pure, or beautiful.

> "First wol I you the name of Seinte Cecilie
> Expoune, as men may in hire storie see:
> It is to sayn in English, Heven's lilie." **

> "That Emelie, that fairer was to seene
> Than is the lilie or hire stalke grene." ††

> "Upon his hand he bore for his delyt
> An eagle tame, as any lily whyte." ‡‡

---

* Political poem, 1400-71.—*Early Eng. Text Soc.*, Vol. IV.
† *Knight's Tale.* ‡ *Romaunt of the Rose.*
§ *Knight's Tale.*
‖ Among the receipts of Bicester Abbey, 19th Rich. II., for lands and tenements " una rosa rubea recept' di Henrico Bowols de Curtlyngton . . . et de uno g'no gariophili rec' de Rog' o de Stodele " . . . &c.—DUNKIN, *Hist. of Bullington and Ploughley.*
¶ " Lillys " and " roses " are the only flowers mentioned on the gardeners' rolls of Norwich Priory.
** *The Second Nonne's Tale.*
†† *Knight's Tale.*
‡‡ *Ibid.*

The yellow flag and purple iris are sometimes indiscriminately spoken of as lilies. In the old medical MS. already referred to, the lilie "that waxit in ʒerdis" is described as white as any milk, and the three other kinds of the field and wood, were yellow, "like saffron," and one "blue purple"; but these are also spoken of as "gladdon" and "greos." Other flowers were brought in from the fields and woods, and perhaps improved by cultivation. The geranium of the flower garden in the Middle Ages was the wild cranesbill, or small herb Robert. The wild scabious and poppy were in the place of the showy annuals and biennials of our gardens of to-day. But many indigenous plants would make no mean show, such as cowslips, daffodils, primroses, foxglove, mullein, St. John's worts, gentian, oxalis, mallow, corncockle, yarrow, campion, centaury, or honeysuckle, all of which we know were grown. There were corners, too, where a peony or tall hollyhock or monkshood flowered, or shaded nook filled with the glossy leaves of the hartstongue, or a portion of the long bed was made bright with pinks and columbines, or sweetly scented with lavender, rosemary, or thyme. In describing the flowers of a garden in Chaucer's time, we must not forget what he called

> "The daysie or elles the eye of day
> The emperise and flour of floures alle."

It found its way into the trimmest gardens; the greenswards and arbours were "powdered" with daisies. To quote Chaucer again:—

> "Home to my house full swiftly I me sped
> To gone to rest, and early for to rise
> To scene this floure to sprede, as I devise
> And in a little herber that I have
> That benched was on turves fresh y grave
> I bad me shoulde me my couche make."

Though a daisy plant is supposed to spoil the most velvety turf, yet none would see it banished from our gardens, and all agree in loving the little flower with the poet who said,

> "Si douce est la Marguerite."

The gardens that were described by Chaucer, although intended for ideal ones, were no doubt but faithful pictures of the gardens of his day, seen through his poet's eye. The garden, "ful of braunches grene," in which Emely was walking when she was watched by the imprisoned knights, was such as might be seen beneath many a feudal castle wall.

GARDEN, FROM FLEMISH MS. OF THE ROMAN DE LA ROSE, LATE FIFTEENTH CENTURY. HARL. 4425.

"The grete tour, that was so thikke and strong,
Which of the castel was the cheef dongeoun
    *   *   *   *
Was evene joynant to the gardyn wal."

We have in history a counterpart of this garden of romance, that of Windsor Castle. When James I. of Scotland was there, in captivity, his solace was writing verse, and he has left us this most charming picture of the garden beneath his prison window:

> "Now was there made, fast by the Towris wall,
> A garden fair;—and in the corners set
> An arbour green, with wandis long and small
> Railèd about, and so with treès set
> Was all the place, and Hawthorne hedges knet.
> That lyf was none walking there forbye
> That might within scarce any wight espy.
>
> "So thick the boughes and the leaves green
> Beshaded all the alleys that there were,
> And mids of every arbour might be seen
> The sharpe greene sweet Juniper
> Growing so fair with branches here and there,
> That as it seemed to a lyf without,
> The boughes spread the arbour all about.
>
> "And on the smalle greene twistis sat
> The little sweet nightingale, and sung
> So loud and clear, the hymnis consecrat
> Of loris use, now soft, now lowd, among,
> That all the gardens and the wallis rung
> Right of their song."

CHAPTER IV.

" And all was walled *that* wone *thout* it wid were
With posterns in pryuytie to pasen when hem list
Orchelardes and erberes eused well clene."
Pierce the Ploughman's Crede, c. 1394.

BEFORE proceeding any further with the history of gardening, it will be as well to pass in review the literature on the subject relating to the periods which have been traversed. The knowledge of herbs and flowers in Saxon times, and for several centuries later, was all learnt from classical authors. The works of Theophrastus, Dioscorides, Galen, Pliny, and Apuleius, formed the basis of Saxon plant-lore. The *Herbarium* of Apuleius (who lived about the fourth century, A.D.) was founded on the works of Dioscorides and Pliny, and it is chiefly through Apuleius that these earlier writers were known. This herbal was translated into Anglo-Saxon, and must have been a very popular book, for no less than four MSS. of it exist, which is a large proportion out of the scanty remains of books of such early times.* The names of plants which are to be found in these MSS. are most interesting, and are useful for the identification of the names used in later herbals. Another

---

* Translations are to be found in Cockayne, *Leechdom and Wortcunning of Early England*, 1864, notes in *Early-English Plant Names*, Earle, 1880—original MSS. Cotton Vitellius ciii. Brit. Mus. date circa 1000-1066. Trinity College, Cambridge, O. 2. 48, 14th century. Also in Harleian 815, Liber Medicinalis. (Harleian 5066, *Herbarium Saxonici*. Thus described in the Catalogue, is not in the MS. thus numbered, and a note to say it was not there in 1804 is signed D.)

good list of herbs in Anglo-Saxon is to be found in Ælfric's *Grammatica*.* This includes most of the simple herbs then known, with the Latin equivalents. The Latin is not always correctly translated, the name of some common native flower being sometimes substituted for a plant which was unknown to the writer.

The earliest writers on this subject in England, were churchmen; Alexander Necham, Abbot of Cirencester, and Bishop Grosseteste, of Lincoln. They both studied at the University of Paris, and thus had an opportunity of seeing for themselves the state of horticulture abroad. Their writings only touch incidentally on gardening. Grosseteste† (b. cir. 1175, d. 1253) wrote on many subjects; he was skilled in medicine, and had a knowledge of the virtues and properties of plants. The works attributed to him are so numerous, that it is scarcely possible that all can have come from his pen, but everything which bore his name continued to be read, and referred to, for more than two centuries after his death. Therefore his works on husbandry must have had considerable influence on horticulture. Palladius's work, *De Re Rustica*, written at some early date, probably in the fifth century, was the foundation of nearly all English writings on husbandry, for several centuries, and most of them, that of Grosseteste included, were merely translations, or adaptations, of this work. *De Re Rustica* is in fourteen books. The first is introductory, the following twelve are devoted in turn to each month of the year, the fourteenth to grafting. Various impossible recipes were thus passed on by men who took no trouble to investigate the truth of their assertions. In the fifteenth century, Grosseteste was as much believed in, as he had been in the thirteenth, although gardening was practised all this time, and something much more accurate could have been written. These works ‡ contain chiefly

---

* *Vocabularies in a Library of National Antiquities.* Wright, 1857. MS. Brit. Mus. Cotton Julius A ii.

† *See* Sam Pegge, *Life of Robert Grosseteste.* 1793. p. 308.

‡ Sloane MS. 686. "The tretyse off housbandry that Mayster Groshe ~de~ made that whiche was Bishope of Lycoll he translate this booke out off frensche in to English."

instructions on such fanciful subjects as these: "To make apples grow without any core"—"To colour apples growing on the tree"—"To make cherries grow without stones"—and many such impossibilities.

Necham, who lived at the same time as Grosseteste, was a more original writer. He was born in 1157, passed the early part of his life at St. Albans, and was made the director of the school belonging to the Abbey at Dunstable; by 1180 he was a distinguished professor at Paris University, returned to Dunstable about 1186, but soon after left the Benedictines of St. Albans, and joined the Augustines of Cirencester, was there elected Abbot in 1213, and died in 1217. Necham's "De laudibus divinæ Sapientiæ," a poem in ten parts, devotes many lines to the praise of various flowers and fruits. The seventh book is on the excellence of such herbs as betony, centaury, plantain, wormwood; the eighth is about fruits—cherries, peaches, medlars, and so forth. He does not, however, confine his praises to English productions, but sings of terebinth, cinnamon, and spices, and fruits which he had probably never seen in their natural state. In like manner, his description in his other work, *De Naturis Rerum*, of what a "noble garden" should be, is drawn from imagination, as many plants, quite unfit for culture in the open air in this country, or even in Europe, are included in the list of what the garden should contain. This is easily accounted for, as Necham, like others of his time, borrowed freely from classical writers. "The garden,"[*] he writes, "should be adorned with roses and lilies, turnsole, violets, and mandrake; there you should have parsley and cost, and fennel, and southernwood, and coriander, sage, savory, hyssop, mint, rue, dittany, smallage, pellitory, lettuce, garden cress, peonies. There should also be planted beds with onions, leeks, garlick, pumpkins, and shalots; cucumber, poppy, daffodil, and acanthus ought to be in a good garden. There should also be pottage herbs, such as beets, herb mercury, orach, sorrel, and mallows." So far, this is evidently a simple catalogue of what was to be

---

[*] The translation of the names of plants is taken from Wright's edition of Necham's works.

seen in his garden at Cirencester, or any other fair-sized garden of his day. But "medlars, quinces, wardon pears, peaches, pears of St. Regula," are followed by fruits such as oranges, lemons, pomegranates, myrrh, and spices, and other things equally incredible.

Another classical writer of uncertain date was Macer. An author of that name was contemporary with Virgil, but the writer of the Herbal which was translated into many languages must have lived at some later date, as he quotes Galen. It is strictly a herbal giving the medicinal uses of herbs and spices. The old translations are valuable, as giving the English equivalents of the Latin names, and Macer's was such a common hand-book, that anyone planting a herb garden, would try to obtain as many of the plants mentioned by him, as could be found in England at that period. The name of the first translator of Macer is lost in obscurity, but there is a manuscript translation, dated 1373, by John Lelamour, schoolmaster of Hertford,* and several other early translations exist, although the book was not printed until about 1530. One of them is curious, from the additions made by the translator or transcriber, of some plants known to him, and not mentioned by Macer.† He subjoins also some further medical recipes, which indicate more of the usual plants of a herb garden. The following example is the recipe given for curing the pestilence:—"Do take and medele, pimpernoll, sauge, auance seint mary gouldes, tansey sorell' and columbyne, stampe *th*ese VII erbes and drink the ioiuse of hem in ole ale or clene wat*er* and it wole distroie the pestilence be it never so felle."

Further information about gardens is to be gained from other medical works. There is an English fourteenth-century medical poem preserved in MS. in the Royal Library, Stockholm, which contains some graphic descriptions of flowers. With regard to the good qualities of rosemary, the author says:— "Rosmarine is bothen erbe & tre, hot and drie of kende

---

* Sloane, No. 5, Sec. 3.
† MS. circa 1440, in the Library at Didlington.

hys lewys arn eue*r*more grene & neue*r* more falty as techy̆ bokes of fysik and ek bokys of skole of sallerne wrot to ye countess of hernaunde and sche sente ye copy to hyre dowter phelyp qwen of Ingelond."* This, of course, was Philippa of Hainhault, wife of Edward III., and it is interesting to note that there is a MS. in the British Museum,† with the following title : —" Chiburn on the virtues of Ros maryn written at the command of the Countess of Henawd who sent the copy to her daughter Phylyp, Queen of England."

Another medical work, by "the venerable doctor, Master Gilbert Kymer," is a treatise addressed to Humphrey, Duke of Gloucester, entitled *Dietarium de Sanitatis Custodia*. Kymer gives a list of herbs to be put in potage, that the Duke might safely take, also full instructions as to what fruits could be taken before meals and what others after. This list includes, besides the commonest fruits, damsons, strawberries, figs, medlars, and peaches, and also foreign fruits and spices.

We find Palladius still translated in the fifteenth, as he had been in the thirteenth, century. There is no clue to the author of the English version, of which a manuscript dating from about 1420 exists at Colchester; ‡ but the name and work of another translator, of the same date, have been preserved. He was a monk of Westminster, named Nicholas Bollard, and either himself translated direct from Palladius, or transcribed or translated through "Godfrey," the parts of the work on husbandry, relating to grafting, planting, and sowing. Robert Salle also re-issued part of the same work. The MS. in the British Museum, containing the work by Salle, ends thus :— " Here endeth the telyng of trees after Godfray upon paladie and her begynneth the tretis of Nicholas Bollard." Then follows the chapter on " the manner of settyng of trees," and grafting, at the end of which it is stated, " here endeth the chapter of the first partie of Godfray upon Paladie de Agricultura." Another MS. of the fifteenth century known as the Porkington Treatise, has a few pages devoted to grafting

---

* *Archæologia*, Vol. XXX.   † Sloane, No. 7. Sec. 5.
‡ Printed E. Eng. Text Soc., ed. by S. T. H. Herrtage.

agrymony saueray tyme rose & spynage
Petrosyll columyne auence & borage
ffenel sothirynwhod waimot & rybwort
herbe Ion herbe robert herbe walter & wallwort
heyhowe poliody parcell & osery
Gromel modewose hyndehill & betony
Glidy valeryan stichys & spreutwort
Cerueyll lydesony lekatt lyly & lyliall wort
Honysop egremoyne honysoke & burnell
Centory horsel adderstong & bygould
Henbane orment wyldetefyll & styhellwort
Wey brede gwith stichys chesunbolden byster
Aueroyn langedibet radystle samyde & seueny
Prymke violet coltthyppe and lyly
Carsyndilles stralbers and nedelwort
langebefe toteshyne tansey & febullwort
Ostry neptes horssollnd & flos campy
dsodyll redmay pynepole & oculy
Rose ryde rose whyte frystone & pympnell
holyhocke coryllender peony if if wold
all thus herbys by seynt ani felt
wold be sette in the moneth of Auerell
ffurther mo wul y not go
But kep of herbys wol y go

Of the kynde of saferolbne

Of saferolbne we mote telle
he shal be copie fayre & lkelle
Saferolbne wul haue wtout leffyng
Welle er y made wel wyth dyng
ffor serthe if they shal bere
pey wold be sette in y moneth of September
a pos day oy for seynt mary day it myzte
orgate next wolk y after p mote y the
wk a Romy p shall ham sette
that p sydys by kep be blont & meto
tyloo puche deps they most sette be
and thus sype mu in yff you turnd to me

Explicit hic liber cui vocatur
Anglice Mayster Ion gardener

and planting of trees which contain almost the same matter as those already cited, with a few additions. The author gives all the usual recipes for making fruit grow without stones, and so on, but he tells also how to graft a vine and a red rose on a cherry, and how to make the fruit turn blue by boring " an hole in the tre nize the rote" and putting in "good asure of Almayne;" also, he says rose hips, or "pepynes," as he calls them, should be sown in February or March, "and dew heme welle with water" "iff thou wolt have many rosys in thy herbere." *

The earliest known really original work on gardening, written in English, is a treatise in verse by " Mayster Ion Gardener," of which a unique manuscript exists in the Library of Trinity College, Cambridge.† It is contained in a small volume of miscellaneous manuscript matter, which was given to the College by Roger Gale, in 1738. This copy was apparently written about 1440, but the poem is probably of earlier date. From the evidence of the language, it appears that the author was Kentish, and from the mistakes of the copyist, it would seem that he was unfamiliar with some of the words which were becoming obsolete at the time he wrote. The existing title, "The Feate of Gardening," is evidently added by a later hand. Nothing definite is known of the author of this poem. He may have been a professional gardener, or he may merely have assumed the name, as symbolic of the craft, just as Langland wrote under the name of Piers Ploughman. We certainly find John a very common Christian name among the gardeners of the period. This treatise is a great step in advance of earlier writers. It is so thoroughly practical, that the directions it contains might be followed with successful results at the present day. It is unencumbered by superstitions, then so prevalent, and quite free from fanciful receipts. The poem contains 196 lines, consisting of a prologue

---

* Porkington MS., the property of W. Ormsby Gore, published by the Warton Club in 1855, under the title of Early English Miscellanies, ed. by G. O. Halliwell, F.R.S., &c.

† Printed in the *Archæologia*, Vol. LIV., with a glossary by myself.

and eight divisions, under the following headings:—" Off settyng' and Reryng' of Treys "—" Of graffyng' of Treys "— " Of cuttyng' and settyng' of Vynys "—" Of settyng' and sowyng' of Sedys "—" Of sowyng' and settyng' of Wurtys "—" Of the kynd of Perselye "—" Of other maner Herbys "—" Of the kynde of Saferowne." This work is invaluable, as it gives incontrovertible evidence of the plants then actually to be found in an English garden, and the way in which they were cultivated, and is, of course, infinitely more worthy of belief on this subject than any translated work. The only other available sources we have for information on this point are the early cookery-books, in some of which the herbs suitable for a garden are enumerated. The following is the list of plants mentioned in John Gardener's poem:—

Plants from "The Feate of Gardening."

Adderstong (*Ophioglossum*).
Affodyll (*Narcissus Pseudo-narcissus*).
Auans (*Geum urbanum*).
Appyl (*Pyrus Malus*).
Asche tre (*Fraxinus excelsior*).
Betony (*Stachys Betonica*).
Borage (*Borrago officinalis*).
Bryswort (*Bellis perennis*).
Bugull (*Ajuga reptans*).
Bygull (*Chrysanthemum segetum*).
Calamynte (*Calamintha officinalis*).
Camemyl (*Anthemis nobilis*).
Carsyndylls (cress. and lily ?)
Centory (*Centaurea nigra*, or *Erythræa Centaurium* ?).
Clarey (*Salvia Sclarea*).
Comfery (*Symphytum officinale*).
Coryawnder (*Coriandrum sativum*).
Cowslippe (*Primula veris*).
Dytawnder (*Lepidium latifolium*).
Egrimoyne (*Agrimonia Eupatoria*).
Elysauwder (*Smyrnium Olusatrum*).
Feldwort (*Gentiana*).
Floscampi (*Lychnis*).
Foxglove (*Digitalis purpurea*).
Fynel (*Fœniculum vulgare*).
Garleke (*Allium sativum*).
Gladyn (*Iris*).
Gromel (*Lithospermum officinale*).
Growdy swyly (*Senecio vulgaris*).
Halsel tre (*Corylus Avellena*).
Hawthorn (*Cratægus Oxyacantha*).
Henbane (*Hyoscyamus niger*).
Herbe Ion (*Hypericum perforatum*).
Herbe Robert (*Geranium Robertianum*).

Herbe Walter (cannot identify).
Hertystonge (Scolopendrium vulgare).
Holyhocke (Althæa rosea).
Honysoke (Lonicera Periclymenum).
Horehound (Marrubium vulgare).
Horsel (Inula Helenium).
Hyndesall (? "Ambrosia." Teucrium scorodonia?)
Langbefe (Helminthia echioides Echium vulgare).
Lavyndull (Lavandula vera).
Leke (Allium Porrum).
Letows (Lactuca sativa).
Lyly (Lilium candidum).
Lyverwort (Anemone Hepatica).
Merege (Apium graveolens).
Moderwort (Artemisia vulgaris).
Mouseer (Hieracium Pilosella).
Myntys (Mentha).
Nepte (Nepeta Cataria, or a turnip).
Oculus Christi (Salvia Verbanaca).
Orage (Atriplex hortensis).
Orpy (Sedum Telephium).
Ownyns and Oynet (Allium Cepa).
Parrow (mistake for Yarrow).
Pelyter (Parietaria officinalis).
Percely (Petroselinum sativum).
Pere (Pyrus communis garden varieties).
Peruynke (Vinca major & minor).
Primrole (Primula vulgaris).
Polypody (Polypodium vulgare).

Pympernold (Poterium Sanguisorba).
Radysche (Raphanus sativus).
Redenay (Red Ray Lolium perenne).
Rewe (Ruta graveolens).
Rose (Rosa, red and white).
Rybwort (Plantago lanceolata).
Saferowne (Crocus sativus).
Sage (Salvia officinalis).
Sanycle (Sanicula europæa).
Sauerey (Satureja hortensis).
Scabyas (Scabiosa).
Seueny (Brassica alba).
Sowthrynwode (Artemisia Abrotanum).
Sperewort (Ranunculus Flammula).
Spynage (Spinacia oleracea).
Strowberys (Fragaria vesca).
Stychewort (Stellaria Holostea).
Tansay (Tanacetum vulgare).
Totesayne (Hypericum Androsæmum).
Tuncarse (Lepidium sativum).
Tyme (Thymus Serpyllum).
Valeryan (Valeriana officinalis).
Verueyn (Verbena officinalis).
Violet (Viola odorata).
Vynys and Vyne tre (Vitis vinifera).
Walwort (Sedum acre).
Warmot (Artemisia Absinthum).
Waterlyly (Nymphæa alba or Nuphar luteum).
Weybrede (Plantago major).
Woderofe (Asperula odorata).

Wodesour (*Oxalis Acetosella*).
Wurtys or Wortys (*Brassica oleracea*).
Wyldtesyl (*Dipsacus Fullonum*, or *sylvestris*).
Ysope (*Hyssopus officinalis*).

List of herbs at the beginning of a book of cookery recipes, fifteenth century. Sloane MS. 1201.

Herbys necessary for a gardyn by letter.

A. Alysaundre (*Smyrnium Olusatrum*), Avence, Astralogia rotunda (*Aristolochia*), Astralogia longa, Atta, Arcachaff (*Angelica?*), Artemesie mogwede, Annes (*Pimpinella Anisum*), Archangel (*Lamium album*).

B. Borage, Betes (*Beta vulgaris*), Betyñ, Basilican (*Ocymum basilicum*), Bugle, Burneti.

C. Cabage, Chervett, Carewey, Cyves, Columbyn, Clarey, Colyaundr', Colewortę, Cartabus, Cressez, Cressez of Boleyñ, Calamyntę, Camamytt, Ceterwort (*? Ceterach officinarum*).

D. Daysez, Dytayñ, Daundelyoñ, Dragaunce (*Arum Dracunculus*), Dylle.

E. Elena campana (*Inula Helenium*), Eufras (*Euphrasia officinalis*), Egrymoyñ.

F. Fenett, Foothistell, Fenecreke (*Trigonella Fœnum-græcum*).

G. Gromett, Goldez (*Calendula officinalis*), Gyllofr' (*Dianthus Caryophyllus*), Germaundr'.

H. Hertez tonge, Horehound, Henbane.

I. Isope, Iertin, Iryngez (*Eryngoes?*), herbe Ive (*Plantago*).

K. Kykombre, yt. bereth apples (*Cucumis sativus*).

L. Longdebeff, Lekez, Letuce, Love ache (*Levisticum officinale*), Lympons, Lylle (*lilium*), Longwortz (*Pulmonaria officinalis*).

M. Mercury, Malowes, Mynt, Mageroñ, Mageroñ gentyle, Mandrake, Mylons.

N. Nept, Nettett rede, Nardus capistola.

O. Orage, Oculus Christi, Oynons.

P. Persely, Pelytor, Pelytor of Spayñ, Puliatt royatt (*Mentha Pulegium*), Pyper white, Pacyence (*Rumex patentia*), Popy whit', Prymerose, Purselane, Philipendula.

Q. Qvyncez.

R. Rapes (*Brassica Napus*), Radyche, Rampsons (*Allium ursinum*), Rapouncez (*Campanula Ranunculus*), Rokettę (*Hesperis matronalis*), Rewe.

S. Sauge, Saverey, Spynache, Sede-wale (*Valeriana pyrenaica*), Scalaceh (? *Sinapis arvensis*), Smalache (*Apium graveolens*), Sauce alone (*Erysimum Alliaria*), Selbestryne, Syves (*Allium Schœnoprasum*), Sorell, Sowthistell, Skabiose, Selia, Stycadose (*Lavandula Stæchas*), Stanmarch (? *Smyrnium Olusatrum*).

T. Tyme, Tansey.

V. Vyolettę, Wermode, Wormesede (*Erysimum cheiranthoides*), Verveyñ.

Of the same Herbes for Potage.

Borage, Langdebefe, Vyolettę, Malowes, Marcury, Daundelyoñ Avence, Myntę, Sauge, Percely, Goldes, Mageroñ, Fenell, Carawey, Rednettyll. Oculus Christi, Daysys, Chervell, Lekez, Colewortes, Rapez, Tyme, Cyves, Betes, Alysaundr', Letyse, Betayñ, Columbyñ, Alla, Astralogia rotunda, Astralogia longa, Basillican, Dylle, Deteyñ, Egrymoñ, Hertestong, Radiche, White pyper, Cabagez, Sedewale. Spynache, Coliaundr' Foothistyll, Orage, Cartabus, Lympons, Nepte, Clarey, Pacience.

Of the same Herbes for Sauce.

Hertes tonge, Sorell, Pelytory, Pelytory of Spayñ, Detey, Vyolettę, Percely, Myntę.

Also of the same Herbes for the copp.

Cost, Costmary, Sauge, Isope, Rose mary, Gyllofr', Goldez, Clarey, Mageroñ, Rue.*

Also of the same Herbes for a Salade.

Buddus of Stanmarche, Vyolette flourez, Percely, Redmyntę, Syves, Cresse of Boleyñ, Purselañ, Ramsons, Calamyntę, Prime Rose buddus, Dayses, Rapounses, Daundelyon, Rokette, Red nettell, Borage flourez, Croppus of Red Fenell, Selbestryñ, Chykynwede.

Also Herbez to Stylle.

Endyve, Red Rose, Rose mary, Dragans, Skabiose, Ewfrace.

---
* Rue is added in fainter ink.

Wermode, Mogwede, Betayñ, Wylde Tansey, Sauge, Isope, Ersesmart (*Polygonum Hydropiper*).

Also Herbes fo[r] *Savour and beaute.*

Gyllofr' gentyle, Mageroñ gentyle, Basyle, Palma Christi, Stycadose, Meloncez, Arcachaffe, Scalaceley, Philyppendula, Popyroyall, Germaundr', Cowsloppus of Jerusalem, Verveyñ, Dyll, Seynt Mar' Garlek.*

Also Rotys for a gardyñ.

Persenepez, Turnepez, Radyche, Karettes, Galyngale, Tryngez, Saffroñ.

* " Seynt Mar' Garlek " is added by another hand.

CHAPTER V.

"Sure gates, sweete gardens, stately galleries
Wrought with faire pillowes and fine imageries ;
All those (O pitie !) now are turned to dust
And overgrowne with black oblivious rust."
SPENSER, *Ruins of Time.*

TOWARDS the end of the fifteenth century fresh influences were brought to bear on the national life, and numerous changes set in. The marriage of Edward the Fourth's sister with the Duke of Burgundy, and through that alliance the increased intercourse with Flanders, led to many alterations in social life. The comparative peace which followed the termination of the Wars of the Roses encouraged a new style of domestic architecture, and comfortable red brick houses succeeded the old castles. The gardens were no longer of necessity confined within the embattled castle walls. The houses in the new style were not built on the tops of hills, but usually on lower lying ground, and were surrounded by a moat. There was some little space within the moat devoted to a garden, or a few plants were placed in the courtyard. The prolonged peace diminished the necessity of keeping all property within the protecting lines of the moat, and thus the custom came in of having gardens beyond it. With this additional space—for there was frequently more room inside the moat than there had been within castle walls, even if the garden were not made outside—there was more scope for play of fancy, and before long several changes in design came in.

One of the first innovations was the railed bed—flower-beds

enclosed by low fences of trellis-work. These trellis railings came into fashion just before Tudor times, but they remained in vogue for many years. When, in 1533, Henry VIII. made great alterations in the gardens of Hampton Court, flower-beds of oblong form were made in the King's new garden. They were surrounded by rails painted green and white—the Tudor colours —as may be seen in the original picture of Henry VIII., from which the illustration on page 89 is taken. In the Hampton Court expenses, 1533, numerous entries refer to the purchase of these rails.

"Paid to [Henry Blankeston, of London, painter] for the like painting of 96 flat pownchens with white and green, and in oil, wrought with antyke

RAILED FLOWER BED, FROM FRENCH MS. OF THE ROMAN DE LA ROSE. C. 1450.
BM. EGERTON 2022.

a both sides bearing up the rails in the said garden at 12d. the piece, £4. 16s.— Also paid to the same for like painting of 960 yards in length of rail in the said garden with white and green, and in oil, price the yard 6d., £24."*

These items are repeated with variations; the posts and rails were painted "white and green in antyke oiled colours," and "flat posts" occur in the place of "flat pownchens."

Another novelty introduced in the first years of the Tudor period, and soon a conspicuous feature in all gardens, was topiary

* Exchequer, Treasury of Receipt, Miscellaneous Books, No. 237. This is a large book of Expenses at Hampton Court, 24th Henry VIII.

work, "opus topiarum," that is to say, quaintly cut trees and shrubs. This art, although new in England, was of very ancient origin, having been known to the Romans. But it is not until this date that it is mentioned as being practised in England. The new idea found great favour in this country, and much time and trouble were expended in producing these monsters in trees, and the taste remained in fashion for more than two centuries. Leland, in his *Itinerary*, in the early years of the

THE MOUNT, ROCKINGHAM

sixteenth century, mentions a place where striking specimens of the work might be seen: "at Uskelle village, about a mile from Tewton, is a goodly orchard with walks opere topiario;" and at "Wreschill Castle" he also describes an orchard with "mounts opere topiario writhen about in degrees like turnings of cokilshells to come to the top without payne." This leads me to speak of yet another peculiarity which was much developed about this time, the "mount," like this one at Wressel Castle, where Leland saw the cut trees. In the thirteenth century there were made in some of the monasteries "mounds" of earth against the garden-walls, to enable the inmates to peer over them into the outer world. During the following centuries, " mounds " or " mounts," of simple construction, were frequently to be found in the gardens, but in Tudor times, the "mount" became a much more important accessory than formerly. They were usually made of earth covered with fruit or other

OLD YEW WALK AND MOUNT ROCKINGHAM.

trees. Mounts were generally thrown up in " divers corners " of the orchard, and were ascended by "stairs of precious workmanship," or a spiral path planted on either side with shrubs, cut in quaint shapes, or with sweet-smelling herbs and flowers. At Rockingham, there remains a specimen of one form of mount. A great terraced-mound of earth, covered with turf and a few trees, is raised against a part of the high wall which surrounds the garden and behind which the keep formerly stood.

From the top of this the eye ranges across the garden with quaintly cut yew-trees, over a magnificent view of the open country beyond; thus the mount served in early times as a "look out" or watch tower. If the garden or orchard happened to be situated in a park, and herds of deer browsed close to its walls, the mount then became useful as a point from which one "myghte shoot a bucke."* The top of the mount was often surmounted by an arbour, either of trellis-work and creepers, or a more substantial building. Probably the finest specimen of this kind of ornament was the "mount" at Hampton Court, and from various sources we can form a very good idea of what it was like. It was situated at the southern end of the "King's New Garden," which was made in 1533, at which time a gardener named Edward Gryffyn superintended the work. The mount was made on a brick foundation, as there were payments made to "John Dallen of London, bricklayer," for "laying of 256,000 of brick upon the walls about the new garden, betwixt the King's lodgings and Thames, and the foundations of the mount standing by Thames, taking for every 1000 14d., by convention £14. 18s. 8d." The earth was then raised and planted with quicksets. The sum of 54s. 8d. was paid to Lawrence Vyncent and John Gaddisby of Kyngston, for four loads of quicksets, every load containing thirty hundred sets of them "to set about the mount by the King's new garden." Another entry refers to the purchase of "ash poles to make rails to bind the quicksets," and "two bundles of wylly roddes to bind" them; and "three pear trees to set in the mount." The most elaborate part of the mount was the arbour. The "South arbour" seems to have been the one on the mount, but mention is also made in the accounts of a west arbour, which was apparently very similar, as the same things were bought for both, and payment made to "John a gwylder smith" "for 300 of broddes serving for the fretts in the roof of the south herber at the mount 12d. the 100, 3s.," and to Galyon Hone, the King's glazier, several sums were paid, of which the following is a sample. "Item in the mount in the garden

* Lawson, *New Orchard*.

GARDEN WITH A GALLERY, FROM "THE SECOND BOOKE OF FLOWERS, FRUICTS, BEASTES, BIRDS, AND FLIES," 1650.

48 lights, every light in the upper story containing $4\frac{1}{2}$ foot, in the nether story every light containing $4\frac{1}{2}$ foot 3 inches, which amount in all (to) 211 foot at 5d. the foot, £4. 7s. 11d." This gives one some idea of how large the arbour was, and how carefully it was made. It appears, furthermore, from the accounts, that the "south herber" was connected with the west one by a gallery running along the wall, which was made of wooden poles and trellis-work. Such galleries were marked characteristics of late fifteenth and early sixteenth century gardens and designs for them are found in some old works; the best of these being in the *Hortus Floridus* of Crispin de Pas (or Passe) which was translated into English in 1615. They existed in Hampton Court before Henry VIII. made his alterations there, and are thus referred to in Cavendish's metrical life of Wolsey.

> " My galleries were fayre, both large & longe
> To walk in them when that it liked me beste
> * * * * *
> With arbours & alleys so pleasant & so dulse
> The pestilent airs with flavours to repulse."

I do not know of a single example of a gallery or arbour, of this description, in existence. They were made of perishable material, such as wood-trellis planted with creepers, vines, roses, or honeysuckle, therefore even those which were not pulled down purposely, must have been long ago destroyed by time. And what is also much to be regretted is, that few, if any, examples are to be found in English illuminated books, although plenty of pictures occur in foreign MSS. of this period, especially French and Flemish. The scarcity of English examples is no doubt partly owing to the destruction of religious books at the time of the Reformation. They are found chiefly in the calendars at the beginning of missals, or Books of Hours, where the miniature for the month of May is frequently a garden, or the garden of the day is introduced, in the illustration of some sacred subject. The gallery ran along the outer wall of the garden, the wall forming one side, posts of wood in a series of arches the other, while the pathway between the wall and the posts was covered in, either with creepers and wood-work, or something more substantial, and affording better shelter. Sometimes the gallery followed the wall round three sides, but it seems to have been the more usual custom to have it on one side only, and it frequently afforded a sheltered walk from the house to the arbour or mount.

Edward Stafford, Duke of Buckingham, during the first years of the sixteenth century, began to lay out very extensive gardens at Thornbury, in Gloucestershire, but he was accused of treason, and hurried to the scaffold, before carrying out his plan. Among the State papers of the time, May, 1521, there is a survey of his lands, and the following extracts appear in it, under the heading of "gardens," and are illustrative of the fashion of galleries. "On the south side of the inner ward [of the castle] is a proper garden, and about the same a goodly gallery conveying above and beneath from the principal lodgings, both to the chapel and parish church. The utter (*outer*) part of the said gallery being of stone embattled, and the inner part of timber covered with slate. On the east side of the said castle or manor, is a goodly garden to walk in, closed with high walls, embattled. The conveyance thither is by the gallery above and beneath, and by other privy ways. Besides the same privy garden is a large and a goodly orchard, full of young graffes well loaden with fruit, many roses and other pleasures. And in the same orchard, are many goodly alleys to walk in openly. And round about the same orcharde is conveyed on a good height other goodly alleys with roosting places, covered thoroughly with white thorne and hasel. And without the same, on the utter part, the said orchard is enclosed with sawin pale (*sawn palings*) and without that ditches and quickset hedges." ... "From out of the said orchard, are divers posterns in sundry places at pleasure to go and enter into a goodly park newly-made." The house and gardens were left to fall into ruins, after Queen Elizabeth's time, and not a trace of the old garden remains.*

Another example of an arbour or "roosting-place," was one made for Elizabeth of York. "10 July 1502 Item payed to Henry

---

* The outer castle wall alone remained, and it was rebuilt, and the present gardens laid out about fifty years ago, by the father of the present owner, Mr. Stafford Howard.

GARDEN HOUSE, JOSELLA.

Smith clerc of the Castle of Wyndsor for money by him payed to a certain labourer to make an herbour in the little park of Wyndsor for a banket for the Queen iiijs. viijd." Again, in the eighteenth year of Henry VII., five shillings were paid for making an arbour at Baynarde's Castle, in London.*

The ordinary arbour was still like those described in earlier times by Chaucer, with a turfed seat, and trellis covered with climbing plants. One is thus spoken of by a poet of the Tudor period †:—

> " The clowdis gan to clere, the myst was rarifiid
> In an herber I saw, brought where I was,
> There birdis on the brere sange on euery syde :—
> With alys ensandid about in compas
> The bankis enturfid with singular solas
> Enrailed with rosers, and vinis engrapid ;—
> It was a new comfort of sorowis escapid."

Other resting-places were arranged along the garden-walls, in the form of shady nooks and corners with grass banks to serve as seats, such as that of which More, in his *Utopia*, makes mention, when he writes :—" We all went to my house, and entering into the garden, sat down on a green bank, and entertained one another in discourse." The arbour or garden-house was sometimes of brick, or stone, built like a turret into the wall ; an early example of arbours like this exists at Loseley, in Surrey. There were originally four houses, one at each corner of the garden-wall, and three of these remain. Another interesting garden of this date is at the Palace, Hadham, in Hertfordshire, which, for many hundred years, belonged to the Bishops of London. It was also the dwelling-place of Katherine, widow of Henry V., after her marriage with Owen Tudor, and it was here that Edmund, father of Henry VII., was born. The garden at the present day is surrounded on two sides by a wall, while the other side is protected by a high yew hedge, three yards thick.

At the beginning of the sixteenth century, a new flower-bed was adopted, as well as the straight-railed beds. This was the "knotted bed," or knets. They were laid out in curious and complicated geometrical patterns. By the year 1520, the

---

* Wardrobe Accounts.   † Skelton, *Garlande of Laurell*.

style was in common use, and most of our English gardens could boast of some kind of novel knotted bed. Cavendish writes of Hampton Court, it was "so enknotted it cannot be expressed." The earth in the knots was either raised a little, being kept in its place by borders of bricks and tiles, or, as was more often the case, it was on the same level as the paths, and then the divisions were made with box, thrift, and so on. Generally, the beds were planted inside their thick margins, with ornamental flowers or small shrubs, somewhat as "carpet

A proper knot to be cast in the quarter of a Garden, or otherwise, as there is full content roome.

KNOT FROM THE GARDENER'S LABYRINTH.

beds" are now laid out; but, sometimes, instead of plants, they were filled with variously coloured earths. In the household accounts of the Duke of Buckingham, in 1502, there is an entry of 3s. 4d. being paid to "John Wynde, gardener, for diligence in making knottes in the Duke's garden." And in the same year, among the accounts of the fifth Earl of Northumberland, a gardener is mentioned as being employed to "attend hourly in the garden for setting of erbis, and clypping of knottes, and sweeping the said garden cleaner hourly." The designs of these knots were very varied. They were either geometrical patterns, or fanciful shapes of animals; the intricate geometric designs being evidently the most popular, as they occur most frequently in books. (See illustration.) The other style is described in the following poem*:—

> "Then we went to the garden glorious
> Like to a place, of pleasure most solacious:
> With Flora painted and wrought curiously
> In divers knottes of marveylous greatnes
> Rampande lyons, stode by wonderfly
> Made all of herbes, with dulset swetenes
> With many dragons, of marveylous likenes
> Of diuers floures, made full craftely
> By Flora couloured, with colours sundrye."

The following are some of the flowers that were cultivated in these knottes, or in the borders, in Tudor times, that are mentioned by contemporary writers:—Acanthus, asphodel, auricula, bachelor's buttons, amaranthe, or "blites," cornflowers. or "bottles," cowslips, daffodils, daisies, "French broome," gilliflowers (3 varieties), hollyhock, iris, jasmine, lavender, lilies. lily of the valley, marigold, narcissus (yellow and white), pansies. or heartsease, peony, periwinkle, poppy, primrose, rocket. roses, rosemary, snapdragon, stock gilliflowers, sweet william, wall-flowers, winter-cherry, violet, and besides these, other sweet smelling herbs, such as mint and marjoram.

Having now gone through some of the principal features of a Tudor garden, the railed beds, knottes, the mount, arbours, and galleries, let us consider further, not only what gardens were made, but what happened to the old gardens in existence during the first part of this period. We have seen, in an earlier chapter, something of the position held by the monastery gardens throughout the land. Now we have reached the years of the Reformation, and so far as this great movement affected gardens, we must glance at its progress. The work of the visitation and then the suppression of the monasteries was begun in 1534. The greater ones were first attacked, and the lesser ones followed. The work was carried on rapidly; in the

---

* *The Historie of Graunde Amoure and la bell Pucell, called the Pastime of Pleasure.* By Stephen Hawes. Ed. 1554.

northern district in 1536, eighty-eight monasteries were reported on in a fortnight :* 202 were suppressed or surrendered between 1538-40. At the time of the Dissolution there were over seven hundred religious houses scattered all over the kingdom. We cannot say that each of these possessed a garden, as some were in towns, in spaces too confined, and some Orders did not devote any of their attention to agriculture. The Benedictines and Cistercians predominated in numbers, and they were, for the most part, large landowners, farmers of their own land, and skilled in horticulture. But of the gardens which surrounded Fountaines, Jervaulx, or Netley, Glastonbury, St. Albans, or Whitby, and many another fine abbey and stately priory, nothing remains. In some instances there is mention made of the gardens by the officers of the Crown, who made the visitations and appropriated everything of value. At Oxford, they regretted that the Austin Friars had felled all their trees, but the Franciscans had "good lands, woods, and a pretty garden." The Cistercians of Waverley were very poor at the time, and the Abbot was granted leave "to survey his husbandry whereupon consisteth the wealth of his monastery." Few traces of old monastery gardens are left. At Westminster there was a fine garden, celebrated for its damson trees, and a garden by the Infirmary, where the sick monks could take the air. Part of this remains in the garden belonging to the College, but some portion of it was built over at the beginning of the last century, when the new College buildings were erected. When Elizabeth came to the throne, she sent for Abbot Feckenham, who had been reinstated in the Abbey of Westminster during Mary's reign. He was planting elms in his garden when he received the summons, and finished his work before he would attend on the Queen. The Abbot ended his days in captivity, and his abbey was soon after transformed into a College, but some of his elm trees, or their successors, remain to this day.

 That which has most often survived destruction, to find a place in a modern garden, on the site of some old cloister, is the fish-pond, although, strictly speaking, it did not always form part of a monastery garden. But it was found useful, and has

* Gasquet, *Henry VIII. and Eng. Mon.*

frequently been spared even by the landscape gardener, who would rather alter than destroy it. At Cirencester, the present parish church is a fine building, but the abbey church beside it, in times past, was so infinitely larger, as quite to eclipse it. Yet now the abbey church and adjoining buildings have so completely disappeared, that almost the only trace of monastic times, in the grounds of the house, built on the same spot, is a small piece of water, the remains of the old fish-ponds. At Hurley-on-Thames the monks' fish stews are still in existence, while at Bisham Abbey, only a mile distant, the garden is surrounded on three sides by a moat, also a relic of monkish days. At Hackness, in Yorkshire, the monks' ponds have been transformed into the present lake, but at Newstead Abbey, Nottingham, the monks' stew, overshadowed by old yews, is untouched, and the eagle pond (see illustration, page 29) there, is also undoubtedly a relic of the Black Friars, a brass eagle lectern having been found in its depths, full of valuable deeds relating to the monastery, there hidden by the friars at the time of their dissolution. At Hatton Grange, in Shropshire, on the site of a cell, Buildwas Abbey, the ponds also remain as originally made by the monks. There are four pools, still bearing their old names—the Abbot's, Purgatory, Hell, and the Bath Pools. They are in sequence, separated by broad dams of earth, and are dug deep into the ground, with steep banks. Thus although the original gardens have vanished, the monastery lands were granted to the great families of the day, and since they passed into secular hands, stately houses have been built, and beautiful gardens, though of a totally different character, have been made, and now adorn what once were the precincts of the old abbeys and priories. Woburn, Welbeck, Burghley, Syon, Battle, Beaulieu, Ramsey, Audley End, and many others, are among the number.

The Earl of Surrey made extensive gardens round the house he built on the site of St. Leonard's Priory, near Norwich, which he called Mount Surrey. About this time the closing of some of the common lands caused some considerable riots, and in 1549 all the trees in the appleyards at Mount Surrey were destroyed by the rebels, and used for making tents and huts. This was one of the most important gardens laid out on the site of a religious house, and it was not until a succeeding generation, when the taste for gardening was still more universal, that many others of the new proprietors followed this example.

We have already had occasion to refer to Hampton Court, in describing the characteristics of Tudor gardens. There are such full accounts left of the expenses incurred in making these gardens, both under the direction of Cardinal Wolsey and of Henry VIII., that, although we do not know the exact plan, a very fair idea of what they were like may be gathered. The land which Wolsey covered with the building, gardens and park, consisted of two thousand acres. In the south-west corner of this ground stood the old manor house, and round this the Cardinal laid out gardens and orchards, separated by brick walls, and beyond the walls, a park. He retained part of the manor house garden, for it is noted several times as "the old garden." John Chapman was head-gardener at this time, and remained in that position, with a salary of £12 a year, when the King took possession of the disgraced Cardinal's lands in 1529. The gardens were soon after greatly enlarged. A new orchard was made to the north of the old gardens, and pears, damsons, medlars, cherries, apples, cucumbers and melons were grown, and forty-three bushels of strawberries were planted at one time. There was a flower-garden which supplied the Queen with roses, and a kitchen garden, where "herbes for the king's table" were grown. A part of these gardens was destroyed when the new ones were made in 1533. The ground was then manured and carefully measured out into several plots, each surrounded by a brick wall. The largest plot was the King's new garden, the site of which is now called the "Privy Garden." In this there were gravel paths, and little raised mounds with sundials on them, and between the paths, railed beds cut in the grass. The rails were trained with roses, and yew, cypress or juniper trees planted in the centre of each bed: while along the walls were apple, pear and damson trees, and under them "violets, prymroses, sweet williams, gillifer slips, mynt and other sweete flowers," and this garden contained the mount and arbour. Another plot was the "Pond Garden," which merely seems to have contained the ponds, and was only decorated with the "beestes," as there is no mention of flowers being planted in it. There was the "little garden," of which not much is known, except that sixty-seven apple-trees were bought for it from "William gardener of London merchant, at 6d. the piece." The distinguishing features of Hampton Court from all other gardens that we know anything of at this time, were the "beasts" and the "dials." The carved animals holding "vanes," and the brass dials, seem to have been put in every part of the gardens and orchards. The beasts were

PICTURE AT HAMPTON COURT SHOWING THE RAILED BEDS AND BEASTS.

set at intervals along the railed beds, and about the mount and all round the ponds, and the entries concerning them in the accounts are very frequent.*

"Also paid to Bryse Auguston, of Westminster, clockmaker, for making of 20 brazen dials for the king's new garden at 4s. 4d. the piece, £4. 0s. 8d. —For making of bestes in timber for the king's new garden—paid to Edmund More.

* 25 Henry VIII. (1533). Exchequer, Treasury of the Receipts, Miscellaneous Books, No. 238.

of Kyngston, freemason, for cutting, making and carving of 159 of the king's and the queen's beestes standing in the king's new garden at 20s. the piece ... £150."

(1530) "Item dien to Anthony Transylyon, of Westminster, clockmaker, for seven dials of him bought which are bestowed in the privy orchard, at 4s. 4d. the piece, 30s. 4d.—Joiners setting up the bestes upon the posts in the privy orchard, Henry Currer, at 8d. by day, 4s.; John Carpenter, at 6d. by day, 3s. Payments for painting the king's festes (= *bestes*) in the privy orchard ... some holding 'fanes' (= *vanes*) with the king's arms."

(1534) "For gilding and painting of the beasts in the king's new garden— To Henry Blankston, of London (various sums for) 11 harts, 13 lions, 16 greyhounds, 10 hinds, 17 dragons, 9 bulls, 13 antelopes, 15 griffins, 19 leberdes (= *leopards*), 11 yallys (2 jalls occur elsewhere), 9 rams, and the lion on top of the mount, also for the vanes."

(1535) "Item in the said harbers (= *south and west arbours*) is set, 25 badges of the king's and queen's, price the piece 3s. £3. 15s. Item in the same harbers is set, 8 arms of the king's and queen's, price the piece 4s. 32s. Paid to Harry Corrant, of Kingston, carver, for making and entayling 38 of the kinge's and queene's beastes in freestone, bearing shields with the kinge's arms, and the queene's, that is to say four dragons, six tigers, 5 greyhounds, 5 harts, 4 badgers, serving to stand about the ponds in the pond yard, at 26 shillings the pece. £49. 8s."

The fountain in the "pond garden" at the present day, is probably a survival of the "pond yard," in which so many beasts were placed. In Henry the Eighth's time they were supplied with water in rather a curious way, as there are entries in the accounts of charges for "labourers ladyng of water out of ye Temmes to fyll the pondes in the night tymes."

There were several other royal gardens, and items with reference to things bought for them, or gardeners' wages, occur in the Privy Purse expenses of Henry VIII. for 1530-32, and Princess Mary, 1536-37. Greenwich is frequently mentioned in these accounts, and it seems to have been one of the favourite summer resorts of Henry, and his daughter. The payments were chiefly made to the head-gardener, named Walsh, for labourers' wages for "weding and delving," and "ordering in the garden." The gardens had probably been laid out when the palace was built by Humphrey, Duke of Gloucester, early in the reign of Henry VI., when it went by the name of "Placentia," or "Plaisance." The head-gardener there in 1519 was Lovell, and he received 60s. 8d. yearly. A little later we find him transferred to the Richmond garden, and his salary raised to £3 a quarter. He supplied the King's table "with damsons, grapes, filberts, peaches, apples, and other fruits, and flowers, roses, and other sweet waters."

There seem to have been two gardens at Beaulieu, or Newhall, the "smalle gardin," and "the grete." The small appears to have been the kitchen-garden, and furnished the "king's table" with "herbes and rootes, and strawberries, artichokes, lettuces, cucumbers, and sallet herbes." The keeper of the great garden in 1532 was one John Rede.*

The gardens within the walls of the Tower of London and at Baynarde's Castle, were kept up in Henry the Eighth's time. Frequent entries in the accounts show that there were royal gardens at Wanstead, where Robert Pury was gardener (1532),† Westminster, Waltham, Woodstock, and Oatlands, but they were probably not on so grand a scale as the more favourite resorts of the King. Windsor received less attention than the other royal gardens during this reign. The gardens at Windsor have now so completely changed, that even the site of the old garden cannot be identified with certainty. There is an account by an eye-witness of Louis de Bruye's reception, in 1472, by Edward IV. at Windsor. They go out hunting, and return late in the evening. "Bey that tyme yt was nere night yett the king showed hym his garden & vineyard of pleasure & so turned into the Castel agayne." This garden and vineyard probably remained unaltered in Henry the Eighth's reign, as we find no mention of changes being made there. The gardens at York Place, the Whitehall of later times, had been laid out by Wolsey with great taste and care, and this place, like Hampton Court, was also given over to the King.

Towards the end of his reign, Henry VIII., having completed his alterations at Hampton Court, turned his attention to laying-out and beautifying the grounds at Nonsuch, near Ewell, in Surrey.‡ He purchased the lands of Cuddington, in 1538, and there built a palace:—

"Which no equal has in art of fame
Britons deservedly do Nonsuche name."

* State Papers, Henry VIII. R. O.   † Ibid.
‡ Minister's Accounts, 31-32 Henry VIII., No. 10. Sir Ralph Sadler, steward of the manor, received 4d a day for the custody of "Gardinorum, Pomariorum et ortorum."

Another contemporary writer, describing the place, says of it : "The Palace itself is so encompassed with parks, full of deer, delicious gardens, groves ornamented with trellis work, cabinets of verdure, and walks so embowered by trees, that it seems to be a place pitched upon by Pleasure herself to dwell in along with Health."* Henry VIII. never quite completed Nonsuch, but it was held for a time by Henry FitzAlan, Earl of Arundel, who continued to carry out the King's designs. Queen Elizabeth, Anne, Queen of James I., and Henrietta Maria, all paid visits to the place, but did not stay there for long. The parliamentary survey of the palace and gardens, made in 1650, shows there were several walled gardens, divided by thick thorn hedges, also alleys, a wilderness, and privy garden, and a large kitchen-garden. There was also a terrace in front of the house, and a "handsome bowling-green." The whole was rather Italian in style, with many fountains and statues. Charles II. gave the place to the Duchess of Cleveland, who pulled it down, and the destruction of this once magnificent palace was completed by her grandson, the Duke of Grafton, who cut down the trees and destroyed the park.†

While such progress was being made in the decoration and laying-out of the flower-garden, the fruit and kitchen-gardens were not altogether neglected. Besides such fruits as were already in common use, others were introduced, and those indigenous in the country, were improved. The strawberry was largely planted, and carefully cultivated : –

> "If frost do continue take this for a lawe
> The strawberies look to be covered with strawe
> Laid overly trim, upon crotchis and bows
> And after uncovered as weather allows." ‡

From the following verse, in September's husbandrie, it is clear where the strawberry plants were procured :—

> "Wife unto thy garden and set me a plot
> With strawbery rootes of the best to be got.
> Such growing abroade, among thornes in the wood
> Wel chosen and picked proove excellent good."

* Nichols, *Progress of Queen Elizabeth.*
† *Camden's Britannia.* Ed. Gough, 1806.
‡ Tusser, *Five Hundred Pointes of Good Husbandrie.*

It was not only for humble folk that wild strawberry roots were gathered, for, in the oft-quoted Hampton Court Accounts, we find several entries of money paid for strawberry roots, brought from the woods for the King's garden.

The raspberry had until this period been more or less ignored, and even now seems not to have been very generally grown. Turner,* in 1548, says of "Rubus ideus in Englishe raspeses or hyndberies .... growe most plentuously in the woddes in east Freseland .... they growe also in certayne

APRICOT TREES ON OLD GARDEN WALL, LITTLECOTE.

gardines of Englande." He also says of them, "The taste of it is soure." The gooseberry, which does not appear in earlier gardens, was now grown. It was planted in some of Henry

---

* "Emptions of strowbery roots violettes and primerose roots for the new garden—also paid to Ales Brewer and Margaret Rogers for gathering of 34 bushels of strowberry roots, primerose and violettes at 3d. the bushel, 8s. 6d. Item to Matthew Garrett of Kyngston for setting of the said routes and flowers by the space of 20 days at 3d. the day, 5s."

the Eighth's gardens in 1516. Turner calls it "a groser bushe, a goosebery bushe," and says of it "It groweth only that I have sene in England, in gardines, but I have sene it in Germany abrode in the fieldes amonge other bushes." This passage is curious, as the subject has frequently been discussed, whether the gooseberry is an indigenous plant in this country. Tusser tells us that they are to be planted in September :—

> "The Barbery, Respis, and Gooseberry too
> Looke now to be planted as other things doo.
> The Goosebery, Respis, and Roses, al three
> With strawberies vnder them trimly agree."

The greatest addition to the number of cultivated fruits was the apricot, which was certainly introduced before the middle of the sixteenth century, probably by Henry the Eighth's gardener, Wolf, about 1524. Turner mentions it in both his works under Malus Armeniaca, and gives Abrecok, or Abricok, as the English name, though he maintains that "an hasty peche is a better and a fitter name for it. But so that the tre be well knowen, I pase not gretely what name it is knowen by." The reason he gives for his name, is that the fruit ripens so much earlier than the peach. The word apricot implies the same idea, being derived from the Latin præcoqua, or præcocca. He says, in 1548, "We have very fewe of these trees as yet," and in 1551, "I have sene many trees of thys kynde in Almany, and som in England." In the beautiful old garden at Littlecote, in Berkshire, there are two apricot trees which still bear fruit, supposed to have been planted when the tree was first introduced into this country.

Tusser, 1573, gives a list of fruits to be set or removed in January, and it includes Apricots, or Apricocks, as he calls them.

The following is his list :—

1. Apple trees of all sorts.
2. Apricocks.
3. Barberies.
4. Boollesse, black and white.
5. Cheries, red and black.
6. Chestnuts.
7. Cornet plums.*
8. Damsens, white and black.
9. Filbeards, red and white.
10. Goose beries.

* = cornel plum = cornel cherries.

EARLY TUDOR GARDENS.

11. Grapes, white and red.
12. Greene or grasse plums.
13. Hurtillberies.*
14. Medlars or marles.
15. Mulberie.
16. Peaches, white and red.
17. Peares of all sorts.
18. Perare plums,† black and yellow.
19. Quince trees.
20. Respis.
21. Reisons.
22. Small nuts.
23. Strawberies, red and white.
24. Seruice trees.
25. Walnuts.
26. Wardens, white and red.
27. Wheat plums.
28. Now set ye may
   the box and bay,
   Haithorne and prim,
   for clothes trim.

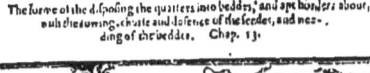

FROM GARDENER'S LABYRINTH.

We cannot prove that red currants had a place in gardens before this time, as they are never mentioned as such; even Gerard, in 1597, does not give them under that name, but describes them as a very small kind of gooseberry without "prickles," of a perfect red colour. But it seems that some sort of currant is intended by "Reisons" in this list.

* = *whortleberries.*
† = *pear-plum.*

Tusser goes on in December's husbandrie to describe how the trees should be planted in the orchard :—

> "Good fruit and good plentie doth well in the loft
> then make thee an orchard and cherish it oft:
> For plant or for stock laie aforehand to cast
> but set or remoove it er Christmas be past
> Set one fro other full fortie foote wide
> to stand as he stood is a part of his pride."

We do not find many other changes in the orchards. Wardens still held a prominent place among pears, and costards among apples. The peach had not improved. Turner speaks of trees abroad, and goes on to say, "The peche is no great tre in England that I could se—the apples are soft and flesshy when they are rype something hory without." Among the Privy Purse expenses of Henry VIII., Mr. Long's gardener is specially mentioned as giving a present of peaches to the King, who at various other times received gifts of cherries, apples, pears, wardens, quinces, medlars, damsons, filberts, and melons.

It was only the large landowners who indulged in a garden specially set apart for flowers and pleasure. The garden of every small manor and farm-house in the kingdom was essentially for use. Fitzherbert, in his *Book of Husbandry*, 1534, enumerates the general duties of a wife, among which he does not forget the garden: "And in the beginning of March or a lyttel afore, is tyme for a wife to make her garden, and to gette as many good sedes and herbes as be good for the potte and to eate, and as ofte as nede shall requyre, it must be weded, for els wedes wyl ouergrowe the herbes." These herbs were much the same as in the previous century, but a few are mentioned in writings of this date, which have not appeared on earlier lists, such as asparagus, melons, taragon, horse-radish, and artichokes, which are first introduced about this date, and grown in the royal gardens. Tusser devotes several lines in his poem to beans and peas. In January—

> "Good gardiner mine
> Make garden fine
> Set garden pease
> and beans, if ye please."

And again,
> "Dig garden, stroy* mallow, now may ye at ease
> And set (as a dainte) thy runciuall pease."

Also
> "Sowe pease (good trull†)
> the moone past full
> Stick bows a rowe
> where runciuals growe."

> "Set plentie of bows among runciuall pease
> to climber thereon, and to branch at their ease."

These quotations show that runcival peas were a favourite dainty. They were a large kind of pea, and the name is supposed to be derived from Roncesvalles, in the Pyrenees. Tusser also gives directions for picking beans—

> "Not rent off, but cut off ripe beane with a knife
> For hindering stalke of hir vegetive life
> So gather the lowest, and leauing the top
> Shall teach thee a trick, for to double thy crop."

In the ordinary course of things, little would have to be bought for a garden, as seeds would be saved, and plants divided and exchanged among friends, year by year.

> "Good huswifes in sommer will saue their owne seedes
> against the next yeere, as occasion needes
> One seede for another, to make an exchange
> With fellowlie neighbourhood seemeth not strange."

Consequently, in old account books we do not find many entries for things bought to stock the garden. But the making so many fine new gardens must have created a demand for plants with which to furnish them. The large quantities of things bought for the newly laid-out gardens could only have been supplied by regular nurserymen, and market gardeners. For instance, such amounts as five hundred rose trees, six hundred cherry trees ‡ at 6d. per hundred, could hardly have been grown in private gardens.

We had a glance at the fruit and vegetable market of

---

* Expression often used probably for the sake of rhythm. = = *weed, or destroy, wild mallow, a common weed.*

† = *good girl, or lass.*

‡ Hampton Court Account.

London in Edward the Second's reign,* and with the great advances in gardening since that time, it is most probable that the market had also increased, and the market gardeners multiplied. Then, as now, the great place for market gardens was the immediate vicinity of London, but some were planted even in the heart of the town, as the following quotation shows:—
" About the latter part of the reign of Henry VIII., the poor people of Portsoken Ward, East Smithfield, were hedged out, and in place of their homely cottages, such houses builded as do rather want room than rent, and the residue was made into a garden by a gardener named Cawsway one that serveth the market with herbs and roots." †

The largest supply of fruit trees came from the orchard at Tenham, in Kent. The history of its establishment is related in a curious and rare pamphlet, entitled, *The Husbandman's Fruitful Orchard.* 1609. The author is unknown, but the epistle to the reader is signed " thy well-willer N.F." ‡ " One Richard Harris of London, borne in Ireland Fruiterer to King Henry the eight fetched out of Fraunce great store of graftes especially pippins, before which time there were no pippins in England. He fetched also out of the Lowe Countries, cherrie grafts and Peare graftes of diuers sorts : Then tooke a peese of ground belonging to the king in the Parrish of Tenham in Kent being about the quantitie of seauen score acres : whereof he made an orchard, planting therein all those foraigne grafts. Which orchard is and hath been from time to time, the chiefe mother of all other orchards for those kinds of fruit in Kent and diuers other places. And afore that these said grafts were fetched out of Fraunce and the Lowe Countries although that there was some store of fruite in England, yet there wanted both rare fruite and lasting fine fruite. The Dutch and French finding it to be so scarce especially in these counties neere London, commonly plyed Billingsgate and diuers other places, with such kinde of fruit, but now (thankes bee to God) diuers gentlemen and others

---

* *See* page 40.
† Stowe, *Survey of London.* Ed. 1598. p. 139.
‡ Imprinted for Roger Jackson, London.

taking delight in grafting . . . have planted many orchards fetching their grafts out of that orchard which Harris planted called the New Garden."

When Drayton wrote his *Polyolbion*, in 1619-22, the orchard must still have been flourishing, as he alludes to it thus,

" Rich Tenham undertakes thy closet to suffice with cherries."—*Song XVIII*.

This orchard is supposed to have produced cherries which sold for £1000 in the year 1540[*]; an immense sum for those days, and it seems an exaggeration when compared with the ordinary prices of cherries, found in the household books about this date; for instance, " Item 9th Julye 1549, 2 lbs. cherrys at my Ladye's comandemente IVd.," and again, " 27th Julye 1549, 4 pond of cherrys IVd."[†] It is difficult to arrive at the ordinary prices given for garden produce. They must, of course, have varied with the seasons, and the quality of the fruit. The difficulty of conveying fruit to market must have kept up the price. One gardener might have great abundance of a certain fruit, while at no great distance a high price was being paid for like wares, but owing to the difficulties of communication, he would be unable to take advantage of this market for his goods. But that they made as much profit as they could, and were not always fair in their dealing, the following Law proves:—
" 2 & 3 Edward VI. c. 15.—Forasmuch as of late divers sellers of victuals not contented with moderate and reasonable gain . . . . have conspired and covenanted together, to sell their victuals at unreasonable prices—butchers, brewers, bakers . . . . costermongers, or fruiterers, £10 fine or twenty days imprisonment and bread and water for his sustenance, second offence £20 and the pillory, third offence £40 or pillory and ears cut off."

The increase in the number of orchards seems to have rendered their legal protection necessary, as another very curious act was passed:—37th Henry VIII. c. 6, sect. 3.— . . . . " Any person maliciously, willingly or unlawfully, after the said

---

[*] Johnson, *Hist. English Gardening*, 1829, p. 56. Philips' *Companion to the Orchard*. Ed. 1821, p. 79.
[†] Le Strange, *Household Books*.

first May (1545) cut or cause to be cut off the ear or ears of any of the King's subjects otherwise than by authority of the law, chance-medley, sudden affray, or adventure : (6) or after the said day maliciously, willingly or unlawfully, bark any apple trees, pear trees, or other fruit trees of any other person or persons (7) that then every such offender and offenders shall not only lose and forfeit unto the party grieved treble damages for such offence or offences, the same to be recovered by action of trespass, to be taken at the common law, but also shall lose and forfeit to the King's Majesty and his heirs, for every such offence X £ sterling in the name of a fine."

Saffron continued to be largely used and grown for the market, and sold at a high price. In the accounts of the Monastery of Durham, "Crocus," or saffron, is of frequent occurrence. In 1531 half a pound was bought in July; the same quantity in August and in November, a quarter of a pound in September, and a pound and a half in October: these items give us some idea of the consumption. In 1539-40 the saffron was bought from Thos. Freeman, of Doncaster, and of a merchant from Cambridge, to the latter, for six and a half pounds of "crocus" £7. 8s. was paid. In 1538 it was bought at "Braydforth fayre." Although it was not cultivated at all in the north, and, as the above quotations show, had to be imported from the Eastern counties, saffron commanded almost as high a price in that part of the country. At Hunstanton, in Norfolk, on "March 26th, 1536, one ounce of saffron cost 8d. and old saffron 12d. the ounce."*

It was a profitable crop, and Tusser, who lived in the Eastern counties, warns the husbandman not to forget it :—

> "Pare saffron plot
> Forget it not
> His dwelling made trim
> look shortly for him
> When harvest is gone
> then saffron comes on
> A little of ground
> brings saffron a pound." †

---

* Le Strange, *Household Books*.
† *Five Hundred Pointes of Goode Husbandrie.* - August.

The work in gardens of all sizes seems to have been superintended by one head-gardener, who had the charge of the buying and selling and planting of the garden stuff; but the actual manual cultivation was done by labourers hired by the day, and not by a permanent staff. The post of head-gardener in any of the Royal gardens was quite an important position.* The wages were from about £12 per annum, and all the money for the payment of labourers passed through the head-gardener's hands. The labourers received 6d., 4d., or 3d. a day, or even 2d. a day if they were given food. The weeding was usually done by women, and 3d. or 2d. a day was the ordinary wage.†

Garden tools have not changed much since the earliest times. The spade and rake we now use are much the same

---

* 1532.—"Also paid by the hands of the forsaid Edmund Gryff(yn) (head-gardener), for digging, gathering, and sorting of the said trees, 12d. Also paid to the said Edmund Gryff(yn), for carriage of the forsaid apple trees, 15d."

1530.—"A gardener at 6d. a day."

1530.—"To John Hutton, for making and levelling of beds in the king's new garden, and raking of the same, by the space (of) 12 days at 4d. a day, 4s."—*Hampton Court Accounts.*

May 8th, 1540.—"To Claaston, for mowyng of the garden at Hunstanton, 2d." September, 1543.—"For dyggen in the garden, 4d."

December 10th, 1549.—"2 ffellowes for helping in the garden for oon week, 2s. 6d."—Le Strange, *Household Books.*

1530.—"Paid to four gardeners for four days—March 18th, 2s. 8d."—.1 *Book of Receipts and Expenses of Cardinal's College Oxford.*

† 1530.—"5 labourers and 15 women weeders in the garden and the orchard;" again, "20 women weeders, 2 labourers, and 2 mowers"—a list of the names of the weeders follows, and the men received 4d. per day, the women 3d.—*Hampton Court Accounts.*

April 23rd (1530).—"Paid to two women rooting up unprofitable herbs (extirpantibus herbas inutiles) in the garden for three days, 16d."

June 6th.—"Paid to Margaret Hall, cleansing the garden, 3d."

June 23rd.—"Joan Fery, working for three days, 10d."

August 19th.—"Paid to Agnes Stringer, working for two days with a half, 7d."

Several more entries of women gardeners follow these: "Paid for bread and drink and herrings and other things (for) the gardeners, all women, as appears by the book of expenses of the second term in the seventh week, 2s. 1¾d."—*Cardinal's College, Oxford.*

"3 whemen for wedyng, 6d."—Le Strange, *Household Books.*

as those of Tudor days. Tusser, in the following passage, enumerates the tools then in use *:—

"Now set doo aske watering with pot or with dish
new sowne doo not so, if ye doo as I wish
Through cunning with dible, rake, mattock and spade
by line and by leuell, trim garden is made."

We know the cost of these tools from various accounts. The prices ranged from 4d. to 1s.†

Probably many of the tools were home-made. Fitzherbert, in 1534, in his *Book of Husbandry*, devotes a paragraph to

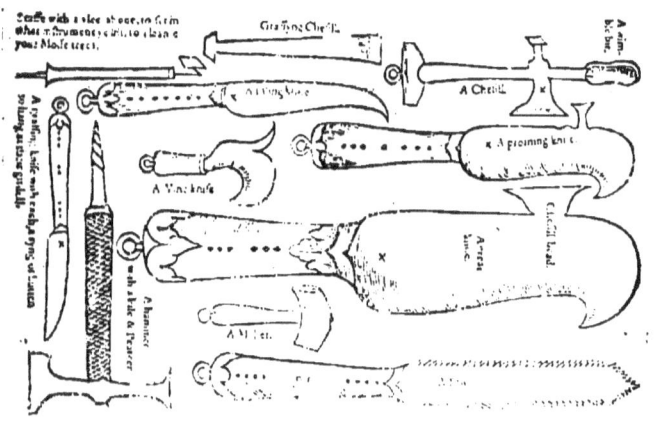

TOOLS USED IN GRAFTING.

showing "howe forkes and rakes shulde be made." He says that they should be prepared in the winter, "when the

* *Five Hundred Pointes of Good Husbandrie*.
† Hampton Court, March, 1533. Item for three iron rakes serving for the King's new garden at 6d. the piece—18d. Item for a hatchet serving for the said garden, 6d. Item for three new knives to shred the quicksets in the new garden at 3d. the piece, 9d. Item for six pieces of round line to measure and set forth the new garden, 12d. Item for two cutting hooks, 2s. Item for two cutting knives, 4d. Item for two rakes, 16d. Item for two chisels, 6d. Item for a graffing saw, 4d. The price paid for a spade at Hunstanton, in Norfolk, on July 7th, 1538, was 8d., and on December 1st, in the same year, 5d. and "for a hattchett, a rake and a parying yearne (= *paring-iron*) for the garden, 10d. March 11th, 1543."—LE STRANGE, *Household Books*.

housbande sytteth by the fyre, and hath nothynge to do than may he make theym redye, and tothe* the rakes with dry wethywode, and bore the holes with his wymble,† bothe aboue and vnder, and drive the tethe vpwarde faste and harde, and than wedge them aboue with drye woode of oke . . . They be most comonly made of hasell and withee." Fitzherbert also gives a list of the tools used for grafting: "a graffynge-sawe . . very thynne, and thycke-tothed," "a graffing-knyfe, an inch brode with a thycke backe, to cleue the stock with all," "a mallet to dryue the knyfe and thy wedge in-to the tree," "a sharpe knyfe to pare the stockes heed, and an other sharpe knyfe to cutte the graffe cleane." " Two wedges of harde wood or elles of yren."

While the husbandman was working in his garden, or making his tools, the housewife busied herself with the preparation of conserves of fruit, and distilling and making decoctions from almost every herb that grew. This business was of such importance that a room was in most houses set apart for the purpose. We have a survival of this custom in the "still room" of modern days. One of Tusser's "five hundred pointes" is "good huswifelie Physicke," in which these stanzas occur:—

> "Good aqua composita, vinegar tart
> Rose water and treakle to comfort the hart.
> Cold herbs in hir garden for agues that burne
> that ouer strong heat to good temper may turne.
>
> Get water of Fumentorie, Liuer to coole
> and others the like, or els lie like a foole
> Conserue of the Barberie, Quinces and such
> with sirops that easeth the sickley so much."

In 1527, a certain printer, " Laurens Andrewe," translated and issued a work entitled, *The vertuose Boke of Distyllacyon of the Waters of all manner of Herbes*, translated from the German of " Jerome of Brynswicke " (Brunswick). It is illustrated throughout with quaint woodcuts, and extraordinary recipes, which, if followed by the housewife, must have added horrors

* = *tooth*. † = *an auger*.

to illness, and perhaps have done her friends and relations more harm than good. Among the plants she is recommended to use are "yellow lillies, floure de luce purpure, periwinkle, house-leek, red and white roses, Solomon's seal, woodbine, peony, marigold, besides all the herbs, such as dill, burnet, dandelion, and fruits, cherries, quinces, peach-leaves, apples, and nuts."

The Household Books of the fifth Earl of Northumberland (1502), contain the following list of "herbes to stylle." "Borage, columbine, buglos, sorrel, cowsloppes, scabious, wild tansey, wormwood, endyff, sauge, dandelion, and hart's tonge." Many herbs in every garden were grown solely for this purpose, and these sweet waters were used in cooking as well as for medicine. A neighbourly gift of distilled herbs was often exchanged, and it is not uncommon to find records of such presents as "sweet waters," "rose water," or "syrup of roses," being accepted by the wealthy from their poorer friends. Similar offerings of flowers or fruit were no less frequent. The Parson of Titteshall sent the Squire of Hunstanton a present of pears and apples, "his boye" receiving a penny for bringing them. On another occasion "wenches," from the same parish, brought him red roses.[*] The Bishop of Norwich sent the Duke of Buckingham a dish of cherries, and one May day "four maydens of Kanisham brought presents of hawthorne to my Lord's Grace, being in his orchard."[†] One feels tempted to pause to entwine a pretty story round these four maidens of Kanisham. Without much strain on the imagination, and with the help of these simple records, it is easy to conjure up delightful visions, and to picture many a fascinating scene of homely country life in Tudor times.

[*] Le Strange, *Household Books* (1540).
[†] Duke of Buckingham's Household Accounts.

CHAPTER VI.

> "Like a banquetting house built in a garden,
> On which the spring's chaste flowers take delight
> To cast their modest odours."
> MIDDLETON, *Marriage*.

THE reign of Elizabeth was a golden era in English history, and abounded in men of genius. Among the many branches of art, science, and industry, to which they turned their attention, none profited more from the power of their great minds, than did the Art of Gardening. Bacon's *Essay on Gardens* is familiar to everyone. Lord Burghley was the patron of Gerard, one of the greatest of English herbalists, and to Sir Walter Raleigh we owe the introduction of our most useful vegetable, the potato.

About this time the persecution of the Protestants on the Continent drove many of them to find a safe refuge in England. They brought with them some of the foreign ideas about gardening, and thus helped to improve the condition of Horticulture.

The Elizabethan garden was the outcome of the older fashions in English gardens, combined with the new ideas imported from France, Italy, and Holland. The result was a purely national style, better suited to this country than a slavish imitation of the terraced gardens of Italy, or of those of Holland, with their canals and fish-ponds. There was no breaking-away from old forms and customs, no sudden change. The primitive mediæval garden grew into the pleasure garden of the early Tudors, which, by a process of slow and gradual development, eventually became the more elaborate garden of the Elizabethan era. What is meant now by a "formal" or "old-fashioned" garden, is one of this type; but, as genuine and unaltered Elizabethan gardens are rare, it is generally the further development of the same style a hundred years later, which is known as a "formal old English garden."

The garden of this period was laid out strictly in connexion with the house. The architect who designed the house, designed the garden also. There are some drawings extant by John Thorpe, one of the most celebrated architects of the time, of both houses and the gardens attached to them. The garden was held to be no mere adjunct to a house, or a confusion of green swards, paths, and flower-beds, but the designing of a garden was supposed to require even more skill than the planning of a house: "men come to build stately sooner than to garden finely:—as if gardening were the greater perfection."* Sir Hugh Platt's opinion† seems to have been the exception that proves the rule, as most other writers were particular in describing the correct form for a garden, but he writes:—"I shall not trouble the reader with any curious rules for shaping and fashioning of a garden or orchard—how long, broad or high, the Beds, Hedges, or Borders should be contrived. . . . Every Drawer or Embroiderer, nay (almost) each Dancing Master, may pretend to such niceties; in regard they call for very small invention, and lesse learning."

In front of the house there was usually a terrace, from which the plan of the garden could be surveyed. Flights of steps and broad, straight walks, called "forthrights,"‡ connected the parts of the garden, as well as the garden with the house. Smaller walks ran parallel with the terrace, and the spaces between were filled with grass plots, mazes, or knotted beds. The "forthrights" corresponded to the plan of the

* Bacon, *Essay on Gardens*.
† *Floraes Paradise, or Garden of Eden*. 1st ed., 1608.
‡ . . . "here's a maze trod indeed,
   Through forthrights and meanders . . .
                                    Tempest, act iii. scene 8.

building, while the patterns in the beds and mazes harmonized with the details of the architecture. The peculiar geometric tracery which surmounted so many Elizabethan houses, found its counterpart in the designs of the flower-beds. "The form that men like in general is a square,"* and this shape was chosen in preference to "an orbicular, a triangle, or an oblong, because it doth best agree with a man's dwelling."† This square garden was usually enclosed by a high brick or stone wall. "He hath a garden circummured with brick."‡ The picture which does duty both in Thomas Hill's *Gardener's Labyrinth*, and in his *Art of Gardening*, shows a square garden with a paling round it. Another illustration, which appears three times in the *Gardener's Labyrinth*, gives a brick wall; while, in a third, the garden is enclosed by a hedge. The custom of covering the walls with rosemary was "exceedingly common in England."§ At Hampton Court rosemary was "so planted and nailed to the walls as to cover them entirely." Gerard‖ and Parkinson both refer to the custom of planting against brick walls. In the North of England, according to Lawson, the garden-walls were made of "drie earthe," and it was usual to plant "thereon wall-flowers and divers sweet-smelling plants."

Bacon has a more magnificent plan:—"The garden is best to be square, encompassed on all four sides with a stately arched hedge. The arches to be upon pillars of carpenter's work, of some ten foot high, and six foot broad, and the spaces between of the same dimension with the breadth of the arch." This "fair hedge" of Bacon's ideal garden was to be raised upon a bank, set with flowers, and little turrets above the arches, with a space to receive "a cage of birds";—"and over every space between the arches, some other little figure, with broad plates of round coloured glass, gilt, for the sun to play upon." It is not likely that such fantastical ornaments to a hedge were usual, though it reminds one of the arched

* Lawson, *New Orchard*, 1618.   † Parkinson.
‡ Shakespeare, *Measure for Measure*, act iv. scene 1.
§ *Hentzner's Travels*. 1598.
‖ Gerard is spelt Gerarde on the engraved title of his herbal, but he signs the Preface without the e.

arcades, already referred to, and does not seem to be at all a new idea of Bacon's. Thomas Hill* discusses the various modes of fencing round a garden. A paling of "drie thorne" and willow he calls a "dead or rough inclosure." He refers to the Romans for examples of the alternative of digging a ditch to surround the garden, but "the general way" is a "natural inclosure," a hedge of "white thorne artely laide: in a few years with diligence it waxeth so thicke and strong, that hardly any person can enter into the ground, sauing by the garden-door; yet in sundrie garden groundes, the hedges [are] framed with the privet tree, although far weaker in resistance, which at this day are made the stronger through yearly cutting, both aboue and by the sides." He gives a quaint method for planting a hedge. The gardener is to collect the berries of briar, brambles, white-thorne, gooseberries and barberries, steep the seeds in a mixture of meal, and set them to keep until the spring, in an old rope, "a long worn roape . . . being in a manner starke rotten." "Then, in the spring, to plant the rope in two furrows, a foot and a half deep, and three feet apart. . . . The seedes thus covered with diligence shall appeare within a month, either more or less"—"which in a few years will grow to a most strong defence of the garden or field." These old gardeners had great faith in all their operations, and but rarely in their works do we find any allusion to possible failure!

Yews were much employed for hedges, but more for walks and shelter within the gardens, than to form the outer enclosure. In the larger gardens there were two or three gates in the walls, well designed, with handsome stone piers surmounted with balls or the owner's crest, and wrought-iron gates of elaborate pattern; or else there was one fine gate at the principal entrance, the rest being smaller and less pretentious, merely "a planched gate,"† or "little door." The main principle of a garden was still that it should be a "garth," a yard, or enclosure; the idea of such a thing as a practically unenclosed garden had not, as yet,

---

* *Gardener's Labyrinth*, 1608.
† *Measure for Measure*, act iv, scene 1.

entered men's minds. But because the garden was surrounded with a high wall, and those inside wished to look beyond, a terrace was contrived. As in the Middle Ages, we find an eminence within the walls, as a point from which to look over them; so at the period we have now reached, the restricted view from the mount did not satisfy, and to get a more extended range over the park beyond and the garden within, a terrace was raised along one side of the square of the wall. "I have seen a garden," says Sir Henry Wotton, "into which the first access was a high walk like a terrace, from whence might be taken a general view of the whole plot below." De Caux, the designer of the Earl of Pembroke's garden at Wilton, made such a terrace there "for the more advantage of beholding those platts."* Another is described at Kenilworth in 1575, "hard all along by the castle wall is reared a pleasant terrace, ten feet high and twelve feet broad, even under foot, and fresh of fine grass."† The terraces, as a rule, were wide and of handsome proportions, with stone steps either at the ends or in the centre, and were raised above the garden either by a sloping grass bank, or brick or stone wall. At Kirby, in Northamptonshire, a magnificent Elizabethan house, now rapidly falling into decay, all that remains of a once beautiful garden, "enrich'd with a great variety of plants,"‡ is a terrace running the whole length of the western wall of the garden. It is now planted with potatoes, and the garden it overlooked is merely a meadow. The lines in Spenser's *Ruins of Time* might have been written on this garden had he but seen it in its present state.

> "Then did I see a pleasant paradize
> Full of sweete flowers and daintiest delights,
> Such as on earth man could not more devize;
> With pleasure's choyce to feed his cheerful sprights.
> Since that I sawe this gardine wasted quite,
> That where it was scarce seeméd anie sight;
> That I, which once that beautie did beholde,
> Could not from teares my melting eyes with-holde."

---

* *Le Jardin de Wilton.* De Caux, 1615.
† Robert Laneham, Letter describing the Pageants at Kenilworth Castle, 1575. Extract in *Praise of Gardens.* Sieveking, 1885.
‡ Morton, *Natural History of Northamptonshire,* 1712.

At Drayton, an Elizabethan house in the same county as Kirby, there is a wide terrace against the outer wall of the garden with a summer-house at each end, as well as a terrace in front of the house, and other examples exist.

The "forthrights," or walks which formed the main lines of the garden design, were "spacious and fair." Bacon describes the width of the path by which the mount is to be ascended as wide "enough for four to walk abreast," and the main walks were wider still, broad and long, and covered with "gravel, sand or turf."* There were two kinds of walks, those in the open part of the garden, with beds geometrically arranged on either side, while sheltered walks were laid out between high clipped hedges, or between the main enclosure wall and a hedge; there were also the "covert walks," or "shade alleys," in which the trees met in an arch over the path. Some of the walks were turfed, and some were planted with sweet-smelling herbs. "Those which perfume the air most delightfully, not passed by as the rest, but being trodden upon and crushed, are three—that is, burnet, wild thyme and water-mints; therefore you are to set whole alleys of them to have the pleasure when you walk or tread." † It appears from a passage in Shakespeare, 1 *Henry IV.*, act ii. scene 4, that camomile was used in the same way. Falstaff says, "For though camomile, the more it is trodden on the faster it grows; yet youth, the more it is wasted, the sooner it wears."

In contrast to this the "closer alleys must be ever finely gravelled and no grass, because of going wet." ‡ Thomas Hill § writes, the "walkes of the garden ground, the allies even trodden out, and leuelled by a line, as either three or four foote abroad, may cleanely be sifted ouer with riuer or sea sand, to the end that showers of raine falling, may not offend the walkers (at that instant) in them, by the earth cleauing or clagging to their feete." Parkinson also has something to say about walks: "The fairer and larger your allies and walks be, the more grace your garden shall have, the lesse harm the

---

* Lawson, *A New Orchard*. 1597. ‡ Bacon, *Ess iv*.
† Bacon, *Essay*. § *Gardener's Labyrinth*.

herbs and flowers shall receive, by passing by them that grow next unto the allies sides, and the better shall your weeders cleanse both the bed and the allies."

The hedges on either side the walks were made of various plants—box, yew, cypress, privet, thorne, fruit trees, roses, briars, juniper, rosemary, hornbeam, cornel, "misereon," and pyracantha. "Every man taketh what liketh him best, as either privet alone or sweet Bryar, and whitethorn interlaced together, and Roses of one, two, or more sorts placed here and there amongst them. . . . Some plant cornel trees and plash them or keep them low to form them into a hedge; and some again take a low prickly shrub that abideth always green, called in Latin Pyracantha." Of the cypress, Parkinson writes: "For the goodly proportion this tree beareth, as also for his ever grene head, it is and hath beene of great account with all princes, both beyond and on this side of the sea, to plant them in rowes on both sides of some spatious walke, which, by reason of their highe growing, and little spreading, must be planted the thicker together, and so they give a pleasant and sweet shadow." Gerard, writing of the same plant, says: "It groweth likewise in diuers places in Englande, where it hath beene planted, as at Sion, a place neere London, sometime a house of nunnes; it groweth also at Greenwich and at other places; and likewise at Hampstead in the garden of Master Waide, one of the Clarkes of hir Maiesties Priuy Counsell." *

Many of the walks and alleys were "shadowed over with vaulting or arch-hearbes." † Bacon thus explains the object of making "these pleached alleys," or "covert" walks. "But because the alley will be long, and in the great heat of the year or day you ought not to buy the shade in the garden by going in the sun through the greene (you ought) to plant a covert alley upon carpenter's work, about twelve foot in height by which you may go in shade into the garden." The "thick-pleached alley," in which Antonio saw Don Pedro and Claudio walking, in *Much Ado About Nothing*, was one of this sort. The word "pleach,"

---

* Thomas Hill, *Gardener's Labyrinth*.
† Ibid.

or "plash," or "impleach," is from the French "plesser," from "plexum," to plait, infold, or interweave. It is used by Shakespeare, not only for cut and intwined trees, as in this case, but also for braided hair, "their hair with twisted metal amorously impleach'd," in *A Lover's Complaint,* and for arms enfolded, "with pleacht armes, bendinge down," in *Anthony and Cleopatra.*

The plants used to form these shady walks were willows, limes, wych-elms, hornbeam, cornel, privet or whitethorn, also "the great maple or sycamore tree cherished in our land only in orchards, or elsewhere, for shade and walks." . . . "It is altogether planted for shady walks, and hath no other use with us that I know."* The alley remaining at Hampton Court is of wych-elm. At Theobalds these trees were chiefly used in those alleys where "one might walk twoe myle in the walkes before he came to their ends." At Drayton, in Northamptonshire, there are two fine specimens of pleached alleys, and the gnarled stems of the wych-elms forming them, bear testimony to their age. The covert walks were sometimes made with a trellis of wood-work, planted with creepers, as we have seen in earlier times, "made like galleries," "covered with ye vine spreading all over, or some other trees which more pleased them." †

Mounts still formed an important accessory to the garden. Bacon, who, it must be remembered, was "speaking of those (gardens) which are indeed princelike," describes the mount. "I wish," he says, "in the middle, a fair mount, with three ascents, and alleys enough for four to walk abreast: which I would have to be perfect circles, without any bulwarks or embossements: and the whole mount to be 30 feet high, surmounted by a fine banquetting-house with some chimneys neatly cast." Such banqueting-houses were often made merely for some special occasion, and decorated with ivy and evergreens, to give them the appearance of permanency. This was an age that delighted in pageants, and what more fitting background for their display than the beautiful gardens that this same love of

---

* Parkinson, *Paradisus.* † Hill, *Gardener's Labyrinth.*

PLEACHED ALLEY AT DRAYTON.

display was creating and developing. When any pageant or "revells" took place, additions were made to the arbours or banqueting-houses in the garden, to accommodate the guests. In June, 1554, " certaine banqueting-houses of Bowes (= *boughs*) and other devices of pleasure," were to be made at Oatlands, and Sir Thomas Cawarden, as "Master of the Tents and Toyles," received a royal command to superintend their erection as he had " good experience heretofore in lyk things."* The following extracts show some of his past experiences, both what he had to do, and the cost of carrying it out.† "4th year of Edward VI.— Banketing-houses 2, the one in Hyede Parke conteynenge in length 57 feet and in bredth 21 feet of assize with a halpace staier (step for daïs) conteining the bredth the one way 60 foote and the other way 30 foote and over the same a type or turret garnished. One other house in Marybone Parke conteyninge in length 40 foote the same adjoined framed, made and wrought of tymber, brick, and lyme, with their raunges and other necessary utensyles therto insident, and to the like accustomed. And 6 standinges whereof were in either of the said parkes. 3 all of tymber garnished with boughes and flowers every (one) of them conteynenge in length 10 foote and in bredth 8 foote * * * Employed on the above works for 22 days, at all hours a space to eat and drynke excepted." Carpenters and bricklayers 1d. the hour. labourers ½d. the hour—plasterers 11d. a day, painters 7d. and 6d. a day. " Charges for cutting boughs in the wood at Hyde Park for trimming the banquetting-house, gathering rushes, flags, and ivy." . . . " Taylors for sewing the roof, &c.: basket makers working upon windows.—Total cost. £169 . 7 . 8."

In Stow's *Annals* another of these banqueting-houses is described. It was made in 1581, at Whitehall, " for certaine Ambassadors out of France." It was round, being 332 feet in circumference, and was built on the S.W. of the palace near the river. Over the canvas roof, painted like clouds, "this house was wrought most cunningly with ivy and holly, with pendants made of wicker rods garnished with bay, rue, and all manner of strange flowers garnished with spangles of gold . . . beautiful

---

* MSS. belonging to M. More Molyneux, Loseley, Surrey.   † Ibid.

Fig. 1. N. 1 SCHOOL

with teasons (= *festoons*) made of ivy and holy, with all manner of strange fruits, as pomegranates, oranges, pompions, cucumbers, grapes, with such like spangled with gold, and most richly hanged."

Of course, such banqueting-houses were only made on State occasions, and could only be afforded by the wealthy. The mount in an ordinary garden was surmounted by an arbour of the plainest description. It may have been a great convenience as a point from which a good view could be secured, especially in a garden not sufficiently grand or large to have a raised terrace; but in these more modest gardens, unless planted with flowering plants and creepers, a mount cannot have been a beautiful object. There is such a mount still to be seen at Boscobel. Nothing could be plainer than this: and it is probably a good sample of the mounts I am speaking of, although it cannot be so early as Elizabethan times. It was most likely made when the house was built, about 1620, and it was in its present state when Charles II. hid in the oak-tree hard by. The battle of Worcester was fought on Wednesday, September 3rd, 1651. The Saturday following, Charles spent in hiding in the "Royal Oak," at Boscobel, and the next day, "His Majesty, finding himself now in a hopefull security, spent some part of this Lord's-day in a pretty arbor in Boscobel garden, which grew upon a mount, and wherein there was a stone table and seats about it. In this place he pass'd away some time in reading, and commended the place for its retiredness." *

The mount was not always a circular lump standing out in the garden; it appears that it was still sometimes banked up against the outside wall. Bacon describes this kind also: "At the end of both the side grounds," he writes, " I would have a mount of some pretty height, leaving the wall of the enclosure breast-high, to look abroad into the fields." The erections placed on the top of mounts did not do away with the use of other arbours in less exposed places in the garden. Some "arbour

---

* *Boscobel, or the History of His Sacred Majesties most miraculous Preservation after the Battle of Worcester, 3 Sept., 1651.* By Thomas Blount, 1660; reprint, 1822. The illustration on page 117 is from this work.

o'ergrown with woodbines," [*] or "pleached bower where honey-suckles ripen'd by the sun, forbid the sun to enter," [†] were sure to be found in a secluded spot. "You may," writes Thomas Hill,[‡] "make the herbes either straight or runing up, or else vaulted or close over the head, like to the vine herbers now adaies made. And if they be made with juniper-wood, you neede to repair nothing thereof for ten years after: but if they be made with willow poles, then must you new repaire them euery 3 yeare after. And he which will set Roses to run about his herber, or beds round about his, must set them in Februarie. . . . And in the like manner you may doe, if you will sowe that sweet tree or flower named Jacemine, Rosemary, or the Pomegranate seedes, unless you had rather decke your herbers comelier with

BOSCOBEL IN 1660.

vines." We learn some of the other plants used for arbours from Parkinson. "The Jacimine, white and yellow, the double Honeysocke, the Ladies' Bower, both white and red and purple, single and double, are the fittest of outlandish plants to set by arbours and banquetting-houses that are open both before and above, to help to cover them, and to give both sight, smell, and delight." The "Ladies' Bower" is Clematis Vitalba, or "traveller's joy," and some five foreign species of Clematis.[§] Kidney beans

---

[*] Fletcher, *Faithful Shepherdess*.
[†] *Much Ado About Nothing*, act iii, scene 1.
[‡] *Art of Gardening*.   [§] *Paradisus*, page 392.

were also employed. They "do easily and soone spring up, and growe into a very great length; being sowen neere vnto long poles fastened hard by them or hard by arbors and banquetting places."*

Parkinson describes a curious arbour made in a lime tree. That tree, he says, "is planted to make goodly arbors, and summer banqueting-houses, either below upon the ground, the boughs serving very handsomely to plash round about it, or up higher for a second above it, and a third also." He goes on to explain the "goodliest spectacle that ever" his eyes beheld, was at Cobham, in Kent, where an arbour was made in this way, boards to tread on were laid on the first series of boughs 8 feet from the ground, the stem again kept bare of branches another 8 or 9 feet, and a second lot of branches plashed to form the roof of the middle, and the floor of yet a third arbour, and stairs arranged to mount up to it; the arbour, he says, would hold "half a hundred men at the least."† The following lines in Spenser's *Faerie Queene* perhaps convey to us a more vivid impression of an Elizabethan arbour than the tumble-down, or overgrown, remains of one, in the corner of some, perchance, neglected garden, could possibly do:—

> "And over him Art, striving to compare
> With Nature, did an arbour green dispread,
> Fram'd of wanton ivy, flow'ring fair,
> Through which the fragrant eglantine did spread
> His prickling arms, entrail'd with roses red
> Which dainty odours round about them threw;
> And all within with flow'rs was garnished,
> That, when mild Zephyrus amongst them blew,
> Did breathe out bounteous smells and painted colour shew."
> —Book II., Cantos V. 29.

The maze was another feature which now became prominent in many gardens. "There be some that set their mazes with Lavender, cotton spike, majoram and such like,‡ or Isope and Time, or quickset, privet, plashed fruit trees." Lawson gives directions for making mazes, and says, "When they are

---

* Gerard's *Herbal*, page 1141.
† *Paradisus*, page 610.   ‡ Thomas Hill.

well formed of a man's height, your friend may perhaps wander in gathering berries till he cannot recover himself without your help." Thomas Hill gives two designs for mazes, though he says they are not "for any necessary commoditie in a garden, but rather" . . . . "that who so listeth having such room in their garden may place the one of them . . in that void place . . . . that may best be spared for the only purpose to sport in them at times." Many people, on the mention of the word maze, will at once think of the well-known example at Hampton Court,

MAZE.

which affords so much amusement to thousands of Londoners, and holiday-makers; but that was not laid out till a very much later date, probably in the year 1700.

Trees cut in fantastical shapes were frequently to be found between the hedges, dotted about and arranged so as to form vistas and walks. Bacon advises in "ordering of the ground within the great hedge" . . . . that "it be not too busy or full of work," or, as we should say, not too elaborate, and he adds, "I, for my part, do not like images cut out in juniper or other garden stuff—they be for children. Little low hedges round like welts, with some pretty pyramids, I like well, and in some places, fair columns."

The idea that cut trees were generally yews is very prevalent, and the remains of topiary work in old gardens still in existence confirm this impression. All the cut trees in the garden at Heslington, near York, are yews. This

HESLINGTON.

garden was laid out soon after the house was built, about 1560. The quaintly-rounded hedge at Rockingham, and the hedges and trees at Erbistock, are two examples of the cut yews of this date. But in the books of the period, other shrubs are spoken of more favourably than yews. It seems, therefore, that it is only because the yew is a slow grower, a sturdy tree, and an evergreen, that more yews than other shrubs have survived. Parkinson says of the "use of the yew," "It is found planted both in the corners of orchards and against the windows of houses, to be both a shadow and an ornament, it being always green." But of the privet he writes, " Because the use of this plant is so much, and so frequent throughout all this land, although for no other purpose but to make hedges or arbours in gardens, &c., whereunto it is so apt, that no other can be like unto it, to be cut, lead, and drawn into what forme one

EXAMPLE OF TOPIARY WORK IN COTTAGE GARDEN, HADDON.

will, either of beasts, birds, or men armed or otherwise: I could not forget it, although it . . . . be an hedge bush." "Your Gardiner," writes Lawson in 1618, "can frame your lesser wood to the shape of men armed in the field, ready to give battell: or swift-running Grey Hounds to chase the Deere, or hunt the Hare. This kind of hunting shall not waste your corne, nor much your coyne." Rosemary also was "sette by women for their pleasure, to grow in sundry proportions, as in the fashion of a cart, a peacock, or such by things as they fancy."*

Flowers were planted in borders along the walks and hedges, "thin and sparingly, lest they deceive the trees"† (*i.e.* rob the trees of nourishment), but the principal receptacles for flowers were "open beds," called "open knots," in contradistinction to the complicated knots. The most practical gardeners did not look with favour on the "curiously knotted garden,"‡ although all books of this period give designs for knots. Parkinson has a page of designs merely to "satisfy the desires" of his readers; he himself considered "open knots" more suitable for the display of flowers. There was not any room left for planting other things between the lines of thyme, thrift, hysop, or whatever the intricate pattern was carried out in. Sometimes the design was simply drawn out in coloured earths, a practice of which Bacon disapproved:—"As for the making of knots or figures with divers-coloured earths .... they be but toys, you may see as good sights many times in tarts." The more simple knots were usually bordered with box, a practice which seems to have been introduced by French gardeners. Parkinson calls it "French or Dutch Box," and recommends it "chiefly and above all other herbs," as it was not so liable to overgrow the beds and distort the pattern, as "Thrift, Germander, Marjerome, Savorie," &c., and did not suffer so much from "the frosts and snows in winter," or the "drought in summer." Lavender cotton (*Santolina chamæcyparissus*), a new importation, was also used, and "the rarity and novelty of this herb being for the most part but in the gardens of great persons. doth cause it to be of greater regard."§

If herbs or box were not used for bordering, "dead material" was the alternative, such as lead, either plain or "cut out like unto the battlements of a church," or oak boards, or tiles, or the shank-bones of sheep, "stuck in the ground, the small end downwards, which will become white, and prettily grace out the ground." Another plan was to use "round

---

* Barnaby Googe's *Husbandry*, 1578. Translation of *Conrad of Heresbach*.
† Bacon.   ‡ *Love's Labour's Lost*, act i. scene 1.   § Parkinson.

whitish or blewish pebble stones"—this method Parkinson puts last in his list, "for it is the latest invention . . . . and maketh a pretty handsome shew." It seems strange that such a simple thing as stones for edging should not have been thought of before. Within these edgings, the "open knots" were filled with flowers, "all planted in some proportion as neare one unto another as is fit for them," which "will give such grace to the garden that the place will seem like a piece of tapestry of many glorious colours." Parkinson divides the flowers to be planted in gardens roughly into two sections, "English Flowers," and "Outlandish Flowers." Among English flowers he names all those we have already noticed as being grown in earlier times, such as primroses, daisies, marigolds, gilloflowers, violets, roses, and columbines, and among outlandish flowers, or "flowers that being strangers unto us, and giving the beauty and bravery of their colours so early, before many of our own-bred flowers, the more to entice us to their delight . . . . are almost in all places, with all persons, especially with the better sort of the gentry of the Land," "namely Daffodils, Fritillarias, Jacinths, Saffron-flowers, Lillies, Flowerdeluces, Tulipas, Anemones, French cowslips or Bears' Ears, and a number of suchlike flowers, very beautiful, delightful, and pleasant."

The number of "outlandish" flowers grown in our gardens was rapidly increasing. All through this period, flowers were coming in, both from the Old and the New World. The following are a few among the best known of these importations:—"The Crown Imperial," both orange and yellow, and varieties of the small Fritillary, then called the "Turkie, or Guinichen flower, or chequered daffodil." The hardy cyclamen (*europœum*); the Lobelia cardinalis, the Passion flower (*Passiflora incarnata*), or "Virgin climer." The Christmas rose. Helleborus niger, niger angustifolius and vernalis. The common white lilac, or "pipe tree," and syringa (*Philadelphus coronarius*); also the common cotoneaster and laburnum; several species of martagon lilies; the common yellow jasmine; the sweet-scented marvel of Peru and evening primrose, and the hardy spiderworts; the African marigold, and sunflowers and larkspurs, both annual and perennial: the snowflakes, which were classed with snowdrops as "bulbous violets"! and Ranunculus, "the crowfoot of Illyria" (*R. illyrius*) and asiaticus, also Bachelor's buttons. (*R. plantanifolius flore-pleno* and *aconitifolius*), from the "Alpish Mountains": sweet Sultan, the Centaurea moschata, Dictamus Fraxinella; Balsam impatiens: some species of campanula, and the bright Convolvulus minor (*C. bicolor*).

Several new plants were introduced by the exertions of some of the leading patrons of gardening. Lord Burghley and Lord Carew were the first to try growing oranges in England. Lord Salisbury employed Tradescant to procure new varieties of fruit trees and other plants from abroad. Lord Zouche, also, deserves a foremost place among the encouragers of horticulture. He was the patron of Lobel, and had a fine Physic Garden at Hackney, of which Lobel had the charge. Lord Zouche himself, also brought back plants from abroad. Gerard mentions two in particular. "The small Candy mustard," which grows in "Austria, Candy, Spain and Italy," was brought by him on his return "from those parts." Also the "Thorne apple," the seeds of which he presented to Gerard.

New plants, and new ideas about gardening, were also coming in from France and the Low Countries, with the influx of Protestant refugees. The Huguenots who came to this country were representatives of almost every trade and craft, and especially that of gardening, which greatly improved under the influence of these new-comers, and members of that craft were among those who took out Letters of Denization in 1544. Many of these foreign gardeners settled about Sandwich, Colchester, and Norwich, and greatly improved gardening in those districts. Foreign gardeners were employed by several landowners in the neighbourhood, to alter and lay out their gardens. In 1575, a Dutch gardener was paid 3s. 4d. for "his travayle from Norwich to Hengrave to viewe ye orchards, gardyns and walks," and 40s. was also "paid to the Dutchman for clypping the knotts, altering the alleys, setting the grounde, finding herbs and bordering the same."* It was these foreigners, also, who

* Huguenot Society. *Walloons and their Church at Norwich*. W. T. C. Moens, 1887.

first set on foot the "Florist Feasts," for which Norwich was famed.

In the gardens typical of this age, between the flower-beds, and at intervals along the terrace or beside the walks, lead or stone vases were sometimes placed, either filled with flowers, or merely for ornament. Beautiful examples of lead vases still exist in some old gardens. At Drayton, in Northamptonshire, there are a number of these vases of different sizes throughout the garden. Two may be seen in the illustration on page 113. Other ornaments were not so frequent as in later times; "Great Princes sometimes add statues and such things for state and magnificence, but nothing to the true pleasures of a garden."*

Parkinson says a garden should have "a fountain in the midst thereof to convey water to every part of the garden, either in pipes under the ground, or brought by hand and emptied into large cisterns or great Turkey jars, placed in convenient places." Bacon writes:—"For fountains, they are a great beauty and refreshment: but pools mar all, and make the garden unwholesome and full of flies and frogs. Fountains I intend to be of two natures; the one that sprinkleth or spouteth water, the other a fair receipt of water of some 30 or 40 foot, but without fish, slime, or mud. For the first, the ornaments of images gilt, or of marble, which are in use, do well. . . . Also some steps up to it, and some fine pavement about, doth well. As for the other kind of fountain, which we may call a bathing-pool, it may admit much curiosity and beauty, wherewith we will not trouble ourselves; as, that the bottom be finely paved, and with images; the sides likewise, and withal embellished with coloured glass and such things of lustre, encompassed also with fine rails of low statues." In the ordinary garden the "fair receipt of water" was not so much embellished, being merely a straight pond with stone steps at each corner, the rest of the bank of smooth turf. November 25th, 1595, Sir Thomas Cecil wrote from Wimbledon to Sir William More, of Loseley, saying that "hearing he has

---

* Bacon.

made divers great pools, he begs him to procure one skilful therein, as certain banks he has made that year about a great pool, have given way through unskilfulness of the workmen."*
The pools at Loseley must have been some time in existence, as on December 21st, 1581, Henry Sledd, Queen Elizabeth's fishmonger, wrote to Sir William More, offering to buy some carp out of his pond. He offers from 12d. to 18d. a piece, according to their size, and adds, "Yf I see they be more worthe . . . . I will mend the pryse." †

Of the first kind of fountain there were many examples in the finest gardens at the time when Bacon wrote. Frederick, Duke of Wurtemberg, describes the one he saw at Hampton Court, in 1592 ‡ :—" In the middle of the first and principal court stands a splendid high and massy fountain, with an ingenious water-work, by which you can, if you like, make the water to play upon the ladies and others who are standing by, and give them a thorough wetting." Of this same fountain Norden wrote in 1598, "Queen Elizabeth hathe of late caused a very beautiful fountaine there to be erected in the second court, which graceth the Pallace, and serveth to great and necessarie use; the fountaine was finished in 1590, not without great charge." Another of the same sort was to be seen at Whitehall, and is described by Hentzner, in 1598:—"A jet d'eau with a sundial, which, while strangers are looking at it a quantity of water forced by a wheel which the gardener turns at a distance, through a number of little pipes, plentifully sprinkles those that are standing round." Hentzner also visited Nonsuch, and notices several fountains. In the "privy gardens" were two "that spurt water one round the other like a pyramid upon which are perched small birds that stream water out of their bills." In the "Grove of Diana," was one "with Actæon turned into a stag as he was sprinkled by the goddess and her nymphs," and a "pyramid of marble full of concealed pipes which spurt upon all that come near." The word "jet d'eau"

---

* MS. letter at Loseley, Surrey.   † Ibid.
‡ Translation, 1602—printed in *England as Seen by Foreigners*. By Brenchley Rye, 1865.

is usually used by contemporary writers for such fountains, and seems to point to their introduction from France.

Other pieces of water were introduced into gardens; like the trout stream running through the orchard at Littlecote, or the stream in the Deanery garden at Winchester, where Isaac Walton used to fish. Beddington (in Surrey), which belonged to Francis Carew, was described by Wurmsser von Vendenheyn, in 1610, as "one of the most pleasant and ornamental gardens in England, with many beautiful streams." At Theobalds and Hatfield there was water. At Hatfield * the banks of the stream in what was called the dell, were beautified with flower-beds and sundry arbours and walks, which were connected with the vineyard on the opposite bank by ornamental bridges. The works were designed and carried out by Mountain Jennings, gardener to the first Earl of Salisbury. A Frenchman, named Simon Sturtivant, planned some elaborate water-works, which were never executed owing to the Earl's death in 1612, as also did Soloman de Caux. One jet d'eau, however, from a design of the latter, was made at a cost of £113 and consisted of a marble basin with a statue of Neptune; 310 lbs. of solder were used to cast the figure, which was probably gilded afterwards. De Caux was the designer of the gardens at Wilton, for the Earl of Pembroke, where there were "foure fountaynes with statues of marble in their midle," and "two Ponds with Fountaynes and two collumnes in the middle, casting water all their height, which causeth the moveing and turning of two crownes at the top of the same." Besides this, the river passed through the garden, and was spanned by an ornamental bridge. The latter was removed later on, and the well-known work of Inigo Jones built in its place.

The garden at Theobalds is also described by Hentzner in 1591:—"In the gallery was painted the genealogy of the kings of England; from this place one goes into the garden, encompassed with water, large enough for one to have the pleasure of going in a boat, and rowing between the shrubs;

* From family MSS. belonging to the Marquess of Salisbury.

here are a great variety of trees and plants, labyrinths made with a great deal of labour, a jet d'eau with its bason of white marble, and columns and pyramids of wood and other materials up and down the garden. After seeing these, we were led by the gardener into the summer-house, in the lower part of which, built semi-circularly, are the twelve Roman Emperors in white marble and a table of truck-stone; the upper part of it is set round with cisterns of lead into which the water is conveyed through pipes, so that fish may be kept in them, and in summer-time they are very convenient for bathing. In another room for entertainment very near this and joined to it by a little bridge was a noble table of red marble."

Having now completed the survey of the several features of an Elizabethan garden, terraces, walks, alleys, mazes, mounts, arbours, fountains and streams having been looked at one by one; it only remains to take a glance at it as a whole. The two following descriptions of a garden take in all these details, and are both contemporary, although from two very different sources. One is the description of a stage arranged to represent a beautiful garden, on the occasion of the performance of a " Maske of Flowers," by the gentlemen of Gray's Inn, at Whitehall, upon Twelfth Night, 1613, " being last of the solemnities and magnificences which were performed at the marriage of the Earl of Somerset and Lady Francis, daughter of the Earle of Suffolke, Lord Chamberlaine;"[*] the other is from Spenser's *Faerie Queene*, the lines in which he pictures a perfect garden, a " second Paradise."

The Maske of Flowers.

" The Daunce ended, the lowd musicke sounded. The Trauers being drawne, was seen a garden of a glorious and strange beauty, cast into foure quarters, with a crosse walke and allies compassing each quarter. In the middle of the crosse walke stood a goodly Fountaine, raised on foure

---

[*] This Maske was printed in 1614 by N. D. for Robert Wilson. It is extremely rare, the quotation is made from a perfect copy belonging to Mrs. Rowley Smith, Plawhatch, Bishop's Stortford.

columnes of Silver. On the toppes whereof strode foure statues of silver, which supported a bole in circuite containing foure and twenty foote, and was raysed from the ground nine foot in height, in the middle whereof upon scrowles of silver and gold, was placed a globe garnished with 4 golden maske heads out of the which issued water into the bole, aboue stood a golden Neptune in height 3 foote holding in his hand a Trident. The garden walls were of brick artificially painted in Perspectiue, all along which were placed fruite trees with artificiall leaues and fruite. The garden within the wall was rayled about with rayles of three foote high, adorned with Ballesters of Siluer, between which were placed pedestalls beautified with transparent lights of variable colours, vpon the Pedestalls stood siluer columnes, upon the toppes whereof were personages of golde, Lions of golde and Vnicornes of silver. Euery personage and beast did hold a torchet burning that gaue light and lustre to the whole fabrique. Euery quarter of the garden was finely hedged about with a lowe hedge of Cipresse and Juniper; the knottes within set with artificiall flowers. In the two first quarters, were two Piramides, garnished with golde and siluer, and glistening with transparent lights, resembling carbuncles, saphires, and rubies. In every corner of each quarter were great pottes of gilliflowers, which shadowed certaine lights placed behind them, and made resplendent and admirable lustre. The two further quarters were beautified with Tulipaes of diuers colours, and in the middle, and in the corners of the said quarters, were set great tufts of seuerall kindes of flowers receiuing lustre from secret lights placed behind them. At the farther end of the garden was a mount raised by degrees, resembling bankes of earth, couered with grasse; on the top of the mount stood a goodly arbour substantially made, and couered with artificiall trees, and with arbour flowers, as eglantine, honnysuckles, and the like. The arbour was in length three and thirtie foot, in height one and twenty, supported with termes of gold and silver. It was diuided into sixe double arches and three doores answerable to the three walks of the garden. In the middle part of the arbour rose a goodly large turret, and at either end a smaller. Vpon the toppe of the mount, on the front thereof, was a banke of flowers, curiously painted behind, while within the arches the maskers sate vnseene. Behind the garden, ouer the toppe of the arbour, were set artificiall trees, appearing like an orchard ioyning to the garden, and ouer all was drawne in perspective a fermament like the skies in a cleere night. Vpon a grassy seate under the arbor, sate the garden gods, in number twelue, apparrelled in long roabes of greene rich taffata cappes on their heads, and chaplets of flowers. In the midst of them sat Primaura, at whose intreaty they descended to the stage, and marching up to the king, sung to lutes and theorboes."

> " Fresh shadows fit to shroud from sunny ray:
> Fair lawns, to take the sun in season due;
> Sweet springs, in which a thousand nymphs did play;
> Soft-running brooks, that gentle slumber drew;
> High-reared mounts, the lands about to view;
> Low-looking dales, disloign'd* from common gaze;
> Delightful bow'rs, to solace lovers true;
> False labyrinths, fond runner's eyes to daze.
> All which by Nature made did Nature 'self amaze,
> And all without were walks and alleys dight,
> With divers trees enranged in even ranks;
> And here and there were pleasant arbours pight,
> And shady seats, and sundry flow'ring banks,
> To sit and rest the walker's weary shanks."†

---

* = *remote from.*
† *Faerie Queene.* Book IV., c. x., 24.

CHAPTER VII.

> "Whose golden gardens seeme th' Hesperides to mock
> Nor these the Damson wants nor daintie Abricock
> Nor Pippin, which we hold of kernel fruits the king
> The Apple-Orendge, then the sauory Russetting
> The Peare-maine which to France long ere to us was knowne
> Which carefull Frut'rers now haue denizend our owne
>   * * * * *
> The sweeting, for whose sake the Plowboyes oft make warre
> The Wilding, Costard, then the wel-known Pomwater
> And Sundry other fruits of good yet severall taste
> That haue their sundry names in sundry counties plac't."
>   DRAYTON, *Polyolbion.*

THE changes in the kitchen, or "cooks-garden,"* were not so marked as in the "garden of pleasant flowers."† As the flower-garden lay in front of the house, "in sight and full prospect of all the chief and choicest roomes of the house; so contrariwise, your herbe garden should be on the one or other side of the house . . . for the many different sents that arise from the herbes, as cabbages, onions, &c., are scarce well pleasing to perfume the lodgings of any house." This is certainly a change from the gardens of earlier times, when herbs covered more or less the whole area of the average garden, when groundsel was allowed a place with leeks, thyme, and lettuce, and was classed among garden herbs indiscriminately with periwinkles, roses, and violets.

Holinshed (died 1580), describing England in his day, points out that the cultivation of vegetables was greatly increased, and

---

* Letter from Peter Kemp to Cecil, 1561.   † Parkinson.

says that vegetables "have been very plentiful in this land in the time of the first Edward, and after his daies, but in process of time they grew also to be neglected, so that from Henry the Fourth till the latter end of Henry the Seventh and beginning of Henry the Eighth, there was little or no use of them in England, but they remained either unknown or supposed as food more meet for hogs and savage beasts to feed upon than mankind. Whereas in my time their use is not onelie resumed among the poore commons, I meane of melons, pompions, gourds, cucumbers, radishes, skirets, parsnips, carrets, cabbages, nauewes, turnips, and all kinds of salad herbes, but also feed upon as deintie dishes at the tables of delicate merchants, gentlemen and the nobilitie, who make their prouision yearelie for new seeds out of strange countries." Holinshed was writing to extol Elizabeth's reign, and though a faithful chronicler of contemporary events, would be tempted to colour them in order to enhance the glory of the period he was describing. Although vegetables were now more fashionable and more used, still from what we have seen of the gardens of earlier times, it seems incredible that the neglect of them had been so entire as Holinshed would have us believe. Parkinson advises some vegetable seeds to be obtained from abroad, especially melons, but says of many of those on Holinshed's list of seeds to be obtained from "strange countries, Redish, Lettice, Carrots, Parsneps, Turneps, Cabbages, and Leekes . . . our English seede . . . is better than any that cometh from beyond the seas."

A striking proof of the progress gardening was making during this period, was the growing importance of those practising the craft in and around London, until at length, in the third year of King James I., they attained the dignified position of a Company of the City of London, incorporated by Royal charter. In that year all those "persons inhabiting within the Cittie of London and sixe miles compas therof doe take upon them to use and practice the trade, crafte or misterie of gardening, planting, grafting, setting, sowing, cutting, arboring, kocking, mounting, covering, fencing and removing of plantes, herbes, seedes, fruit trees, stock sett, and of contryving the conveyances to the same belonging, were incorporated by the name of Master Wardens,

Assistants and Comynaltie of the Companie of Gardiners of London."* Thomas Young was appointed first Master, and

CHARTER OF THE GARDENER'S COMPANY.

seven years was the term of apprenticeship to the Company.

* From the Original Charter belonging to the Company.

It was hoped that the formation of this Guild would put a stop to frauds practised by gardeners in the City, who sold dead trees and bad seeds "to the great deceit and loss" of their customers. But it appears that these abuses continued to exist, and a second Charter was granted in the fourteenth year of James I., and the Company was invested with further privileges. No person was allowed to "use or exercise the art or misterie of gardening, within the said area, without the licence and consent"* of the Company, nor were any persons who had not served their apprenticeship, and received the freedom of the Company permitted to sell any garden-stuff, except within certain hours, and in such places and markets as were open to other foreigners who had not the freedom of the City. The Company were also permitted to seize any "plants, herbs or roots that were exposed for sale by any unlicenced person and distribute them among the poor of the place where such forfeitures shall be taken." And it was also lawful for any four members of the Guild "to search and viewe all manner of plants, stocks, setts, trees, seedes, slippes, roots, flowers, hearbes and other things that shall be sould or sett to sale in any markett within the Cittie of London and sixe myles about," and to "burn or otherwise consume" all that was found to be "unwholesome, dry, rotten, deceitfull or unprofitable." William Wood was elected first Master under the new charter. There were two Wardens, the number of Assistants was increased to twenty-four, a Beadle was appointed, and the Company was granted a livery. The rights and privileges of the Company were again confirmed by Charles I., in 1635. The arms are a man digging, and the supporters two female figures with cornucopiæ; the crest, a basket of fruit, and the Motto, "In the sweat of thy browes shalt thou eate thy bread." Although licenced by the Charter, to have a Hall in which to assemble, they never appear to have possessed one. The Company for long has ceased to exercise its privileges, but it still exists, and ranks seventieth among the City Guilds.†

All the herbs already in cultivation were retained, mostly on

* Second Charter, 1616, in the possession of the Company.
† There is an account of the Company in Bradley's Treatise on *Husbandry and Gardening*, 1726.

account of their medicinal properties, which were in many cases both varied and comprehensive. For instance, the decoctions of "Blessed Thistle" or *carduus benedictus*, either the leaves ground, or the juice drunk, or the leaves applied outwardly, were supposed to cure deafness, giddiness, loss of memory, the plague, ague, swellings or wounds, the bites of serpents, or mad dogs, and many other complaints. With faith in such a catalogue of its uses, it is not astonishing that the "Blessed Thistle" was cultivated in every garden. Another plant that was grown in all gardens, from the tenth century onwards, was the Mandrake (*Mandragora vernalis* and *autumnalis*). More ridiculous superstitions cluster round this plant than are attached to any other. The roots were supposed to resemble the figure of a man, and to possess certain mystic powers, therefore spurious roots were manufactured in this form, and sold as charms. It was said to shriek when pulled from the ground, and the sound was so horrible that anyone who heard it went out of his mind, or died. Shakespeare refers to this superstition:

"And shrieks like Mandrakes torn out of the earth,
That living mortals, hearing them, run mad."
Romeo and Juliet, act iv. scene 3.

Not only in the Herbals proper, but in almost every practical work on gardening, the "vertues and physic helps" of each flower are enumerated. Thomas Hill devotes four pages to the "physicke helps and worthie secrets of the Colewort," or cabbage. Even Parkinson finds some medicinal use for nearly every plant, and only a few "are wholly spent for their flowers sake";* even of tulips he confesses to have "made trial," and preserved the bulbs in sugar, and found them pleasant. "That the roots are nourishing, there is no doubt .... for divers have had them sent by their friends from beyond sea, and mistaking them to be onions, have used them as onions in their pottage or broth, and never found any cause of mislike, or any sense of evil quality produced by them, but accounted them sweet onions." †

By far the most important introduction into the kitchen garden was the potato. The generally received idea is that the potato

---

* Larkspur, *Paradisus*, p. 278.   † Parkinson, *Paradisus*, p. 77.

was first brought to Europe by Sir Walter Raleigh, from Virginia, but this is doubtful. There have been great discussions among botanists on the subject of its native habitat. That Sir Walter Raleigh and his companion, Thomas Herriott, brought the potato back with them from the New World, in 1585 or 1586, is a fact. But it was also brought to Europe by the Spaniards between 1580 and 1585. The potato has been found in a wild state only in Chili, but, it is probable, that before the arrival of the Spaniards in America, the plant had spread by cultivation into Peru and New Granada. From thence it was most likely introduced, in the latter half of the sixteenth century, into that part of the United States now known as Virginia and North Carolina, and there discovered by Raleigh, unless he found it among the provisions of some Spanish ship captured by him on its way from Chili or Peru. Gerard gives a picture and account of the "potatoe of Virginia " (*Solanum tuberosum*) which " he had received " from that place. The original species still exists in cultivation, in Europe, and differs but slightly from the ordinary varieties now grown. Gerard's description of the flower and root is accurate. He calls it "a meate for pleasure," being " either rosted in the embers, or boiled and eaten with oile, vinegar and pepper, or dressed any other way by the hand of some cunning in cookery." He thus describes the tuber, " Thicke, fat and tuberous, not much differing either in shape, colour or taste from the common potatoes, saving that the rootes thereof are not so great nor long, some of them round as a ball, some ouall or egge fashion, some longer and others shorter, which knobbie rootes are fastened into the stalkes with an infinite number of threddie strings." " The common potatoe," he refers to, is at first sight puzzling, but he really means the Batata or Sweet Potatoe, *Ipomœa Batatas*. The origin of this plant is also a subject of discussion ; America and Eastern Asia both lay claim to it, but the strongest evidence seems to point to its introduction from the New World. Christopher Columbus is supposed to have brought the plant back to Queen Isabella, and early in the sixteenth century it was cultivated in Spain. Both Gerard and Parkinson grew it in their gardens, but as it was always killed by the frost at the end of September, they never saw it in flower. Sweet potatoes were eaten in various ways, roasted, sopped in wine, or cooked with prunes, and conserves were made of them. They were sometimes called Skirrets of Peru. Parkinson names a third plant in his list of potatoes "Potatos of Canada." "We in England, frome some ignorant and idle heads, have called them Artichokes of Jerusalem, only because the root being boiled is in taste like the bottom of an Artichoke head." "This plant has no similitude . . . with an artichoke . . . neither came it from Jerusalem or out of Asia, but out of America."* None of these authors make any attempt to account for Helianthus tuberosus being called "Jerusalem," but it can be explained, as the plant is a kind of sunflower or "Girosole," of which latter word Jerusalem is a corruption. Goodyer gives the history of its first introduction.† "In anno 1617 I received two small roots thereof from Master Franqueuill of London . . . the one I planted, the other I gave to a friend, mine brought mee a pecke of roots, wherewith I stored Hampshire." Of the use of these Parkinson writes, "The Potatos of Canada are by reason of their great increasing, grown to be so common here with us at London, that even the most vulgar begin to despise them, whereas when they were first received among us they were dainties for a Queen, but the too-frequent use, especially being so plentiful and cheap, hath rather bred a loathing than a liking of them." Goodyer also classes them as "meat more fit for swine than men."

Both the ordinary artichoke (*Cynara Scolymus*) and the cardoon (*Cynara Cardunculus*) were grown, but the latter were never as popular in England as they were abroad, probably because "we cannot yet find the true manner of dressing them, that our country may take delight therein."‡ The artichokes grown in England were considered the best, and plants were exported to Italy, France, and the Low Countries.

Greater attention was paid to the culture of melons. All gardening books give instructions for growing them, apparently without great success, for Parkinson is honest enough to say, "Muske melons have been begun to be nursed up, but of late

---

* Johnson's Edition of Gerard's *Herbal*, 1633.   † Ibid.
‡ Parkinson.   "1 oz. of Cardone" seed in 1761 cost 1s.—*Household Accounts, Stonor.*

dayes in this Land, wherin although many have tried and endeavoured to bring them to perfection, yet few have attained unto it." The seeds were planted in April, in a hotbed, and carefully covered with straw; when they had sprung up they were given an hour's sun in the morning, and re-covered, then, when they "have gotten four leaves," are planted on a well-manured sloping bank in a sunny sheltered place, and covered with a pot, or some shelter, until they were well grown. Sir Hugh Platt writes, "When your mellons are as big as Tennis balls, then if you nip off at a joynt, all the shoots that are beyond them, the mellons will grow exceeding great." He also gives a direction learned from "Mr. Nicholson Gardiner." "Lay your young Mellons upon Ridge-tiles to keep them from the ground, and for reflection," and he tells us that the seed should be steeped in milk for twenty-four hours before sowing. Parkinson says the best seed came from Spain, and not from France, but some seed was saved in England. Gerard saw some good melons at the "Queene's House at St. James," grown by Master Fowle, and also "great plenty" at Lord Sussex's at "Bermondsey by London." It was usual to eat them with pepper and salt, and "to droun them in wine for feare of doing more harme."* These "musk-melons" are *Cucumis melo*, the same as are now termed melons, and they were "of a russet colour and green underneath . . . . . deep furrowed and ribbed . . the inward substance is yellow, which only is eaten."† "Melons or pompions," include pumpkins and gourds of all kinds. These were eaten especially by the poorer classes, cooked in various ways. Parkinson says they eat as "a dainty dish," pompions, the seeds taken out and filled with pippins, and baked altogether.

Vegetables then did not have at all the same relative value as nowadays; some which are now scarcely grown, such as skirrets, holding a prominent place, while others were not so much valued. The heading of a chapter in Hill's *Gardener's Labyrinth* will illustrate this fact. "What care and skill is required in the sowing and ordering of the Buckshorne,

* Parkinson. † Ibid.

Strawberrie and Mustarde Seede." Buckshorne is *Plantago coronopus*, and was largely used in salads, " especially in sallets in the sommer time, although the same have no apt succour nor taste." The strawberry, Hill continues, requires " small labour, but by dilligence of the Gardener, becommeth so great, that the same yeeldeth faire and big Beries as the Beries of the Bramble in the hedge. . . . The Berries in sommer time, eaten with creame and sugar, is accounted a great refreshing to men, but more commended, being eaten with wine and sugar." Mustard was grown only for the seeds, not for the use of the seedlings in salad. The seed pounded with vinegar was eaten " with any grosse meates, either fish or flesh." * Hill gives a long list of complaints it will cure. " The juice taken diuers mornings fasting doth procure a good memorie." He recommends it to be dropped into the eyes to remove dimness of sight, one would have thought rather to ensure an opposite effect. The powder of the seeds taken as snuff " marvelously amendeth the braine"!

Nauewes and turnips, though spoken of separately, seem to be one and the same thing, as Hill says of them—" The propertie many times of the ground dooth alter the Nauewe into a Turnup, and the Turnup into a Nauewe." He recommends poppies to be " sowne in the beddes among colewortes," which does not speak well for the cabbages. Beans were still largely grown by the poorer classes, but kidney beans, of which Gerard depicts eight sorts, two from America, were " a dish more oftentimes at rich men's tables than at the poor." Peas were sown at midsummer for autumn use, and also in August and September for the following spring. Dried peas were used at " sea for them that go long voyages." The rouncial was still much grown, also the green and white hasting, called so because of its earliness. The following were also popular varieties : the sugar pease, the spotted, the gray, the pease without skins, and the Scottish or tufted, or the rose, and the early French, "which some call the Fulham Pease, because those grounds thereabouts do bring them soonest forward for any quantity, although sometimes they

---

* Gerard.

miscarry by their hast and earliness."* The "Rams ciche" or "ciche pease" (*Cicer arietinum*) was occasionally grown. Turner says he had seldom seen it in England, and Gerard says it "is soun in our London gardens, but not common." This "Chick Pea" never became popular. Miller, writing a hundred years later, says it was much grown in France and Spain, but rarely sown in England.

Any practical gardener, if asked the use of an orchard, would, doubtless, reply that the use is to ensure a sufficient supply of fruit; but Lawson tells us that no one can deny, "that the principal end of an orchard is the honest delight of one wearied with the workes of his lawful calling"; and, again, he speaks from experience, being himself an old man, and says that the orchard "takes away the tediousnesse and heavie load of three or four score years." What a truly magical power must an Elizabethan orchard have possessed! Such an introduction makes one keen to leave the kitchen-garden, and traverse again the flower-garden, on the north-east side of which we should probably find the orchard. It was thoughtfully put there when it was possible, that the fruit trees might help to shelter the more tender plants of the flower garden, and some tall forest trees, "Walnuts, Elms, Oaks or Ashes," were planted at a good distance beyond, to shelter but not overshadow the orchard. "The extent of an orchard was much larger than that of a garden, and it would require more cost, which everyone cannot undergo," to build a brick wall round it. Instead of this, mud walls, wooden palings, or a quickset hedge were substituted. But Parkinson recommends a wall of brick or stone, in spite of the expense, "as the gaining of ground and profit of the fruit trees planted there against, will in short time recompense that charge." "On the south wall your tenderest and earliest fruits, as Apricocks, Peaches, Nectarins, and May or early cherries, should be set on the east and north, and on the west, plums and quinces, spread upon and fastened to the walls by the help of tacks and other means to have the benefit of the immediate reflexe of the

* Parkinson.

CASTLE BROMWICH.

sunne."* This arrangement of the walls was suited only to the southern counties. Lawson writes that in his county (Yorkshire) the best fruit to grow was "Apples, Pears, Cherries, Filberds, red and white Plums, Damsons and Bullaces," and he further adds, as a warning, " we do not meddle with apricockes, nor peaches, nor scarcely with quinces, which will not like in our colde parts." Allusions to the fruit trees trained against a wall occur in the poems and plays of the time. Marlowe mentions cherries on a wall, and Ben Jonson in *Every Man in his own Humour* (act i. scene 1) makes Wellbred write to Edward Knowell, " Leave thy vigilant father alone to number over his green apricots evening and morning on the north-west wall." The idea of thus growing fruit was of recent date. Sir Hugh Platt, writing in 1600, says, " Quinces growing against a wall, lying open to the sun and defended from cold winds, eat most deliciously. This secret the Lord Darcey brought out of Italy, quaere, would this suit of all other fruits?"

In front of the trees trained to the wall or running parallel with the outer hedge, was a path, and this was bordered with a row of low-trained fruit trees, " Cornelian cherry trees plashed low, or gooseberries, curran trees, or the like," or "pippins, Pomewaters or any other sort of apple, planted " all along the side walk. There were arbours at the corners of the walks, and banks of camomile or other sweet herbs on which to rest. The paths were well sanded, and under the trees "green grass kept finely shorn." Between the raspberries and currants beside the path, the ground, says Lawson, should be "powdered with strawberries." In fact, all was done that the orchard might be well ordered, and made fit "for refreshing one's spirits." The arbours were much the same as those in the garden, and like them were often raised on mounts. In such an arbour in his orchard in Gloucestershire, Shallow invited Falstaff to "eat a last year's pippin of my own graffing," with a dish of Leathercoates. The Leathercoat was " a good winter apple of no great bignesse, but of a very good and sharp taste." †

Much care was taken to preserve pippins for a length of

---

* Lawson, *New Orchard*. 1618. † Parkinson.

time. Lawson gives directions for gathering and storing them. "You should have a long ladder of light firre, also a gathering apron like a pocke before you made of purpose, or a wallet hung on a bough, or a basket with a sive bottom . . . . an hooke to pull boughes to you." For storing, apples and pears should be laid "in a drie loft . . . in a heape ten or fourteen days, that they may sweate," they must then be wiped and dried "with a clean and softe cloth," and afterwards laid between layers of straw. Sir Hugh Platt gives a recipe for "apples kept without wrinkles." "Gather not your Pippins till the full moon, after Michaelmas; so may you keepe them a whole yeare without shrinking; and so of grapes and all other fruits."

"Our orchards," writes Holinshed, "were never furnished with such good fruit, nor with such varietie as at the present." The varieties of almost every kind of fruit had been increased by cultivation. The number of apples was "infinite," and as Gerard and Parkinson found it quite impossible to give the names of all the kinds grown in their time, it would be useless to attempt such a catalogue now. Gerard gives woodcuts of the "Pomewater tree," "The Baker's ditch apple tree," "the King of Apples," "The Quining, or Queene of Apples," and "the Sommer" and "Winter Pearmain." Parkinson says of the Queen Apple, two sorts, both "great, fair, red, and well relished," and Ben Jonson thus refers to the same apple:—

> "Only your nose inclines
> That side that's next the sun to the queene apple."

"The golding pippin," Parkinson writes, "is the greatest and best of all sorts of pippins." He gives also, the Summer, French, Russet, spotted and yellow pippins, and adds, "I know no sort of pippins but are excellent, good, well-relished fruits." He is not so lavish in his praise of some of the other sorts of apples, as "The Paradise Apple," "not to be commended," or "Twenty sorts of sweetings and none good." He names several from France, and brackets together "Pome de Rambures, de Capandas and de Calual, as all fair and good apples brought from France." The following are a few names from among those he calls "very good," or "fair," "great," "goodly," and "very well rellished." "Pearmain, Russeting, Broading, Flower of Kent, Davie Gentle, Costards Harvey, Deusan or Apple-John, Kentish Codlin, and Worcester apple." We can gather which were the best known and most popular sorts, from casual references to them in various writings of the period, such as,

> "In July come ginnitings and quadlings."—BACON, *Essay on Gardens*.
> "Ripe as a pomewater."—*Love's Labour's Lost*, act iv. scene 3.
> "I am withered like an old Apple-John."—*1st Henry IV.*, act iii. scene 3.
> "Pippins, caraways and leathercoats."—*2nd Henry IV.*, act v. scene 3.
> "And after pleasing gifts for her purvey'd,
> Queen-apples, and red cherries from the tree."
> *Faerie Queene*, Canto VI., fragment of Book VII.
> "Tho' would I seeke for Queene Apples unrype."
> *Shepheards's Calendar*, June.
> "Not yet old enough for a man, nor young enough for a boy: as a squash is before 'tis a peascod, or a codling when 'tis almost an apple."
> *Twelfth Night*, act i. scene 5.

Cooking apples were baked or roasted and dressed in many other ways, and the choicer varieties were served as now for dessert at the end of dinner.

> "I will make an end of my dinner;
> There's pippins and cheese to come."
> *Merry Wives of Windsor*, act i. scene 2.

"The best sort of apples serve at the last course of the table, in most men's houses of account, where, if they grow any rare or excellent fruit, it is then set forth to be seen and tasted.* Cider was still made in quantities, and the largest orchards were of cider apples, but there was yet another use made of this fruit. The "pulp of apples and swine's grease and Rosewater" was made into an ointment, "used to beautifie the face," "which is called in shops pomatum." †

The Quince, which is now almost entirely neglected, received much attention. Hugh Platt says they "may well be grafted on a medlar" (but not a medlar on a quince, proved by Master Hill). Gerard gives three varieties, Parkinson six, and writes,

---

* Parkinson. † Gerard.

"There is no fruit growing in this land that is of so many excellent uses as this."

The varieties of pears were even more numerous than of apples. Gerard says he knew someone who grew "at the point of three score sundrie sorts of Peares, and those exceeding good: not doubting but if his minde had beene to seeke after multitudes he might have gotten togither the like number of those worse kindes . . . to describe each pear apart, were to send an owle to Athens, or to number those things that are without number." The eight varieties he figures are the following: "the Jennetting, Saint James, Royall, Burgomot, Quince, Bishop, Katherine, and the Winter Peare." The Katherine pear was a popular variety, "known to all," as these lines in "A Ballad upon a Wedding," by Sir John Suckling (1609-1641), testify:—

> "Her cheek so rare a white was on,
> No daisy makes comparison;
> Who sees them is undone;
> For streaks of red were mingled there,
> Such as are on a Catherine pear,
> The side that's next the sun."

The various kinds of "Bon Cretien" were among the best grown. One sort Parkinson mentions as the ten-pound pear or "Bon Cretien" of Syon, "so called because the grafts cost the master so much the fetching by the messenger's expenses, when he brought nothing else." The same pears did not suit all counties alike, some kinds were more grown in one part than another; as for instance, the Arundell and the Robert, which were specially plentiful in Norfolk and Suffolk. Wardens were still reckoned amongst the best cooking pears. Parkinson notes "the pear of Jerusalem being baked it is as red as the best Warden, whereof Master William Ward, of Essex, assured me, who is the chief keeper of the King's granary at Whitehall." A glance down Parkinson's list, containing some sixty-five sorts, some of which are quoted already, shows several names still familiar in the nineteenth century, such as Bon Chrétien, Bergamot, Windsor, and "Pear Gergonell." Several varieties of pears are noted by Lyte in the copy of Dodoens's *Herbal*, now in the British Museum, annotated by him, and marked with the alterations he intended to make in his translation. A list of names of pears in his handwriting is also preserved by his descendants, which shows how much attention he gave to this fruit. Bulleyn, in his work on Health,* mentions a "kind of peares growing in the City of Norwich, called black friers' peare—very delicious and pleasant, and no lesse profitable." "A phisition of the same citye called doctoure Marseilde, said he thought those peares without all comparison were the best that grow in any place of England."

Bulleyn also remarks on the cherries growing in Norfolk. "In the county of Kent be growing great plentye of the fruite. So are there in a towne near unto Norwich, called Ketreinham."† It is probably to the influence of the Huguenots in these two counties, that the improvement in fruit culture—especially of the cherry—is owing. To these foreigners we may also ascribe the advance in hop-growing, which about this time was coming into favour. Several varieties of cherry were grown; the best known were the Flanders or Kentish, the Spanish, "Gascoigne," and Morello, also a variety called "Luke Warde's cherry, because he was the first that brought the same out of Italy." ‡ Parkinson describes thirty-five named varieties. Sir Hugh Platt gives an account of what he calls "a conceit of that delicate knight," Sir Francis Carew, at Beddington, when Queen Elizabeth visited him there. He covered a cherry tree with canvas kept damp, to retard the fruit, only removing "the tent when assured of her Majesties coming, so that she had cherries at least one moneth after all cherries had taken their farewell of England."

The garden or tame sort " of Plummes are of diuers kindes, some white, some yellow, some blacke, some of the colour of a chesnet, and some of a lyght or clear redde; and some great, and some small; some sweet and dry, some fresh and sharpe, wherof eche kinde hath a particular name. The wilde Plummes are least of al, and are called slose, bullies, and snagges." § It is evident from this description that the number of plums had greatly increased. John Tradescant was a great grower of plums, as of all fruit. He ‖ " hath wonderfully laboured to obtain all the

---

* *A newe Book entituled the Gouernement of Healthe.* William Bulleyn, 1558.

† Ketteringham. ‡ Gerard. § Lyte's *Herbal*. ‖ Parkinson.

rarest fruits he can hear of": and also "Master John Millen, dwelling in Old Street, who from John Tradescant and all others that have good fruit hath stored himself with the best only, and he can sufficiently furnish any." Gerard says that the greatest variety of plums was to be found in the garden of Mr. Vincent Pointer, at Twickenham, but he adds that "my selfe is not without some, and those rare and delicate." Mirabelle, or "Myrabolane," were grown. Parkinson gives sixty-one varieties by name, but he does not recommend them all; some are only "reasonable good rellished," others "waterish," and "The Margate plum the worst of an hundred." The list includes some "Mussell" plums, the same as the modern "muscle," so much used for grafting, and Damsons, also "The perdigon, a dainty good plum, early, blackish, and well rellished," doubtless the parent of the Perdrigon violet Hâtif, and others.

The Apricot, which we have seen was introduced in the Tudor period, was grown "in many gentlemen's gardens throughout all England." The "great apricock," and the two Mascolines of Parkinson are types still well known. He distinguishes six varieties in all. The Argier apricock seems rather of the "Musch Musch" type. It was brought by John Tradescant "returning from the Argier voyage, whither he went voluntarily with the Fleet that went against the Pirates in the year 1620."* Sir Hugh Platt gives many hints on the culture of this fruit. He writes, "A grafted Apricot is best, yet from the stone you shall have a fair Apricot." Again: "Mix cow-dung and horse-dung well rotted with fine earth and claret wine lees of each a like quantity, baring the roots of your trees in January, February, and March; and then apply of this mixture to the roots of your Apricot trees, and cover them with common earth. By this means Apricot trees as never bare before, have brought forth great store of fruit. . . . This of Mr. Andr. Hill." Another of his observations on Apricots is worth recording. "Plant an Apricot in the midst of other plumme-trees round about it . . . then in an apt season bore through your plum-trees, and let in to every one of them one or two of the branches of your Apricot tree . . . and lute the holes up with tempered loame: . . and the next year cut off the branch

* Parkinson.

from the Apricot tree. . . . Take away in time all the head of your plum tree . . . . . and so you have gotten many Apricot trees out of one." Later on "And. Hill" is quoted again, and his advice is to plant the trees against an east wall, and to protect them with a "course cloth . . . in the night or in cold weather." Platt also mentions, as rather an unusual thing, that "Sir Francis Walsingham caused divers Apricock trees to be planted against a south Wall, and their Branches to be born up also against the wall, according to the manner of vines, whereby his plumbs did ripen three or four weeks before any other." In 1611, "£100 was paid to William Hogan, keeper of His Magesties still-house and garden at Hampton Court, for planting the walls of the said garden with apricot trees, peach trees, plum trees, and vines of choice fruits." *

Gerard figures four varieties of peach. "The white peach with meate about the stone of a white colour; the red peach with meate of a gallant red colour, like wine in taste and therefore marvellous pleasant: the D'auant peach with meate of a golden colour; and the yellow peach, of a yellow colour on the outside, and likewise on the inside . . . of the greatest pleasure and best taste of all the other of his kinds." He makes no mention of the nectarine, which, however, by Parkinson's time had become well known. Six varieties are described in a chapter to themselves, although he says "they have been with us not many years." He gives twenty varieties of peach, and a woodcut illustrates six of these; two of them are considerably smaller than the apricot on the same plate. Although Platt tells us that a peach grafted on a nut will have no kernel, he cannot quite believe—although he gives the recipe—that a peach tree watered three days running with goat's milk, when beginning to flower, will produce pomegranates. Most of his other observations on their culture are practical and correct. They like, he says, a clay soil, and to be water-logged at the root destroys them. They will grow from stones, and bring forth a "kindly peach," but they thrive best when grafted on a plum stock. Bacon mentions nectarines

---

* Issue Rolls of the Exchequer, James I. By F. Devon, 1836.

as coming in September, along with "peaches and melocotones." Of the latter, Parkinson writes it "is a yellow fair peach . . . and is better rellished than any of them."

The only "curran," so called by Gerard, is the small grape or currant of Corinth, classed with grapes. The red currant is referred to under Gooseberries or Flaberries. Parkinson, however, gives them a chapter to themselves, and explains the difference between them and those "sold at the Grocers." He describes the red, white and black kinds, and says the white are "more desired . . . because they are more dainty and lesse common." Raspberries, white and red, were* eaten "in summer-time, as an afternoon dish to please the taste of the sick as well as the sound." The cornel tree or Cornelian Cherry (*Cornus mas*) was introduced about this time, and found a place in orchards along with barberries, service berries and almond trees.

Before closing this rapid review of the fruit of this period, I must say a few words about vineyards and grapes. Many of the larger gardens had vineyards attached. Barnaby Googe says they were invariably placed on the western side of the garden, and it is curious to note that such is the position of the one mentioned in *Measure for Measure*, act iv. scene 1.

> "He hath a garden circummured with brick,
> Whose western side is with a vineyard back'd;
> And to that vineyard is a planched gate,
> That makes his opening with this bigger key.
> This other doth command a little door,
> Which from the garden to the vineyard leads."

Gerard gives five pictures of what he calls "tame" or "manured" vines. He advises "shavings of horn disposed about the roots, to cause fertility." Parkinson's list includes twenty-three names. He says that Tradescant grew twenty sorts, but "he never knew how or by what name to call them." "The ordinary grape, both white and red, which excelleth crabs for verjuyce, and is not fit for wine with us," was probably what was usually grown in vineyards, the choicer sorts being only found, as these old writers would say, in the gardens of the curious. He has on his list black and white "Muscadine," and the "Frontignack";

---

* Parkinson.

the other names are such as "the claret wine grape," "the Rhenish wine grape." Platt gives several recipes for keeping grapes—in pots covered with sand, the bunch hung up with the end of the stalk stuck in an apple; or he says they can be preserved on the vine by covering the bunches with oiled paper. He constantly refers to the vineyard, and how to "order" and plant it. The way he classes the orchard and vineyard together shows the latter was by no means uncommon : " Master Pointer keepeth conies in his orchard, onely to keepe downe the grasse low ; . . . also in vineyards the use is to turne up the ground with a shallow plough, as often as any grasse offereth to spring, but I think the prevention of grass in orchard and vineyard is much better, if it were not too costly." He maintains that there is no reason why English wine should not be as good as that on the Continent. He attributes the ill-success in England to the bad way the vines were pruned, and he accuses " the extreme negligence and blockish ignorance of our people, who do most unjustly lay their wrongful accusations upon the soil, whereas the greatest, if not the whole, fault justly may be removed upon themselves."

The vineyards attached to the royal gardens at Windsor and Westminster were still flourishing. In 1618 fish-ponds were made in the " vine garden " at Westminster, " for the king's cormorants, ospreys and otters."* At Oatlands, in Surrey, there also appears to have been a vineyard, as payments occur in 1619 for " planting of new and rare fruits, flowers, herbs and trees," in the King's garden there, and " for dressing and keeping the vines." † The first Earl of Salisbury planted a vineyard at Hatfield, on the north bank of the River Lea, on a piece of ground sloping to the south, hedged in with privet and sweet briar. Hatfield had been given to Cecil by James I., in 1607, in exchange for Theobalds, to which the King took a great fancy. This was the second time that Hatfield had changed hands in this way. The manor belonged to the Abbey of Ely before the Conquest, and after Ely became a bishopric, the bishops made their residence at Hatfield, until Henry the Eighth's time. This

* Issue Rolls of the Exchequer, James I. By Devon, 1836.
† Ibid., July 23, 1619.

king also wishing to possess the place, effected an exchange of land with the bishop. Ely was in early times famous for its vines, and doubtless vineyards existed also at Hatfield during the centuries it was Church property, so that when Cecil planted a vineyard it was no new experiment. Mme. de la Boderie, wife of the French ambassador, sent thirty thousand vines to be set in the new vineyard, which are referred to in the following letter to Cecil :*—
. . . " understanding your Lordship's speech yesterday, that you were about to send some present of gratification to Mme. de la Boderye in regard of your vines, Lest your Lordship's bounty which knows the true limitts of honor of it self, should be misledd by my disesteeming the things upon a sodayne when I valued them but att £40 I thought good to let your Lordship know before it be too late that I misreckned myselfe for 20,000 at 8 crowns the thousand, cometh to near £50 sterling, besydes the cariage and besydes, the ambassador sent me word yesterday by his maistr-d'Hostel that there are 10,000 more a coming which he hath consigned to be delivered heer to me for your Lordship's use." As these were more plants than the vineyard would hold, some were kept in a nursery to put later in the place of any that were "defectyve or dying." A few muscat, and other vines, not grown before in England, were brought from Paris, by Tradescant, who was then director of Cecil's garden, and he also received five hundred plants from the Queen of France ; Pierre Collin and Jean Vallet, who probably brought over this present, were permanently engaged to plant and dress the vineyard. This vineyard does not appear to have been kept up for many years, as the last reference to it among the family papers is dated 1638, in which year Lady Hatton sent some vine cuttings.

In spite of the efforts of the writers of the early seventeenth century, vine-culture was never really revived in England, and vineyards gradually ceased to be planted. A few isolated instances occur later on. Brandy is said to have been made at Beaulieu in the last century, and Fairchild, in 1722, had a flourishing vineyard in Hoxton. These were probably nearly the last serious attempts at vine-culture.

* From family papers belonging to the Marquess of Salisbury.

In the writings of this period, we first find ideas for protecting and sheltering delicate plants, which a little later developed into orangeries and greenhouses, and finally into the hothouse and stove. Sir Hugh Platt, especially, in the second part of the *Garden of Eden*, not printed until 1660, frequently mentions the possibility of growing plants in the house, and utilizing the fires in the rooms to force gilliflowers and carnations into early bloom. "I have known Mr. Jacob of the Glassehouse," he writes, "to have carnations all the winter by the benefit of a room that was neare his glassehouse fire." Holinshed, while praising the orchards of his day, says, "I have seen capers, orenges and lemmons, and heard of wild olives growing here," but he does not say how they were preserved from cold. Gerard also describes both oranges and lemons, but he is too honest to pretend that they grow in England. A few oranges, however, were successfully reared in this country. "I bring to your consideration," writes Parkinson, in the treatise on the Orchard, "the Orenge alone, without mentioning Citron or Lemmon trees, in regard of the experience we have seen made of them in divers places, For the orenge tree hath abiden with some extraordinary looking [after it] and tending of it, when as neither of the other would by any means be preserved any long time." "They must," he goes on to say, be kept in "great square boxes, and lift there to and fro by iron hooks in the sides . . . to place them in an house or close gallery in for the winter time . . . but no tent or mean provision will preserve them." Platt suggests that if planted against a concave-shaped wall, lined with lead or tin to cause reflexion, they might "happily bear their fruit in our cold Clymate. Quaere, if these walls did stand so conveniently, as they might also be continually warmed with kitchen fires; as serving for Backs unto your chimneys, if so they should not likewise finde some little furtherance in their ripening."

The experiment of growing lemons was tried by Lord Burghley. There are some interesting letters extant in which the history of the way in which the tree was procured is preserved. Cecil wrote to Thomas Windebank, who was then in Paris, March 24th and 25th, 1561-62, saying he had heard from his son Thomas, that

ORANGE COURT AT BURGHLEY HOUSE. FROM A PICTURE AT BURGHLEY.

Mr. Carew was going to have certain trees sent home, and "I have already an orange tree; and if the prise be not much, I pray you procure for me a lemon, a pomegranate, and a myrt tree; and help that they may be sent home to London with Mr. Caroo's trees; and beforehand send me in writing a perfect declaration how they ought to be used, kept and ordered." The answer to this letter is dated April 8th, 1562, from Paris:— "Sir, According to your commandment I have sent unto you by Mr. Caro's man, with his master's trees, a lemon tree and two myrte trees, in two pots, which cost me both a crown, and the lemon tree 15 crowns, wherein, Sir, if I have more than perhaps you will at the first like, yet it is the best cheap that we could get it, and better cheap than other noble men in France have bought of the same man, having paid for six trees 120 crowns. . . . Well I think this good may ensue by your buying it, that if the tree prosper . . . you will not think your money lost. If it do not prosper, it shall take away your desire of losing any more money in like sort. My Lord Ambassador and Mr. Caroo were the choosers of it." He then gives directions for the "ordering" of the trees, which were to stand out in some sheltered place during the summer, and be lifted into the house for the cold months from September until April. If the tubs were filled up with earth, the plants could remain in them "this two or three year, so heed be taken that the hoops fall not away and that the earth shed not." The lemon "hath been twice grafted and is of four years' growth, and this year he would look for some fruit." How these particular trees flourished, we do not know, but one of the oldest parts of Burghley House is the "Orange Court," a long room with many large windows where the trees were sheltered for the winter.

This is one of the first instances of their importation, but orange and lemon trees were great rarities in this country, until many years later. Lord Carew, referred to in these letters, is said to have had the first trees. On Sunday, August 19th, 1604, James I. gave a banquet at Whitehall to the Constable of Castile. "The first thing the King did was to send the Constable a melon and half a dozen of oranges on a very green branch, telling him that they were the fruit of Spain transplanted to England." ... "The Ambassador then divided the melon with their Majesties."*

James I. made an attempt to promote mulberry-culture, with a view to establishing a silk industry. He imported the trees from France. They had been introduced from Italy into Provence about a hundred years previously, and during the reign of Henri IV. (1589-1610) into the Orleans district. King James in November, 1609, sent a circular-letter to the Lords-Lieutenant of all the counties of England, ordering them to make public the announcement that in the March following a thousand mulberry trees would be delivered at each county town, and all who were able were persuaded and required to buy them, at the rate of three farthings the plant, or six shillings the hundred. He also had a treatise on the cultivation of mulberries published. The King set the example by having four acres planted with mulberry trees, near the palace of Westminster. The large sum of £935 was the cost of walling in the area, and levelling the ground and planting the trees.† Among the MSS. at Hatfield, there is the draft dated 1606 of a patent for the importation of mulberry trees:—the Patentee was to bring in, "only the white mulberry and such as shall be plants of themselves, and not slips of others, and of one year's growth." Each year he was to bring at least a million, which he should cause to be planted and preserved, and he was not to take above a penny for each plant. Cecil, in furtherance of the King's scheme, himself bought five hundred trees from France in 1608, but it is not known where they were planted. In the Exchequer Rolls, under the date 1608, we find £100 for trees and plants for silkworms, and in 1618 "£50 to the keeper of the gardens at Theobalds for making a place for the King's silkworms and providing mulberry leaves." The solitary mulberry trees, so often to be seen in gardens in all parts of England, were probably planted when this effort was made to bring them into notice. But a few trees, still in existence, are even older. The

---

* Translation of a Spanish MS. in the British Museum, printed in *England as Seen by Foreigners*. By Brenchley Rye.

† Issue Rolls, James I. By Devon.

four trees in the West garden at Hatfield were, according to tradition, planted by Queen Elizabeth; one in the garden at Syon House was planted when the place was still a monastery, and at Ribston, in Yorkshire, there is a fine old tree which dates from the time when it was in the hands of the Templars, or of the Knights of St. John of Jerusalem, who succeeded them. Shakespeare twice refers to the fruit :—

> "Volumnia . . . thy stout heart
> Now humble as the ripest mulberry
> That will not hold the handling."
> *Coriolanus*, act iii, scene 2.

He could not with one masterly touch of the pen have described this peculiarity of the fruit, had it not been familiar to him.

The custom of strewing rushes (various species of *Juncus*) on the floor, was very general in the Middle Ages. Frequently we find notes of payments for rushes, such as in 10th of Henry III., 1226, "12d. for hay and rushes for the Baron's chamber," and in the Household Rolls of Sir John Howard, 1464, item "paid to gromes off chamber for reshis 16d." Queen Mary's presence chamber was strewn with rushes, also that of Elizabeth, though she added thereto the luxury of a Turkey carpet. In Princess Elizabeth's accounts, 1551-2, a small sum was entered "to the steward for rushes." The guest chambers were always freshly strewn :—

> "So here a chamber . . .
> * * *
> I shall warande fare strewed
> It should not else to you be showed." *

In the *Taming of the Shrew*, Petruchio, just after his marriage, sends his servant to Grumio to prepare the house for his bride. Grumio arrives late, and in haste calls, "Where's the cook? is supper ready, the house trimmed, rushes strewed, cobwebs swept?" Such had been for long years the custom, but in Henry the Eighth's reign an improvement on the plain rushes became the fashion, and sweet-smelling herbs and flowers were added. By Elizabeth's time this practice was much in vogue. As early as 1516 "flowers and rushes" were purchased "for chambers," for Henry VIII. In 1552, in Princess Elizabeth's

* *Towneley Ministry.*

accounts, there are numerous entries of payments to a certain Thomas Briesly, for "flowers and herbs by him provided for the same purpose." The sum of £10 was paid in 1565 and 1567, to Robert Jones, for providing boughs and flowers for the Council Chamber.* Queen Elizabeth was so fond of having a constant supply of flowers for strewing, that a waiting-woman was appointed with a fixed salary to have flowers always in readiness. So late as 1713 this office had not been abolished, as there is a letter extant in the State Archives addressed to Alice Blizard, who held the post of "herbe strewer to Her Majesty the Queen." Parkinson, writing about what flowers are suitable for laying out knots, says of both Germander and Hyssop, "they must be kept in some form and proportion with cutting, and the cuttings are much used as a strawing herb for houses, being pretty and sweet."

The houses must have been made very fragrant with many herbs and flowers, not only strewn on the floor but placed in vases about the rooms. In the Loseley Accounts in 1556, the item occurs, "a blewe potte for flowers 1d."† Parkinson says of both Yew and Box, they are used "to deck up houses in the winter-time." Not only in pots and vases were flowers to be found, but many were skilfully arranged into little posies, and worn as personal ornaments. Violets made into garlands, posies, and nosegays "are delightful to look on, and pleasant to smell."‡ "Auriculas do seem every one of them to be a nosegay alone of itself . . . . . they are not unfurnished with a pretty sweet scent, which dothe adde an increase of pleasure in those that make them ornaments for their wearing."§ Another curious button-hole was the Fritellaria, which, says Parkinson, was "worn abroad" by the "curious lovers of these delights."

Some flowers had particular meaning attached to them, and were therefore worn on special occasions. Rosemary was borne at funerals :—

"There's rosemary, that's for remembrance,"

said Ophelia, and strange to say, it was also worn at marriages. Anne of Cleves, when she arrived at Greenwich as a bride, wore

* Acts of the Privy Council. New Series, Vol. VII., 1893.
† *Archæologia*, Vol. XXXVI.   ‡ Gerard.   § Parkinson.

"on her head a coronet of gold and precious stones, set full of branches of rosemary." At a rustic wedding witnessed by Queen Elizabeth at Kenilworth, "each wight had a branch of green broom tied on his left arm (for that side is near the heart) because rosemary was scant there."

> "Down with the rosemary and bays,
> Down with the mistletoe;—
> Instead of Holly, now upraise
> The greener box, for show.
> *   *   *
> When yew is out and birch comes in,
> And many flowers beside
> Both of a fresh and fragrant kin
> To honour Whitsuntide.
> Green rushes then, and sweetest bents*
> With cooler open boughs,
> Come in for comely ornaments
> To re-adorn the house."
> HERRICK, *Candlemas Eve*.

Parkinson again refers to the flowers in houses when writing about wall-flowers. "The sweetness of the flowers," he says, "causeth them to be generally used in nosegayes and to deck up houses." The "greater flag" was also used for the same purpose. Plants were grown in rooms also, and Platt gives a long paragraph with suggestions of the best plants to grow, and tells how to water them, and give them air and light. Window boxes, too, were used: "In every window you may make square frames either of lead or of boards well pitched within; fill them with some rich earth, and plant such flowers or hearbs therein as you like best." For the more shady parts of a room he advises rosemary, sweet briar, bay, or germander. And "in summer-time," he continues, "your chimney may be trimed with a fine bank of mosse, . . . or with orpin, or the white flower called 'everlasting.' And at either end one of your flower or Rosemary pots. . . . You may also hang in the roof and about the sides of the room small pompions or cowcumbers pricked full of Barley. . . You may also plant vines without the walls, which being let in at some quarrels, may run about the sides of your windows, and all over the sealing of

---

* A sort of grass. (*Agrostis*.)

your rooms. So may you do with Apricot trees, or other plum trees, spreading them against the sides of your windows."

This great delight in growing flowers for domestic decoration, was a marked feature in English life at this period. A Dutch traveller, Levimus Leminius, a physician and a native, Zierikzee, visited England in 1560. He was charmed with English comfort, and thus writes *:—" Their chambers and parlours strawed over with sweete herbes refreshed mee;—their nosegays finely intermingled with sundry sorts of fragraunte floures, in their bed chambers and privi rooms with comfortable smell cheered me up and entirely delyghted all my senses."

* Translation by Thomas Newton, published in *The Touchstone of Complexions*, 1581 —reprinted in *England as Seen by Foreigners*. Brenchley Rye.

CHAPTER VIII.

> " Bring hether the Pinke and purple Cullambine
> With gellillowers,
> Bring Coronations, and Sops in wine,
> Worne of Paramoures
> Strowe me the ground with Daffadowndillies
> And Cowslips and Kingcups and loved Lillies,
> The pretty Pawnce
> And the Chevisaunce
> Shall match with the fayre flowre Delice."
> SPENSER.

WHILE Henry VIII. was reigning in England, great advances were being made on the Continent in the science of Botany. The Botanic Garden at Padua was founded in 1545, and was quickly followed by one at Pisa. But it was nearly a century later before we could boast of one in England. The rest of Europe was before us too, in Botanical literature. The *Aggregator Practicus di Simplicibus* was probably printed by Schœffer between 1475-80. The *Ortus Sanitatus* was printed in 1485, and was the basis of all the botanical works that immediately followed it. It was also the foundation of the English *Grete Herball*. This book was printed by Peter Treveris, and several editions of it appeared. The first of these is said to have been printed in 1516, but the existence of a copy of this issue seems somewhat doubtful, the earliest edition, of which many copies are extant, being that of 1526. A translation of Macer's *Herbal* was printed about 1530, but it is to William Turner that we owe the first really English Herbal. Herbal literature has perhaps more in common with botanical researches than gardening, but by studying the early Herbals, we gain much knowledge from the side lights they throw on garden history. Turner especially deserves a place in this history, as he did a great work not only for botany, but for gardening. He had a garden of his own at Kew, and mentions some of the gardens of the day in his works. He was born at Morpeth, in Northumberland, between 1510-15. He studied in Cambridge, where he was the friend of Latimer and Ridley. Turner was a Reformer, and twice his books were prohibited and condemned to destruction. He travelled in Italy, Germany, and Holland, and received the degree of Doctor of Medicine in Italy. On his return to England, he held several church preferments. He was Dean of Wells, but he was deprived of his Deanery, and exiled, in Mary's reign, though he was reinstated, for a time, on the accession of Elizabeth, and he died on July 7th, 1568. His *Libellus di Re Herbaria* was printed in 1538, and dedicated to the king. The "Names of Herbes," in 1548, was dedicated to his patron, the Protector Somerset, from whose house at Syon the preface is dated. Syon had been granted to Somerset on the suppression of the Bridgittines in 1539. Throughout the work there are frequent references to the garden there. Turner's *Herbal* was printed in 1551, and the " seconde parte " of the Herbal in 1562.

Thomas Tusser was the author of a well-known work on husbandry. He was born about 1523-5, at Rivenhall, in Essex. In his early years he was trained as a singer, and sang in the choir at St. Paul's. He was afterwards under Nicholas Udall, at Eton, and in 1543 went to Cambridge, and remained there until he came to Court as a retainer of Lord Paget. After ten years of Court life, he retired to a farm in Cattiwade, in the parish of Brantham, Suffolk, on the borders of Essex. It was there that he composed his poem, *One hundred Pointes of Good Husbandrie*, which appeared in 1557. He soon after left that farm, and was moving about for some years—going to Ipswich, West Dereham in Norfolk, Norwich, Fairstead in Essex, London and Cambridge; and died in London in 1580. He enlarged on the first book, and in 1573 the first edition of

Five Hundred Pointes of Good Husbandrie was published. Tusser was good, practical, and simple-minded. In the poem he gives useful hints for the cultivation of a garden, as he touches on gardening among the pointes of husbandrie for each month. The other "pointes" include all departments of farming; besides advice about housekeeping; how to keep Christmas, and how to treat wife, children, servants, and friends; and his counsel on this last point should hold good at the present day, though few would wish to follow all his injunctions on husbandry :—

> "Good friend and good neighbour that fellowlie gest
> With hartilie welcome, should have of the best."

William Bulleyn, a learned physician, wrote a book entitled *The Government of Healthe* (1558). Although devoted to the herbs used in medicine, some curious information on gardening can be gleaned from it.

The history of the Herbals of this period is rather involved, as they were so much copied one from another, and the same plates were used in several works. The authors of every country borrowed freely from ancient writers, especially Dioscorides and Columella. The former was translated into Italian, and published with many additions in 1544, by Mattioli, the learned Italian botanist and physician. Dodoens, another of the great botanists of the sixteenth century, who copied much from Dioscorides, was born at Mechlin in 1517. He published at Antwerp in 1554, *A History of Plants*, written in Dutch, which was translated into French by Clusius (Charles de l'Excluse), and printed at Antwerp in 1557. Henry Lyte translated the work into English from the French of Clusius, and Lyte's version was printed at Antwerp in 1578, the same woodcuts being used for the work in all the three languages. Each of these books went through several editions. Meanwhile, Dodoens greatly enlarged his original, and embodied it in a new work, *Stirpium Historiae*, Pemptades sex: in thirty books. This great Herbal was translated into English by Dr. Priest, who died before he could publish his translation.

Gerard's *Herbal*, 1597, is founded entirely on that of Dodoens, parts of it being exact translations. Gerard professes to have "perused divers Herbals set foorth in other languages," but does not own to having copied so largely as he did. In the second edition of Gerard's *Herbal*, corrected and enlarged by Johnson in 1633, in the Preface to the Reader, this fact is pointed out, and, moreover, Johnson maintains that the translation made by Dr. Priest, which Gerard states to have

L'OBEL. FROM AN ENGRAVING IN THE TYSSEN LIBRARY, HACKNEY.

perished, really came into Gerard's hands, and was largely used by him, Gerard himself not being sufficiently proficient in his knowledge of Latin. "I cannot," wrote Johnson, "commend my author for endeavouring to hide this thing from us." L'Obel and Garret both helped to amend some mistakes in the Latin in the Herbal, while it was going through the press. L'Obel himself was the author of a work on plants, *Stirpium Adversaria* (1570). In this he was assisted by Peter Pena, whose acquaintance he had made while studying at Montpelier. Mathias de Lobel, or L'Obel, was born at Lille, in 1538, and travelled about Europe, and practised medicine both in Antwerp and Delft, before he came to England. For many years he took charge of the garden belonging to Lord Zouche in Hackney, and was made "Botanist to the King" (James I.). The familiar "lobelia" was so named in his honour, by Plumier. The first rudiments of a scientific classification are to be found in his works; which are therefore considered superior to those of Dodoens, who never attempted anything of the kind. He had studied Mattioli, and frequently refers to him, but his work, although esteemed by the learned, being in Latin and never translated, could not become popular, as did the work of his contemporary, Gerard, which was written in English. Gerard's *Herbal* has always maintained a conspicuous position in the literature of flowers, and the second issue, so ably edited by Thomas Johnson, tended greatly to increase the popularity and the value of the work.

John Gerard, or Gerarde, was born in 1545, at Nantwich, in Cheshire, and died in 1607. He was not only a physician, and learned "in simples," but also a practical gardener, and cultivated a physic-garden of his own at Holborn, then a suburb of London, where he lived. The first work he published was a catalogue of the plants in his garden,* which contained nearly eleven hundred kinds, both native and foreign. For twenty years he superintended the gardens of Lord Burghley, and dedicated his great work to this patron. Although the Herbal cannot lay claim to originality, yet Gerard, translator and adapter as he was, has left an indelible mark of individuality on his work. His notes on the localities of flowers are specially characteristic, as also the way in which he mentions his friends from whom he received presents of plants, or information about them.

---

* There is a unique copy of this work in the British Museum, reprinted and edited by B. Jackson.

Instances, such as the following, occur on almost every page:—
"Ciprepedium Ladies' Slipper. I have a plant thereof in my garden which I received from Mr. Garret, Apothecary, my very good friend." "The golden Mothwort or Cudweed (*Helichrysum*) . . . being gathered before they be ripe or withered remaine beautiful long time after, as my selfe did see in the hands of Mr. Wade one of the Clerks of her Maiesties Counsell which were sent him . . . from Padua." "The finger Hart's tongue . . . I found in the garden of Master

GERARD. FROM TITLE PAGE OF HIS "HERBAL," 1597.

Cranwick dwelling at Much-dunmow, in Essex, who gave me a plant for my garden." The friends, such as these, who assisted Gerard, are very numerous, and of most of them nothing further is known than the few words in which Gerard introduces them, such as "a Learned Merchant of London, Mr. James Cole, a louer of plants and very skillful in the knowledge of them." "Mr. Garth, a worshipfull gentleman, and one that greatly dilighteth in strange plants, who very louingly imparted to me" a Solomon's seal received from Clusius. The names of some people, however, occur so frequently, that we can gather more particulars about them. Thomas Hesketh is constantly referred to, as collecting certain plants chiefly in Lancashire and the North of England, and sending specimens to Gerard to grow in his garden. Thomas Edwards, of Exeter, was also a botanist, and collector of English wild flowers. Master Nicholas Lete, Merchant, of London, not only himself searched for flowers, both in England and France, but was so "greatly in loue with rare and faire floures and plants . . . he doth carefully send into Syria hauing a servant there at Aleppo, and in many other countries, for the which, my selfe and likewise the whole land are much bound unto him." One of the plants he brought to this country was a cabbage "with crincly leaues" of a "blewish green." Gerard mentions also his procuring a yellow gillyflower from Poland, showing the extensive range of his collectors. Gerard also had a collector, William Marshall, whom he "sent into the Mediterranean," and who brought him from thence the seeds of the plane-tree, and plants of the prickly pear or " Prickly Indian Fig-tree."

James Garret, we know from other sources also, was a skilful gardener, and especially clever at growing tulips. He was a "learned apothecary of London," and a good Latin scholar, and was generous in imparting knowledge and giving plants to both Gerard and Clusius. It would be tedious to enumerate all the friends referred to by Gerard, as they are very numerous, and the list of these helpful friends could be greatly added to by looking into the 1633 edition where Johnson's acquaintances are as prominent as those of Gerard. It is refreshing to see the way in which these old herbalists wrote to each other, and helped one another. Johnson, even more than Gerard, worked in harmony with other botanists and physicians, and they went expeditions together in search of rare flowers. Johnson wrote some Latin tracts descriptive of some tours he made with friends in the South and West of England, and constantly in the Herbal references to his rambles with other collectors occur. In writing of a kind of grass he says:—" I never found this but once, and that was in the companie of Mr. Thomas Smith and Mr. James Clarke, Apothecaries of London, when riding into Windsore Forest upon search of rare plants."

Thomas Johnson was born at Selby, in Yorkshire, but was himself an apothecary of London, and had a shop on Snow Hill. It was in this shop on Snow Hill that the banana was first exhibited in England. Johnson received the bunch of fruit from Dr. Argent, who got it from Bermuda. Gerard had only seen a pickled specimen sent from Aleppo. Johnson hung the bunch up in his shop until it ripened. He says: "Some have judged it the forbidden fruit; other-some the grapes brought to Moses out of the Holy Land." He was the most eminent botanist of the time, and obtained some distinction as a soldier. He joined the army to fight for the Royalist cause, and died from wounds received at Basing in 1644. The most important of Johnson's friends and assistants was John Goodyer. He noticed for the first time many native plants, and his knowledge of botany must have been very considerable, from the way in which he is referred to by both Johnson and Parkinson. Thomas Glynn, and George Bowles, were two other collectors, whose names should not be altogether forgotten.

Ralph Tuggy is another name not often remembered, and yet, from frequent references to him, he must greatly have helped the progress of gardening. Johnson mentions Tuggy as if he was almost as well-known as Parkinson or the Tradescants, and his garden at Westminster contained many plants then very rare. He was especially famous for his pinks and carnations, and auriculas, and it appears that his widow kept up his garden after his death, which occurred before 1633. Johnson described some eight hundred more plants than Gerard, and added many woodcuts. The total number in the completed Herbal was 2717, and the number of pages in this ponderous folio reached over 1600.

Between the first appearance of Gerard's *Herbal* and the second edition, Parkinson had published his *Paradisi in sole Paradisus terrestris*, the most popular gardening work of this period. Although the medicinal properties are given a place in it, as in all early books on plants, it is quite distinct in character from these other Herbals. The title of the book is a play upon his name, Park-in-Sun's Earthly Paradise, and the quaintness, freshness, and originality of the title is characteristic of the whole book. Parkinson has the power of inspiring his readers with a love of flowers and a feeling for their beauty, and still, after a lapse of centuries, no gardener could fail to be refreshed and stimulated in his art by a perusal of the *Earthly Paradise*.*

PARKINSON, FROM THE TITLE PAGE FOR HIS "PARADISUS."

Parkinson was born in 1567, and, like all the botanists already mentioned, was an apothecary. He lived in London, and was possessed of an excellent garden, and that he had also travelled appears from his works. He was "apothecary to King James," and was made "Botanicus Regius Primarius," by Charles I. He dedicated his *Paradise* to Queen Henrietta Maria. The

* The feelings that the book might inspire in children is very prettily shown in *Mary's Meadow*, by Juliana Horatia Ewing.

exact date of his death is uncertain, but it occurred soon after the publication of his work, entitled *Theatrum Botanicum*, in 1640. This book has more to do with botany than with gardening, and although he describes even more plants than are to be found in Gerard, there is no special improvement in classification, the arrangement being chiefly according to their medical qualities. The French botanists, Jean and Gaspard Bauhin, had brought out their works since the publication of Gerard's *Herbal*, and Parkinson made use of these, as well as of those of L'Obel. The blocks for Parkinson's illustrations were cut in England.* Those for Gerard and Johnson came from abroad, as did also the greater part of Turner's.

The busiest workers and collectors of foreign plants in the time of James I. and Charles I., were the three generations of John Tradescants. The grandfather, a Dutchman, came to England, probably early in the reign of James I. The next John, "the father," was gardener to the first Lord Salisbury, the Lord Treasurer; to Lord Wolton, to the Duke of Buckingham, and, in 1629, was made gardener to Charles I. They all travelled about Europe, the father in Barbary also, and the grandson made a voyage to Virginia. They collected curiosities during their travels, and formed a museum, called "Tradescant's Ark," a catalogue of which was published in 1656, *Museum Tradescanteanum*. When the last Tradescant died, in 1662, he left the museum to Mr. Ashmole, who bequeathed it to the University of Oxford. Besides the museum, at their house in Lambeth they had a good garden, where they cultivated many of the plants they imported. This was visited by the King and Queen, and was the resort of the learned of all classes. The remains of this garden existed in 1749, at which date Sir William Watson wrote a paper describing it, for the Royal Society.† He noticed two very large arbutus trees, which had not suffered from the severe cold of 1729 and 1740, when "most of their kind were killed." "In the orchard" there was "a tree of Rhamnus Catharticus"

---

* For history of woodcuts see *Pulteney's Sketches of the Progress of Botany*, 1790. Chap. 12.

† *Phil. Trans.* Vol. 46, page 160.

(Buckthorn) twenty feet high, and near a foot in diameter."
Watson also mentions a deciduous cypress, "Cupressus americanus acacia foliis deciduis" (*Taxodium distichum*), a tree which the Tradescants introduced. The tulip-tree was also one of their importations. Evelyn thus refers to it: "Poplar of Virginia—I conceive it was first brought over by John Tradescant, under the name of tulip-tree (from the likeness of its flowers), but is not, that I find, taken much notice of in any of our Herbals. I wish we had more of them, but they are difficult to elevate at first."*
Some other plants brought over by them have more fortunately preserved their memory. Tradescant's Daffodil, called, "the great rose daffodil," in Parkinson, is Plenissimus, still described as "the largest and richest yellow of all double daffodils." †
Tradescant's Aster still bears their name, and the Tradescantias, or spiderworts, are a well-known genus. During his travels, Tradescant made purchases for his patron, the first Earl of Salisbury, and some of his original bills are preserved at Hatfield. Many of the items are of interest, showing not only the prices paid for known plants, but also for some new ones, which he was the first to introduce.

The following are extracts from this interesting series ‡ :—

"3 January, 1611—John Tradescant his bill for Routes flowers, seedes, trees and plants by him bought for my Lo: in Holland Bought at Leyden in Holland— For roots of flowers of Roasses and shrubs of strang and rare, £3.— ... Also bought at Harlem in Holland of Cornellis Helin of the Rathe ripe cherry trees 32 at 4s. the peece, £6. 8s.—for flowers called anemones, 5s.—for 16 Province Roses, 8s.—for two mulbery trees 6s.—for the great red currants 6 plants 1s.—for two arbor vita trees 1s.—fortye frittelarias at 3 pence the peece 10s. 5 January 1611—bought at Brussells and in Holland . . . . for the rathe ripe portingall quince on e] tree, 6s.—for the lion's quince tree 3s.—for two great medlar trees of Naples 5s.—for tulipes roots at Harlem at ten shillings the hundred 800, £4.—for on dussin of great blacke curants 1s.—on excedyng great cherye called the boores cherye 12s.—on Aprycoke tree called the whit aprycoke 6s.—also bought of the archedukes gardener called peere vyens 10 sorts 20s.—on chery tree called the Archedukes cherye 12s.—Also bought of

---

* One of the oldest tulip-trees is at Waltham, in Essex. "The largest and biggest that ever was seen, there being but one other, in Great Britain, and that at Lord Peterborough's."—*History of Waltham*, Farmer, 1735.

† Barr's *English Daffodil Catalogue*, 1893.

‡ From the original MS. at Hatfield, by the kind permission of the Marquess of Salisbury.

Mr. John Jokkat for the double Echatega the martygon pompone blanche, the martygon pompong orang coller an the Irys calsedonye and the Irys susyana £2. 5 January 1611—bought in France—Bought at Parrys, on pomgranat tree vithe many other small trees at the root 6s.—on bundell of genista hispayca 2s.—8 pots of orrang trees of on years growthe grafted at 10s. the pece £4. —ollyander trees 6 at halfe a crowne the pece 15s.—Myrtil trees 7 at halfe a crowne the peece 17s. 6d.—two fyg trees in an other baskit called the whit fygs vithe many other rare shrubs give me by Master Robyns 4s.—Also of vynes called muscat two bundals of plants 4s.—Cheryes called Biggandres at 2 the peece 24, £2.—Sypris trees at on shilling the peece 200, £10.—blak mulberry trees at 2s. the peece 17, £1. 14s.—peache the troye 4 trees at 2s. the pece, 8s. (also alberges, malecotton peaches same price) on pot of the dubble whit stok gilliflower and on pot of the other gilliflowers. 3s.

The total sums on these bills amount to £110. 8s. 9d. for plants, and a few shillings for baskets, with padlocks and hampers to pack them in;—the travelling expenses being extra. There is also a note on the first bill of £38. from Sir Walter Cope, evidently for trees bought for him at the same time as Lord Salisbury's. "Master Robyns," referred to by Tradescant, was Jean Robin, a famous French botanist (1550-1629) and first curator of the "Jardin des Plantes." He is frequently mentioned by Gerard as "Robinius of Paris." The genus "Robinia" is named after him.

The tombstone of the Tradescant family is still to be seen in Lambeth churchyard, on the N. E. of the chancel, erected in 1662 by the widow of the younger John. The quaint epitaph runs as follows :—

> "Know, stranger, ere thou pass, beneath this stone
> Lye John Tradescant, grandsire, father, son
> The last died in his spring ;—the other two
> Liv'd till they had travell'd Art and Nature through ;
> As by their choice collections may appear
> Of what is rare, in land, in sea, in air.
> Whilst they (as Homer's Illiad in a nut)
> A world of wonders in one closet shut.
> These famous Antiquarians that had been
> Both gardeners to the Rose and Lily Queen
> Transplanted now themselves, sleep here, and when
> Angels shall with their trumpets waken men
> And fire shall purge the world, these hence shall rise,
> And change this garden for a Paradise."

Sir Hugh Platt was supposed to be the most learned man of his time, in soils and manures. He published a work on that subject in 1594, and also *The Jewel House of Art and Nature*. His work on gardening, which deserves our attention, was printed first in 1600, under the title of *The Paradise of Flora* and again with the addition of a second part in 1660, with the title, *The Garden of Eden*. This last edition appeared some time after Platt's death, and was edited "by a kinsman" of his, Charles Bellingham. "That learned and great observer," Sir Hugh Platt, "knight of Lincoln's Inne, gentleman," had a garden of his own, in London, and an estate near St. Albans, and it also appears, from references in his works, that he passed some time at Copt Hall, in Essex, which belonged to Sir Thomas Henneage. He was intimate with all the chief gardeners of his day, and is most conscientious in giving the credit of any piece of information to the friend from whom he learnt it. Thus he frequently refers by name or initial to Mr. Andrew Hill, Mr. Tavener, Mr. Pointer, of Twickenham, Garret, the apothecary, Pigot, the gardener, Mr. Nicholson Gardiner, and others, all evidently well known to his readers, as authorities on the subject. He recommends various manures for different plants, and for the general improvement of the soil. Fern spread over the earth during the winter, and then dug in— "Ashes of ferns are excellent" and "soot enriches the ground," also shavings of horn. "Onions and bay salt sown together have prospered exceeding well." He is careful to specify the best kind of manure for every plant. On the reverse of the title-page of *The Jewel House of Art and Nature*, he gives a picture of an exceptionally large ear of barley, "grown at Bishop's Hill, Middlesex, in 1594, the ground being manured with sope ashes."

Another plant-lover of this date who deserves to be remembered, is Dr. Penny. Not much is known of his life. He was a physician, and travelled abroad, and also about England, and collected many plants. He was a friend of the most eminent botanists of the day, Clusius, Gesner, Turner, Lobel, Gerard, &c. He must have been well known at the time by the way in which he is referred to by these writers, although now his name is remembered by few. Gerard speaks of him as "Thomas

Pennie, of London, Doctor of Physic, of famous memorie and a second Dioscorides for his singular knowledge of Plants, . . . lately deceased . . . whose death myself and many others do greatly bewail." Johnson refers to him in the same way: " Of famous memorie, a good physician and skilfull Herbalist." He was the introducer of several plants, and was the first to find some of our native species. Clusius named Hyperium balearicum " Pennaei " after him, as he brought it first from Majorca. Geranium tuberosum was also called after him; this plant was brought to England by Turner, who "bestowed it on Dr. Penny," from whom Clusius received it.

Other writers on gardening, of about this time, have been quoted already; but little is known of their lives, beyond what can be gathered from their works. William Lawson, who treats of orchards and fruit-trees, was a north-countryman, and wrote from his own experience. Thomas Hill, or Didymus Mountain, as he sometimes styled himself, published several works, which he did not profess to have composed, but " gathered out of the best approved writers of gardening, Husbandrie, and Physic." * Of some authors we know not even the names. N. F., the writer of a good treatise on fruit in 1608 and 1609, cannot be identified, nor do the initials correspond to any of the many names quoted by other writers, unless Fowle, mentioned by Gerard as the " skilful keeper " of Queen Elizabeth's garden at St. James, and famous for the musk melons he grew there, had a Christian name beginning with N.

* *Gardener's Labyrinth*, 1751.

CHAPTER IX.

" . . . . retired leisure
That in trim gardens takes his pleasure."
MILTON.
" That is the walk, and this the arbour ;
That is the garden, this the grove."
GEORGE HERBERT.

THE period now to be surveyed falls naturally into three divisions. The first, the reign of Charles I.; the second, the Commonwealth; the third, the Restoration. The development of gardening in each of these has its own distinctive character. The current of slow progress in horticulture runs on smoothly, but garden design does not alter much until the third portion of the time. During the Commonwealth, there was a movement towards the improvement of orchards and market gardens, and the reign of Charles II. witnessed a great revival in gardening in all its branches. The early part is merely a continuation of the gardening in the time of James I.; the men whose works have already been quoted were still alive—Parkinson, Johnson, and the Tradescants—and they form a link with the Elizabethan age. Sir William Temple and John Evelyn, whose names are so intimately connected with the garden history of the Restoration, in like manner connect that period with the brilliant days of gardening at the close of the seventeenth century.

Each succeeding generation of gardeners had a very poor opinion of the capabilities of their predecessors, while they thought the excellence of their own gardens could hardly be surpassed. Holinshed maintained that there never were such gardens as those of Elizabeth's reign, but by the middle of the seventeenth century gardening was so much advanced that the early years of Elizabeth were looked back upon as a time of almost primitive horticulture. After a large allowance is made for probable exaggeration, the fact remains that the progress was sufficiently marked to be felt by the writers of the time. Rea, writing in 1665, "to the Reader" of his *Flora Ceres and Pomona*, says his reason for publishing his work was that after "seriously considering Mr. Parkinson's garden of pleasant flowers, and comparing my own collections with what I there found (I) easily perceived his book to want the addition of many noble things of newer choicing, and that a multitude of those there set out, were by time grown stale, and for unworthiness turned out of every good garden." Rea is writing about the pleasure garden, but Hartlib, ten years earlier, writes in the same strain of nursery gardening.

Hartlib, a Pole by birth, settled in England early in Charles the First's reign. During the Commonwealth he received a pension from Cromwell of £100 a year. His *Legacy of Husbandry* is a review of agriculture, and his remarks are most practical. He is strongly in favour of increasing the number of nursery gardens and orchards, and argues chiefly on the ground that gardening improved the land, and would pay well, if properly managed. "Gardening though it be a wonderfull improver of lands as it plainly appears by this, that they give extraordinary rates for land . . . from 40 shillings per acre to 9 pound and dig and howe, and dung their lands which costeth very much . . . yet I know divers which by two or three acres of land maintain themselves and family and imploy other about their ground; and therefore their ground must yield a wonderful increase or else it could not pay charges;—yet I suppose there are many deficiencies in this calling, because it is but of a few years standing in England, and therefore not deeply rooted nor well understood. About fifty years ago, about which time ingenuities first began to flourish in England, this art of gardening began to creep into England into Sandwich and Surrey, Fulham, and other places." He goes on to say that old men in Surrey remembered "the first gardeners" to plant cabbages, coleflowers, and to sow turnips and carrots: "they paid 8 pound per acre yet the gentleman was not content fearing they would spoile his ground because they did use and dig it. . . . Many parts of England are wholly ignorant . . . where the name of gardening and howing is scarcely known. . . . Gardening-ware (unless about London) is not plentiful or cheap. . . We have not nurseries sufficient in this land of Apples, Pears, Cherries, Vines, Chestnuts, Almonds, &c.: but gentlemen are necessitated to send to London some hundred miles for them." Further on, however, he says that "there are many gallant orchards" in Kent, about London, in Gloucestershire, Hereford, and Worcester, and these we know had existed a long time previous to the fifty years he ascribes to them. In Kent and Surrey, he adds, plums usually "pay no small part of the rent."

It was not the Puritan party only who were occupied in the improvements of orchards. One of the great Royalist families took a prominent part in the work. To this day, at Holme Lacy, in Herefordshire, is to be seen the same long green walk flanked with yew hedges, down which Charles I. may have passed, when he stayed with Lord Scudamore, the year which is marked in history by his loss of the battle of Naseby. After the death of the king he had served so faithfully, Scudamore went with the expedition to the relief of the French Huguenots at Rochelle, and on his return to Holme Lacy, occupied himself with planting and grafting apple-trees. He introduced the Red Streak Pippin, from which the choicest sort of cider was made. Ambrose Philips (1671-1749) commemorates this fact in his poem "Pomona." He praises the Musk apple, and adds:—

> "Yet let her to the Red-streak yield, that once
> Was of the sylvan kind, uncivilized,
> Of no regard, 'till Scudamore's skilful hand,
> Improv'd her, and by courtly discipline
> Taught her the savage nature to forget—
> Hence called the Scudamorean plant, whose wine
> Whoever tastes, let him with grateful heart
> Respect that ancient loyal house."

The orchard at Holme Lacy still remains, and the garden now possesses one of the finest walls of "cordon" fruit in the country. Walter Blith, Author of *The English Improver, or a New Survey of Husbandry*, 1649, was another "Lover of Ingenuity," as he styled himself, and he also impressed upon his countrymen the advantages of planting orchards, and urged those in other parts of England to copy what was done in the West of England, and to plant "the Vine, the Plumb, the Cherry, Pear, and Apple," he advises also "the more planting of cabbage, carrot, onion, parsnip, artichoak and Turnep."

These led the way, and other Agriculturists followed this good example, and tried by their writings to give a stimulus to the industry of market-gardening. Ralph Austen, in 1653, wrote a Treatise on Fruit Trees, and dedicated it to Hartlib. The first part of his work, full of arguments in favour of gardening and fruit-culture, based on scriptural authority, and interspersed with texts, is typical of the puritanical style of the times. In another of his works, *The Spiritual use of an Orchard or Garden of Fruit Trees*, this is carried to such excess that there is but little information about gardening, although every process, grafting, transplanting, and so on is compared to some stage in a Christian's life. This puritanical spirit is also apparent in the title of Adolphus Speed's book, in 1659, *Adam out of Eden*, and the rest of the title-page is indicative of the practical side of these writers. It runs thus:—"Shewing Among very many other things, An Approvement of Ground by Rabbies from £200 annual Rent to £2000 yearly profit all charges deducted." But how this feat was to be accomplished it is needless to go into!

During the Commonwealth, gardening was treated from a more practical point of view; what would pay best to cultivate, was considered, and how the soil could be most improved, and made more fruitful. Not many gardens were laid out, and many of the existing ones suffered during the wars, especially the Royal Gardens. Nonsuch and Wimbledon were sold, and a survey made of Hampton Court, with a view to selling it, in 1653, but the order was "stayed until

Parliament" took "further notice," and it was left untouched. The absence of large new gardens is more marked when compared with the numbers which appear to have been laid out after the Restoration.

The progress, during the middle part of this period, was in the culture of economic plants, and not in garden design, or in the flower garden. Many of the old superstitions about plants were exposed. Austen fills several pages in contradicting old-fashioned notions, "Errors discovered," he calls it, such things as writing an inscription on a peachstone or almond, and planting it, expecting the same writing to appear in the ripe fruit of the tree:—"To have all stone fruit taste as yee shall think good lay the stones to soak in such liquor as yee would have them taste of," or "to have red apples, put the grafts into Pikes' blood." He thus sums up these recipes:—"These things cannot be." "Errors in practise," he seeks to correct also, and shows much good sense in his remarks, on planting or moving fruit trees:—"Many remove their trees in Winter or neere the Spring whereas they ought to remove them in September or thereabouts." Another error was "planting trees too neere together; I account 10 or 12 yards a competant distance for Apple-trees or Pear-trees, for Cherry or Plum 7 or 8." Many plant "too old trees in orchards, and neglect to plant their trees in as good or better soyle, then that from which they are removed." He points out some of the writings in which such errors were to be found. "*The Countryman's Recreation*, 1640, is full of these fancies," also in the works of "Didymus" or Thomas Hill, and the *Country Farm*, by Gabriel Platt. The necessity of refuting such errors shows how primitive many gardeners still must have been in their ideas. A small work on fruit trees by Francis Drope in 1672 is free from absurdities; but Adam Speed's book, a few years later than Austen's, is full of errors as apparently ludicrous as those "discovered" by Austen, so gradual is the passage "from darkness to dawn." I need only quote two of his solemn assertions as specimens:—"To make white lilies become red, fill a hole in a lily root with any red colour," and "the roots of roses set among broom, will bring forth yellow Roses." He suggests that sow thistles should be planted, as "they will maintain" calves, lambs, pigs . . . and millions of rabbits, and Jerusalem Artichokes, because they would feed poultry and swine. Some of his remarks, however, are more sensible; for instance, he observes, of potatoes, "they will make very good bread, cakes, paste, and pyes . . . increase of themselves in a very plentiful manner, with very little labour; they will likewise grow and thrive very well, being cut in slices, and so put into the earth."

Vegetable pies and tarts seem not to have been unusual; Markham, in *The English Housewife*, 1637, gives several recipes, one for "spinage tart" flavoured with cinnamon, rose water and sugar; another of spinach, sorell, parsley, and eggs. He gives also long lists of varieties of salads, " Cookery sallats," such as " boyled carrets," radishes skirrets ;—" simple sallats, onions, lettuce, samphire, Beanecods, sparagus or cucumbers," served with oil, vinegar, and sugar; and " compound sallats," which " are usuall at great feasts and upon Princes' tables;" these consist of, "first the young Buds and knots of herbs," such as " Red sage, mints, lettuce, violets, marigolds, spinage " . . . also " cabbage done with cucumber, currants, orange, lemons, olives, figs, and almonds." Carrots were used for adorning dishes, cut into "scutchions, arms, birds or beasts." Lamb and mutton should be garnished, he says, with prunes or currants, and fish with barberries.*

Among the quantities of varieties of fruits of which we have given some examples in a former chapter, Austen gives a selection of the best. He commends among apples, the summer and winter Pearmain, the small pippin, the Harvey, the Queene and the Gilloflour. Out of the four to five hundred sorts of pears, he selects the " Winsor" and " Sommer Bergamot." " But for a constant bearing kind I know none better than the Catherine peare";—" Greenefield excellent . . . will last indifferent well, a great bearer;"—" Choke peare, accounted a speciall kind, for

---

* The price paid for one pound of Barberries in 1618, was 3s.—Le Strange, *Household Accounts*.

Perry, although the peare to eat is stark naught." "Flanders cherries, most generally planted here in England. The Black Hart cherry is a very special fruit." "The best nectarine is the Roman red. But it is very hard to be propagated, as for grafting none take that way, and but few with inoculating, which I conceive is the reason it is dearest of all plants with us:" "The nutmeg and the Newington the best peaches: very large and gallant fruit." "I know of but one kind of Figs that come to ripeness with us: the great Blew-fig, as large as a Catherine Pear. The trees grow in divers gardens in Oxford, set against a south wall, and be spread up with nayles and Leathers." Austen was the greatest authority on fruit-trees in his day.*

The ruthless hand of man has done more to destroy the old gardens of England than the changes of time and seasons. But some vestiges of the gardens of each period still remain to us. Although no "princely" gardens were being laid out during the middle of the seventeenth century, like those of its latter end, many an old manor-house garden may date from about this time. This is not a history of "Gardens," so it is impossible to give anything like a complete list of the beautiful old gardens that are still to be found throughout the length and breadth of England. I must content myself by mentioning a few typical examples, to serve as illustrations of the fashions and plans of each successive century. The garden of Chilham Castle, in Kent, with its terraces, bowling-green, and clipped trees, was laid out in 1631. That of Bilton, in Warwickshire, with its fine holly and yew hedges, was begun in 1623. Bulwick, in Northamptonshire, with terraced slopes, pond, and fine wrought-iron gates, was being laid out at the same time, and finished in 1674. And at Mitford, in Northumberland, although the Manor House (dated 1637) itself is in ruins, the old wall of the garden still encloses a tangle of roses, sweet herbs, and old apple-trees, and a sun-dial, which for 250 years has faithfully marked the hours as they fly. Instances such as these could be found in every county in

---

* A good treatise on fruit in MS., probably written by Joshua Chandler about 1651, is entirely founded on Austen, and parts of it are transcribed from Austen's work, with the omission of his references to Scripture.
The MS. is in the possession of Miss Willmott.

BULWICK.

England. Household accounts give us a few glimpses into the management of such gardens. In the interesting series at Hunstanton, of the Le Strange Household books, such items as the following occur :—

"1628 Nov. 6, for a Bagg to Bring the fruit home in 1s. To a man for digging of flaggs for the Bowling ground 4s. For 65 foote of Oake Bord for the gardin doores 7s.

1629 Paid for dikinging and Hedging of Heacham orchyard, 2 men for 7 dayes a peice 7d. in clearing the garden and digging of it 11s. 8d.

1630 6 wheel Barrowes £1, for a crest for the gardin house end at 2s. 8d. £1. 18s., and for crest for the moate wall 16d. £12. 2s. 5d. and for the gardin entry doorstall 3od, and for crest over that door.

Oct. 16, 1631 for a gardin spade 3s.

1632 To the gardiner for a quarter's wages wanting 2 weekes £2.

1635 For 2 greate gardin basketts 4s.

1637 Pumpes and Pipes for the garden £2. 4s. To a gardener of Creake for slips and seeds, 2s."

What greatly adds to the interest of the Accounts of Hunstanton, is that the part of the garden there referred to within the moat, has been but little altered since that date. The bowling-green is still there, and a square plot of garden with thick low hedges, in front of the house, is hardly changed. The note book of Henry Oxenden, of Barham, Kent, between the years 1638 and 1668,* contains many interesting gardening entries : —

" Feb. 11, 1635 set the hawksbill pares in the garden in Maydeken.

" 1635 planted the cherry garden at great Maydeken.

" Feb. 14, 1652 gave Mr. Barling 4 apple trees and a peare tree, viz. a musk pare tree.

" Feb. 10, 1652 sent my Coz Henry Oxinden the yew tree . . . lent him then my stone rowle.

" Nov. 16, 1647 planted twentie-five peare trees in the garden that is walled about at Great Maydeken witness my sonne Thomas and my sonne Hobart.

" Nov. 1654 tooke up out of the Nursery at Maydeken 1 quince tree, 2 warden trees and 3 other peare trees, and set y^m in Byton, and 1 pear tree against the bake house windore, I allso sete one medlar tree and a nutmeg peach tree in the garden.

" Fb. 19, 1655 grafted one of the best pares Capt. Meriwether hath uppon a tree beside the house at South Barham ; made a crosse upon it : it is to be eaten in Feb.

"1639, hee (Sir Basil Dexivell at Boome) planted his orchard agt. his back dore agt. the Hall.

"Feb. 7, 1647 Lieutenant Hobday planted 10 apple trees, in his orchard next his garden, which I gave him.

"1665 Ms. Adie, Relict of Ed. Ady, new coped the wall round about the gardens and the Greene Court."

The note book of Sir Thomas Hanmer,* about this

HUNSTANTON.

date, contains some memoranda about the fruit trees in his garden at Bettisfield :—

"Against the South wall are one Apricocke from Mr. Rea,† three Apricockes from London, one peache from a French stone, raised at Bettisfield 1660, and two red-heart cherries from Trevallyn. In the corner next to the turf

---

* A memorial of the Parish and Family of Hanmer in Flintshire by John, Lord Hanmer— Privately printed. 1877.

† Author of *Flora, Ceres and Pomona*.

walk one pear from Bowen, I think a bergamot. Against the West wall there, from the south wall to the door, all plums from Colonel Jeffreyes, except one double-flowered cherry, and one morocco plum next the door ; on the other side the door, first a bullen plum, then a Turkey plum, then a king plum, then a Catalonia plum, and a Duke cherry, a cornelian. Against the North Wall these plums from Trevallyn, viz. the Apricocke plum and the orange, and one plum from Colonel Jeffreyes. Against the East Wall in the great garden, may cherries, a carnation cherry, about the middle of the wall, a duke cherry at the end, close by the North Wall, a cornelian cherry from Rea marbled, and a turkey plum from Rea . . . In the little court . . . are three peaches from Mr. Bate, viz. a Morills . . . a Newington . . . then a Persian peach. . . . Against the East Wall of the little garden, beginning from the South Wall, first three peaches raised 1660 at Bettisfield, from French stones, then a peach de Pau, then a Savoy peach."

These little details cannot fail to be of interest. They show how a man, an ardent Cavalier, who had lived through such stirring scenes, turned his attention to his garden to pass away the years of inaction and waiting until the Restoration. He took up gardening not only as a pastime, but really gave his brains to it, as well as his time, and made himself a thorough master of the art, as further notes from his pen show. Another Royalist, who has always been recognized as one of the greatest patrons of gardening, was Lord Capel, son of the Lord Capel who was beheaded in 1648. He was created Earl of Essex in 1661, and died in the Tower in 1683. He made the garden at Cassiobury, which is frequently referred to as one of the most beautiful gardens of the seventeenth century. His brother, Sir Henry Capel, was also a gardener, and introduced "several sorts of fruits from France."* He had a garden at Kew, in it were "curious greens"; it was "as well kept as any about London" and his "flowers and fruits" were "of the best." † Sir Henry was created Baron Capel of Tewkesbury in 1692, hence there is apt to be some confusion in the various allusions to Lord Capel, as both were gardeners. The Earl of Essex seems to have confided the chief care of his gardens to Cooke, a celebrated gardener and author of a work on fruit trees, though, as

* Switzer, *Ichnographia Rustica*, 1718.
† Gibson, *Gardens about London*, 1691.

Evelyn remarks,* "no man has been more industrious than this noble Lord in planting about his seate adorn'd with walkes, ponds, and other rural elegancies. . . . The gardens at Cassiobury are very rare, and cannot be otherwise, having so skilful an artist to govern them as Mr. Cooke, who is, as to the mechanical part, not ignorant in mathematics, and pretends to astrology." Sir Henry does not appear to have had such assistance;—"his garden has the choicest fruit of any plantation in England, as he is the most industrious and understanding in it." †

Another distinguished Royalist and gardener was John Evelyn. His great work on Forest trees does not really come into our subject. It was written for the Royal Society (of which Evelyn was one of the first Fellows) with the idea of being a practical assistance in the planting of trees in parks, woods and forests, and went far beyond the narrow limits of a garden. But gardens are incidentally referred to, as the following extracts show. He urges the hardiness of cedars, and regrets they are not more grown. Perhaps it was at his suggestion that some were planted in the Chelsea physic garden in 1683. The ilex, also, he proves to be hardy by the remains of one in the Privy Garden, Whitehall, "Where once flourished a goodly tree of more than four score years." " Phillyrea is sufficiently hardy, which makes me wonder to find angustifolia planted in cases and so charily set into the stoves among the oranges and lemons." He had " four large round " Phillyreas, " smooth-clipped," in his own garden at Says Court, Deptford.‡ Under Hornbeam, he notices the "admirable" hedges at " Hampton Court and New Park," " the delicious villa of the noble Earl of Rochester." " These hedges are tonsile, but where they are maintained to 15 or 20 feet height . . . they are to be kept in order with a scythe of 4 foot long, and very little falcated, that is, fixed in a long sneed or straight handle, and does wonderfully expedite the trimming." . . These hedges are a great "convenience for the protection of our orange trees, myrtles, and other rare perennials and exotics." The laurel

* Evelyn's Diary.   † Ibid.   ‡ Gibson, *Gardens about London*, 1691.

was so commonly used for the same purpose, that Evelyn says "it seems as if it had only been destined for hedges." Holly for a garden-hedge he also enthusiastically praises:—"Is there under heaven a more glorious and refreshing object of the kind than an impregnable hedge of about 480 feet length, 9 feet high, and 5 feet in diameter, which I can show in my now ruined gardens at Says Court (thanks to the Czar of Muscovy) any time of year, glittering with its armed and varnished leaves." This is quoted from the later edition of the *Silva*, and the "ruin" of the garden refers to the damage done there by Peter the Great, who lived at Sayes Court to be near Deptford during his visit to England (1698). He is said to have amused himself by being wheeled about the garden in a wheel-barrow, over borders and through hedges, regardless of consequences. In his Diary, on June 8th, 1698, Evelyn writes:—"I went to Deptford to see how miserably the Czar had left my house after three months making it his Court. I got Sr. Christr. Wren, the K.'s surveyor, and Mr. London, his gardener, to go and estimate the repairs, for which they allowed £150 in their report to the Lords of the Treasury."

Besides the interest he took in his own garden, Evelyn helped to lay out others. The family seat of the Evelyns, Wotton, in Surrey, he says, was one of "the most magnificent that England afforded, and which indeed gave one of the finest examples to that elegancy since so much in vogue." He, however, helped his brother to carry out various alterations in 1652. With much deference to so distinguished a gardener as Evelyn, at this distance of time we may be allowed to doubt if all his alterations were improvements. There was a "mount" or "mountaine," and a moat within ten yards of the house. This was taken away by "digging down the mountaine and flinging it into a rapid streame . . . filling up the moat, and levelling that noble area where now the garden and fountain is." In 1658 he went "to Alburie (Albury, near Guildford) to see how that garden proceeded, which I found exactly don to the designe and plot I had made, with the crypta thro' the mountain in the park 30 perches in length, such a Pausilippe is no where in England besides. The Canall was now digging and the vineyard planted." This curious cutting through the hill still exists, besides other traces of the old work, and a very fine yew hedge. Again, he shows himself to be the advocate of a holly hedge, in the following extract from his Diary:—"25 Sept. 1672, I din'd at Lord John Berkeleys . . . it was in his new house or rather palace. . . For the rest, the fore court is noble, so are the stables, and above all the gardens, which are incomparable by reason of the inequality of the ground, and a pretty piscina. The holly hedges on the terrace I advised the planting of." Berkeley House, which was burnt to the ground, stood on the site of what is now Hay Hill, Berkeley Square, and Lansdowne House.

In 1664 Evelyn published his *Kalendarium Hortense, or Gardener's Almanac,* a most popular work, which went through a number of editions, and appeared with the last corrected edition of the *Silva,* in 1705, and Evelyn died at the end of the same year. The flowers to be planted and the business to be done in each month is carefully gone through. He gives also a list of the comparative tenderness of flowers, and divides them into three classes, those "least patient of cold," "to be first set into the conservatory or otherwise defended," those "enduring second degree of cold," and accordingly "to be secured in the conservatory," class III. "not perishing but in excessive colds to be last set in or protected under matrasses or slighter coverings." His classifications of some of the plants are rather singular. The first begins well with Acacia Aegyptiaca (= *A. vera*), Aloe American (= *Agave americana*), then Amaranthus tricolor, but the list contains also Styrax Colutea, or bladder senna, and white lilac, which are hardy, while oranges, lemons, oleanders and "Spanish jasmine" (*J. odoratissimum*) are in the second class with the "Suza Iris" (*I. susiana*), "summer purple cyclamen" (*C. europoeum*), and "Digitalis Hispan" (*lutea*). The last list classes together, pomegranate and pine-apples with Eryngium planum and winter aconite.

In Rea's *Flora, Ceres and Pomona,* the approximate size of a garden is given. The dimensions are much more modest than Bacon's "princely garden," eighty square yards for fruit, and thirty square yards for flower-garden for a nobleman, for a "private gentleman 40 square yards fruit and 20 flower, is enough; a wall all round of brick 9 feet high, and a 5 feet wall to divide the fruit and flower gardens, or else pales painted a brick colour. The large square beds to be railed with wooden rails painted, or box-trees or pallisades for dwarf trees." Most of the designs he gives are squares, with T or L shaped beds, fitting into the angles and along the walls of the garden, these borders to be about three yards wide. In the corners of each bed were to be planted "the best crown Imperials, lilies Martagons and such tall flowers, in the middle of the square beds great tufts of pionies, and round about them several sorts of cyclamen, the rest (of the beds) with Daffodils, Hyacinths, and such like. The streight beds are fit for the best Tulips, where account may be kept of them. Ranunculus and Anemonies also require particular beds—the rest may be set all over with the more ordinary sorts of Tulips, Frittilarias, bulbed Iris and all other kinds of good roots. . . . It will be requisite to have in the middle of one side of the flower garden a handsom octangular somer-house roofed everyway and finely painted with Landskips and other conceits furnished with seats about and a table in the middle which serveth not only for delight and entertainment but for many other necessary purposes as to put the roots of Tulips and other flowers in, as they are taken up upon papers, with the names upon them untill they be dried, that they may be wrapped up and put in boxes. You must yearly make your hot bed for raising of choice annuals, for the raising of new varieties of divers kinds. These gardens will not be maintained and kept well furnished without a Nurcery, as well of stocks for fruits as of flowers and seedlings where many pretty conclusions may be practised."

Rea's description shows what great attention was paid to the culture of bulbs, especially tulips, in the average small garden. "Tulip fever" was at its height, and although it never reached such a climax in England as it did in Holland, the flowers were justly popular. Fifty years after the first tulip was seen in Augsburg (1559) the flower was well known and largely cultivated throughout Germany, Holland and England. About seven distinct varieties were grown, and endless variations propagated from them, and the rage for procuring fresh colours became a passion among gardeners. Rea's son-in-law, Samuel Gilbert, in his *Florist's vade mecum*,* gives a plan of a garden for tulips. The beds are divided into squares, and numbered up to fifty, and each division was intended for a distinct variety of tulip.

A present of tulips was much valued, or an exchange was effected among friends, and each new variety carefully treasured. The following notes occur in a pocket-book of Sir Thomas Hanmer: "Tulips sent to Sir J. Trevor 1654 1 Peruchot 1 Admiral Enchuysen 1 of my Angelicas 1 Comisetta 1 Omen 1 of my best Dianas, all very good bearing rootes, sent by my wife from Haulton." "June 1655 Lord Lambert, I sent him by Rose a very great mother-root of Agate Hanmer." This was a tulip grown in his own garden at Bettisfield, its colours were gris de lin, crimson and white. Sir Thomas Hanmer has also left notes on their culture. " Set them in the ground about the full moon in September about four inches asunder and under four inches deep, set the early ones where the sun in the spring may come hot on them. Set the later kinds where the noon sun may not be too fierce on them. Let the earth be mold taken from the fields, or where woodstacks have been, and mix it with a fourth part or more of sand. Make your beds at least half a yard thick of this mold. Tulips live best planted alone, but you may put some anemonies with them on the outside the beds if they be raised high and round. They will come up in December and January, and the early sorts flower in the latter end of March, and beginning of April, the other a fortnight or more after them. Set the mother-roots by themselves, and the young offsets by themselves. The new varieties of tulips come from sowing their seeds, but the seedlings will be five years at least before they bear a flower. Keep old strong roots for seed, of such kinds as have blue cup and purple chives, and are striped with pure white,

---

* Second Edition, 1683.

and carnations or gridelines or murreys. The single colours with blue cups or bottoms, and purple chives will most of them parrach or stripe and will stand two years unremoved when the roots are old."

A further catalogue of the contents of the flower garden at Bettisfield in 1660 is chiefly a list of its tulips. Each bed is mentioned, and every row of bulbs taken separately, and the name of each bulb, as many as thirteen ranks, all carefully arranged. But other flowers also found corners, although not allowed beds to themselves. This was another bed at Bettisfield. "In the middle of this bed is one Double Crown Imperial. In the end are six rows of Iris raised from seed by Rea;—also polyanthuses and daffodils. In the four corners of this second bed are four roots of good anemonies." In one there was a preponderance of Narcissus, all described " Belles du Val narcissi, all yellow." . . " Belle Selmane narcissi, right dear ones," and so on. "The border under the South Wall in the great garden is full of good anemones, and near the musk-rose are two roots of the daffodil of Constantinople from Rea, and a Martagon pomponium." These extracts show that Thomas Hanmer was a friend of the gardener and author Rea. He made a catalogue of choice plants, "yet such as will bear our climate," with short "directions for their preservation and increase, not meddling with their medical qualities," and it is believed that these notes were given to Rea, who made use of them in his book.

Sir Thomas was also a friend of Evelyn and imparted some of his knowledge of plants to him. On August 22nd, 1668, he writes to Evelyn, enclosing him some papers: "They are but common observations, but true ones, and most of the famed secrets for meliorating flowers will not prove so." In 1671 he writes again, this time sending Evelyn some plants:—

"Bettisfield, *Augst.* 21*st*, 1671.

"Sir, I send you herewith some rootes of severall sorts: the bear's ears and some of the anemones and ranunculus are very good, but the tulips (except Agat Hanmer and the

Ariana, and some others) are not extraordinary; indeed, my garden affords not now such varieties of rare tulips as I had formerly; most of my best died the first yeare I came to live at this place, and I have not furnisht my selfe anew, because I thinke neither this ayer nor earth agrees with them. I suppose your flower garden, being new, is not very large, and therefore I send you not many things at this tyme, and I wish the beares eares doe not dry too much before you receave them; they will be a fortnight at least before they come to Deptford, and therefore sett them as soone as may be, and water them well (if it raine not) for three or fower dayes, and plant them not in too hott a sun. I thought once to have ventur'd some gilliflowers, having two years since raised some very good ones from seed (w^h I never did before, nor I thinke never shall againe, because the wett in England hinders the ripening of the seed more than in Holland and Flanders) but there is such store of excellent ones all about London, that I had not the confidence to adventure any to your view;—and I doubted whether being soe long on the way would not kill them. Sir, I wish I were better able to serve you either in these bagatelles or more weighty occasions: I should with great alacrity and satisfaction, I assure you, lay hold on all opportunityes to express myselfe how really I am

"S^r.

"Yo^r affectionate faithfull servant,

"Tho. Hanmer.

"My wife and my selfe humbly present o^r services to your worthy lady, and your selfe, as also to my noble friend Sr Richard Browne. I convey this letter and the box to you by my son Tom Hanmer, who is constantly at his chamber in ffig-tree Court in the Inner Temple, and can send your commands to mee at any tyme. You will find in the box some very good bear's ears seed, which you know better to sow and order than I can direct."

Other flowers mentioned as rarities by Gerard and Parkinson had become very generally known. Among the lilies this is noticeable:—"The red lily (*L. canadense rubrum*) is a flower so vulgar, every countrywoman can form an idea of it in a stranger's head, by their rustick descriptions. . . . Next comes martagans, a rambling flower onely fit for flower pots or chimneys, and to be planted in by borders or under hedges." * Carnations were still popular flowers:—"*Caryophyllus hortensis* called July flowers, and are indeed summer glory as Tulips the pride of the spring. . . . the nobler sorts which are called Dutch July flowers or more vulgarly carnations raised from seeds in the Netherlands and other parts adjoining to the sea, and thence conveyed to us." †

The sensitive plant, *Planta Mimosa*, the sensible or humble "plant," was a new acquisition in Charles the First's time. The seeds were "yearly brought out of America." ‡ This would be one of the tender annuals, for which the hot bed would be prepared. Another plant grown in this way was Tobacco, "Sow on a hot bed as early as you can after Christmas," writes Sharrock, "then plant under South Wall or otherwise with hedges or fences of Reed to be defended from sharp weather." § Jacoboea marina (= *Sprekelia formosissima*) came from N. America, in 1658. Jasmine (= *odoratissimum*) from Madeira about the same time, and many other plants were introduced.

So much is done to encourage the improvement of flowers nowadays, by Shows, Competitions and Prizes, that it is difficult to realize that the efforts made in that direction long ago were spontaneous. The earliest record I have noticed of encouragement of the growth of flowers (except of course gratuities for presents of flowers, at a much earlier date) is mentioned by Pulteney,‖ "Mr. Ray informs us that the people of Norwich had long excelled in the culture and production of fine flowers, and that in those days (c. 1660) the florists held

---

* Gilbert, *Florist's Vade Mecum*.   † Ibid.   ‡ Rea.
§ The first description of Tobacco in English appeared in 1580 in a work entitled *Joyfull News from the Newfound World*, translated from the Spanish of Monardus by J. Frampton. There is an account " of the Tobacco and of his great vertues " and a woodcut of the plant.
‖ *Sketches of Botany*, 1797.

their annual feasts, and crowned the best flower with a premium as a present."

The introduction of foreign tender plants led to the gradual growth of conservatories and hothouses. In a previous chapter I noticed some hints Sir Hugh Platt gave for the protection of delicate plants during the winter. In the second part of his work, first printed in 1660, he not only thinks of protection, but has also a feeble idea of forcing, an art which did not develop until many years later. He writes, "Quaere, If pease beans pompeons musk mellons, and other pulse seeds, put in small pots ... and placed in a gentle stove or some convenient place aptly warmed by a fire and then sown in March or April would they come up sooner?" Again he says, "why not utilize a kitchen fire planting them (*i.e.* apricots or vines) near a warm wall, or brewers, diers, soap boilers or refiners of sugar, who have continual fire, may easily convey the heat of steam of their fires (which are now utterly lost) into some private room adjoining wherein to bestow their fruit trees."

Attention was now turned to growing oranges, and the houses built for the shelter of these trees are the earliest kind of conservatory. Very far removed from the modern glass structure, they were like large rooms with big windows and a stove or open fire to warm it in the coldest time, or " in default of stoves or raised hearths you must attemper the air with pans of Charcole." * The oranges were planted in cases, and were lifted out to adorn the garden during the summer months, but were "committed betimes into the conservatory." No garden was complete without its "collection of choice greens." Already in the time of Charles I. there existed several orangeries. At Wimbledon, the favourite resort of Henrietta Maria, was one of the finest examples. The orange garden was laid out "in four knots," bordered with box, and turfed squares with walks round them. In this the oranges stood out in tubs in the summer time, and there was a garden house in the orangery, where the trees, forty-two in number, were stored for

* Rea, *Flora, Ceres and Pomona*, 1665, also Sharrock.

the winter. These trees were valued, when the Parliamentary survey was made prior to selling the place, at £420. The survey of these grounds forms a very complete picture of a garden of this date, the various terraces, trees, walks, summer-houses and everything it contained being carefully described and valued.*

After the Restoration, conservatories became more general, and are noticed by several of the writers of the time. There were houses built for the reception of "tender greens" at the

ORANGERIE AND CANAL, EUSTON. FROM A SKETCH BY EDMOND PRIDEAUX, C. 1716.†

Oxford Botanic Garden, and later on at Chelsea Physic Garden. The gardens of Essex House in the Strand possessed a fine collection " of choicest greens," under the care of John Rose, one of the most celebrated gardeners of that day. His

* Printed in *Archæologia*, Vol. X, 1789. Reprinted in an Appendix to this volume from original MS. in the Record Office, Parliamentary Survey, No. 72.

† In the possession of Charles Glynn Prideaux Brune Esq.

treatment of plants in cases is thus quoted by Rea : —" In spring and autumn you must take some of the earth out of the cases, and open the rest with a fork or other fit tool . . . fill up again with rank earth two parts dung well rotted." That orange-trees, however, were still considered a great novelty, the following extract from Pepys' Diary will show:—" 25 June 1666. Mrs. Pen carried us to two gardens at Hackney (which I every day grow more and more in love with) Mr. Drake's one, where the garden is good, and house and prospect admirable, the other my Lord Brooke's, where the gardens are much better, but the house not so good nor prospect good at all. But the gardens are excellent, and here I first saw oranges grow, some green, some half, some a quarter, and some full ripe, on the same tree,

ORANGERIE AT CHISWICK. FROM AN ENGRAVING BY ROCQUE, 1736.

and one fruit of the same tree do come a year or two after the other; I pulled off a little one by stealth (the man being mightily curious of them) and eat it, and it was just as other little green small oranges are—as big as half the end of my little finger. Here were also a great variety of other exotique plants, and several Labyrinths, and a pretty aviary." He visited this garden on a former occasion, May 8th, 1654, and says of it :—" One of the neatest and most celebrated in England," but either the oranges were not there then, or he did not see them.

Gardeners seem to have understood that a certain amount of air was necessary for plant life, but I think they by no means realized the power of light. Sharrock, writing on the subject, comes to the conclusion that " the coldness and briskness of the free air . . . produces verdure," and to prove this, he takes for example flowers shut in rooms, the leaves of which become paler, and the "whiting the leaves of Artichokes, Endive, Mirrhis Cichory, Alexander, and other plants, which is done by keeping them warm without the approach or sentiment of the cool fresh aire." It is to be wondered how they got delicate plants to live by sheltering them in dark places during the winter months. "Some defend their Mirtles, Pomegranates and such other tender Plants, either by houses made of straw like Bee hives, or of boards (with inlets for the sun by casements, or without them) Litter of Horse Stables being laid in very cold weather about the houses of defence."

Le Nôtre was invited to England by Charles II., and it has generally been believed that he accepted the invitation, and that St. James's Park, as well as alterations at Hampton Court and Whitehall, were made from his designs, and under his direction. In 1661 a certain Adrian May was appointed by Royal Warrant, "supervisor of the French gardeners employed at Whitehall, St. James, and Hampton Court, to examine their bills, &c., and see that they have due satisfaction." This shows it is a fact that Frenchmen were employed, if not the great Le Nôtre himself, Perrault, or some of his pupils. Switzer in 1718 mentions Perrault's visits to England, but says nothing of the coming of Le Nôtre. Jean de la Quintinye, who was the great French gardener and fruit grower, as Le Nôtre was their chief garden architect, certainly visited England, and gave hints to and corresponded with the principal Englishmen of rank who followed the fashion, and were lovers of gardens. His works were translated by Evelyn and London and Wise, and were quite the standard books in England, and his illustrations of the manner of grafting and pruning, are admirable. Rose, who was considered the best practical gardener of his time, was sent by the Earl of Essex to study at Versailles, and on his return he was appointed Royal Gardener by Charles II. Thus the French influence was strong in England, and grand gardens, belonging to the largest houses of the nobility, not old-fashioned manor-house gardens, were laid out in the French style.

One good reason why it was in large gardens only that this style was adopted, was, that to carry out such vast ideas as those of Le Nôtre, space was required. The trees were planted in longer, larger, bolder avenues. There were wide paths and terraces, adorned with statues, and fountains and cascades. All French pictures of gardens show also numbers of seats, and arbours of stone, with a background of trellis-work, or closely-clipped trees, in the form of alcoves and arches. The semi-circular garden at Hampton Court was also laid out during the reign of Charles II. under the direction of Le Nôtre. He designed the avenues, and the canals which were "near completed" in 1662. The gardens were somewhat altered a few years later. In the time of Charles II. there was a large central fountain, with syrens and statues, by Farrelli, which was removed under William III., besides twelve smaller fountains. The work was begun soon after the Restoration, when Charles returned fresh from having seen the glories of Versailles, spent large sums of money, perhaps with some idea of rivalling the magnificence of Louis XIV. Among the fountains were laid geometrical beds and plots of grass, each with a conical-shaped yew in the centre. Some of these yews, no longer clipped into stiff forms, still remain.

One of the French gardeners who helped to carry out the alterations at Hampton Court was Beaumont, who was the designer of Levens in Westmorland, though the work he did there is certainly not in the style of Le Nôtre. At Levens there is a portrait of him with this inscription on it, " M. Beaumont, gardener to James II., and Colonel James Grahme. He laid out the gardens at Hampton Court and at Levens." Colonel Grahme was a staunch adherent of James II., and after the Revolution of 1689, for political reasons, found it safest to live in the North, on the estate he had lately purchased, and it was during his time, and under the direction of Beaumont, that the gardens assumed the form which they retain almost unaltered to this day. They are, therefore, a most perfect example of the Dutch type of garden of this period. One feature which was apparent in every garden of this date, was the bowling-green or alley, which had come into fashion

LEVENS. FROM A PICTURE BY GEO. S. ELGOOD.

a hundred years earlier. At Levens there still remain some of the bowls with the Bellingham crest, and as Colonel Grahme bought the place from the Bellinghams in 1687, the bowling-green must have existed some years previously. Many examples of old bowling-greens still remain:—there is a very fine one at Chilham Castle, in Kent, 207 ft. long and 126 ft. wide, also good examples at Cusworth and Bramham, in Yorkshire, Holme Lacy in Herefordshire, at Powis Castle and many other places. They were of various forms and sizes, and there was generally a raised bench or terrace on one or more sides of the open green, frequently with a pavilion from which the spectators looked on at the game, while the bowling-alley, on the contrary, was completely hidden by overshadowing trees. A bowling-green at Warwick Castle is thus described in 1673:—"Within the gate . . . is a fair Court, and within that, encompassed with a pale, a dainty bowling-green, set about with laurel, firs, and other curious trees,"* and in 1681 the Duke of Norfolk's garden near Norwich is described by the same writer, Thomas Baskerville: "Taking a boat for pleasure to view this city by water, the boatman brought us to a fair garden belonging to the Duke of Norfolk, having handsome stairs leading to the water, by which we ascended into the garden, and saw a good bowling-green, and many fine walks." In all his journals, Baskerville notices the public bowling-greens at all the small towns, and attached to many of the inns he stayed at. Thus, of Pontefract Castle, he writes, "of which now only remains the platform and stump of the bottom of the wall 2 or 3 yards above ground, but yet it is handsome, because employed to fine gardens and a bowling-green, where you may have for your money good wine," also at Bedford "the ruins of an old castle, containing within it a fine bowling-green." Among others he notes Saffron Walden, "a very good bowling-green without the town," and of Watton, a small town in Norfolk, he says there is little remarkable, save a fine new bowling-

---

* Thomas Baskerville's Journal MSS. of the Duke of Portland, Hist MSS. Report 13.

green at the "George Inn." These pieces of good turf must have added much to the beauty of the gardens, and in the small towns served as a public garden and recreation ground.

Every garden also contained one or more sundials. They formed, as a rule, a centre to the design, and were in themselves a fitting ornament to a garden. The sundial has frequently survived destruction, when all other traces of an old garden have been obliterated. At Exton, in Rutlandshire, the old sundial stands in front of the house which was burnt down,

SUNDIAL, EUSTON, WITH THE ARLINGTON ARMS, ABOUT 1671.

almost the only vestige of the garden which formerly lay in front of its windows. In some dials the owner's coat of arms was used to form the style, or in others the motto of the family was inscribed round the dial, which is often a great help in fixing the date of the construction. Occasionally an entire garden was laid out like a sundial, the figures being planted in box or yew. There is a good example of one after this design at Wentworth Stainborough, which was made in 1732, in which the letters are of box and the style of yew.

Gardeners from all times have had great difficulties to contend with, in the extirpation of garden pests. Their minds were chiefly exercised in devising schemes for keeping down the moles. When Queen Elizabeth paid a visit to Theobalds, and Lord Burghley prepared a Masque in her honour in May, 1591, speeches were recited before her, composed by George Peele, describing the processes of making the garden, and comparing its beauties to the virtues of the Queen. The first speech was that of the "Molecatcher," which began thus:—
"I cannot discourse of knots and mazes, sure I am that the ground was so knotty that the gardener was amazed to see it, and as easy had it been, if I had not been, to make a shaft of a cammock * as a garden of that croft."† The ordinary mole-catchers were paid by the number of moles they caught, "usually 12d. a dozen for all the olde moles they catch, and 6d. a dozen for younge ones. Now as for those who send purposely for a mole-catcher to gette a single mole in a howse, garden or the like, they will seldom take lesse than 2d. and sometimes 3d. for her if they gette her, because they have payment onely for those they catch and if they misse the lose is theires."‡ The farmer, Henry Best, in the East Riding of Yorkshire, who made these notes, has also left the account of what he paid himself to the mole-catchers. In "1628, April 28, paid to John Pearson for killing moules in the carre one and a haif dozen olde ones 13½d., two dozen young ones 6d.," and so on. Several curious recipes for killing moles are found in old gardening books. Sharrock gives the following "Remedies against Moles"§:—"By watering moles are drowned or driven up into so narrow a compass that they may be easily taken. Mr. Blith relates one spring, about March, a mole-catcher and his boy in about ten dayes time, in a ground of 90 acres, took 3 bus[hels] old and young. Among Mr. Speed's notes there are these receipts:

* = a crooked tree.
† Dramatic and Poetical Works of R. Greene and G. Peele. By Dyce, 1861.
‡ Rural Economy in Yorkshire, 1641. Surtees Society, 1857.
§ An Improvement in the Art of Gardening. By Robert Sharrock, 3rd Ed., 1694.

Take red herrings and cutting them in pieces burn the pieces on the molehills, or you may put garlicke or leeks in the mouths of their Hill, and the moles will leave the ground. I have not tryed these ways, and therefore refer the reader to his own tryal, belief or doubt."

For the destruction of other garden pests many equally fanciful remedies were in vogue. Lawson recommends to pick off all caterpillars with the hand, "and tread them under foot." "I like nothing of smoake among my trees," he says; "unnaturall heates are nothing good for naturall trees." He enumerates the things necessary for keeping the garden free from " beasts," "besides your out strong fence, you must have a fayre and swift greyhound, a stone-bowe, gunne, and if neede require, an apple with an hooke for a Deere, and a Hare-pipe for a hare," and against blackbirds, bullfinches, and other small birds, " the best remedy here is a stone-bow, a peece." No survey of the garden would be complete, without mention of the bees, whose hives were to be found in them all, and the management of which was considered a necessary part of a gardener's duties, and writers on gardening subjects generally devoted a chapter to bees.*

One memorable event in the time of Charles I. was the formation of the first Botanical Garden in England, at Oxford, in 1632. This was just a hundred years after the establishment of the earliest in Europe, that at Padua. Henry, Earl of Danby, founded and endowed it; he gave five acres of land, also built greenhouses, and a house for the gardener. The fine gateways, bearing a date and inscription in praise of the Founder, were designed by Inigo Jones. Jacob Bobart, a German, from Brunswick, first had charge of it, and he was succeeded by his son, also Jacob.

The marshes for bog plants, to be seen at Kew and elsewhere at the present day, which are the admiration of lovers of a "wild garden," are no new thing. Bobart had one at Oxford, which is thus described by Robert Sharrock.† "The Artificial Bog is made by digging a hole

---

* Thomas Hill, *The right ordering of Bees*.
† *An Improvement to the Art of Gardening*. 3rd Edition. 1694.

in any stiff clay, and filling it with earth taken from a bog
. . . of this sort, in our garden here in Oxford, we have
one artificially made by Bobart, for the preservation of Boggy
plants, where being sometimes watered, they thrive for a
year or two as well as in their natural places." A catalogue
of the garden, which contained some 1600 species and varieties,
was published by Bobart in 1648. Of these nearly six hundred
were native plants. The catalogue is a tiny book, and no
space is given to describe the flowers. It is merely a list
of names, the first part Latin-English, the second English-
Latin. The list contains among trees "Abies mas," "male
Firretree," "Arbutus," "Strawberry tree," "Arbor Judae,"
"Judas tree," "Ash tree," &c. Among the flowers are about
twenty sorts of Roses, including "York and Lancaster,
Provence, Austrian and Cinnamon, 11 violas, 9 clematis, 7
Colchicum and 9 crocus, double and single peony, 4 foxgloves,
10 Lychnis, Campian, Bee orchis, orchis serapius," &c. The
list also contains "Nicotiana, English Tabacca," "Yucca,
Indian Bread," "Stinging nettle," and 4 sorts of moss, "cup,
club, hard sea, and tree mosse."* The plant names follow
each other in alphabetical order, quite regardless of any
classification. The first attempt to separate indigenous from
foreign plants was made by William How, in his work entitled
Phytologia Britannica (1650).

Although this is not a history of the progress of Botany, that
science is so intimately connected with gardening, that some
references to it cannot be left out, for how could the immense
number of plants now cultivated, be understood or identified, if
it were not for systematic classification? The two great pioneers
in this work are John Ray and Robert Morison. Their relative
merit has been the subject of some discussion. Both began to
work out a system about the same time. Ray gave an outline of
his classification in 1668, in the tables in Bishop Wilkins's *Real or
Universal Character*. Morison's first ideas are embodied in his

---

* A second and enlarged edition was published in 1658, with the
co-operation of Philip Stephens and William Brown, both botanists of
Oxford. It is a great improvement on the first, and makes frequent
reference to Gerard and Parkinson.

work, *Hortus Blesensis*, 1669, and further developed in his *Plantarum Umbelliferum*, 1672, and his *History of Plants*,* 1680. Ray's complete system, shown in his *Methodus Plantarum*, did not appear until two years later, his *Synopsis* in 1690, and the revised *Methodus* in 1703. Morison professes to have worked out the system entirely from Nature, but Ray, with perhaps more honesty, owns his indebtedness to Caesalpinus and other foreign writers, and even to Morison. It was Ray who first separated the Monocotyledons from Dicotyledons, and thus laid the basis of the "Natural System" now universally followed. Ray (1628-1705) was the son of a blacksmith, near Braintree, in Essex; he was educated at the Grammar School there, and in 1644 went to Cambridge, where he soon showed his love of natural history, and especially of botany, and published his catalogue of plants round Cambridge in 1660. He travelled much about England, and also spent three years abroad with his friend, also a naturalist, Francis Willoughby. In 1667 he was made a Fellow of the Royal Society, and contributed many writings to their "transactions." He settled near his native place in 1679, and there passed the remainder of his life in study, and the production of his great works on Natural History and Botany. Morison (1620-1683) was a native of Aberdeen. Being a staunch Royalist, when the war broke out he joined the army, and on the failure of the King's cause went to France. There he studied Natural History, and became so distinguished a botanist that he was appointed Curator of the fine gardens of the Duke of Orleans at Blois, in 1650. Charles II., on his Restoration, invited Morison to return to England, and gave him the supervision of the Royal Gardens. In 1669 he was appointed Professor of Botany, at Oxford, with the degree of Doctor of Physic, and there he lectured and laboured at his *Historia Plantarum Oxoniensis*, until his death, caused by an accident, in 1683. The systems evolved by these two men differed from those of all preceding Botanists; inasmuch as they were the first to classify plants

---

* *Plantarum Historiæ Universalis Oxoniensis*, pars secunda. The first part was never published. 1680.

according to some real likeness in the fruit or flower, and not merely from similarity of habit or place of growth. Morison divided herbaceous plants into fifteen classes; Ray into twenty-five, and trees and shrubs into eight. These systems, which paved the way, so to speak, for Jussieu, Robert Brown, and others, came at a time when they were most needed. From East and West, from the Old World and from the New, plants were pouring in yearly in increasing numbers; and the necessity of arranging these newly-acquired treasures, was the foremost task of Botanists.

CHAPTER X.

> "When lavish art her costly work had done,
> The honour and the prize of bravery
> Was by the garden from the Palace won."
> COWLEY.

A GOOD idea of the number of gardens existing in England in the time of William and Mary may be gathered from the diary of Celia Fiennes,* who travelled on horseback through the country. In every county, and at almost every stage of her journey, she mentions or describes some garden more or less notable. The fountains, or "waterworks," were perhaps the most characteristic feature in the larger gardens, and of these she gives many elaborate descriptions. At Chatsworth there were fountains innumerable, one a willow tree "which rains from each leaf," and there is one bason in the middle "of one garden that's very large and by sluces besides the images severall pipes plays out ye water: about 30 large and small pipes altogether, some fflush it up that it ffrothes like snow." At Wilton there was a grotto with pipes concealed apparently all round and over the roof, which sent forth a sort of shower bath which "washes ye spectators." Again, at Bradby, Lord Chesterfield's house, "In one garden there are 3 fountaines wherein stands great statues. Each side on their pedistalls is a dial, one for ye sun, ye other a clock wch by ye water worke is moved and strikes ye hours, and chimes ye quarters, and when they please play Lilibolaro on ye chimes. All this I heard when I was there."

* *Through England on a side-saddle in the time of William and Mary.* 1888.

These waterworks, introduced as we have already seen in Tudor times, were now very much in vogue. The ideas for them came from abroad, both from France and Holland. The fountains at Versailles, and other places in France, are too well known to require notice. But waterworks of quaint forms and surprise-arrangements were typical of Dutch gardens, and William of Orange brought these into popular favour in this country, together with many other Dutch fashions. In 1621, Lord Chaworth in his diary* remarks on the "verie fyne gardens" surrounding the house of the Infanta Isabella, in Brussels, "wherein are ye most varietie of the best waterworks of ye world." The gardens at Boughton, Northamptonshire, were laid out during this reign, when the house was rebuilt by Ralph, first Duke of Montague. They were very extensive, covering over a hundred acres, and were remarkable for the "sumptuous waterworks." There was the "parterre of statues, the parterre of Basins and the water parterre, wherein is an octagon basin whose circumference is 216 yards, which in the middle of it has a jet d'eau, whose height is above 50 feet, surrounded with other smaller jet d'eaus. . . . The Canal at the bottom of all, is about 1500 yards in length in four lines falling into each other at right angles. At the lower end of it is a very noble Cascade . . . adorned with vases and statues. The Cascade has five falls. The perpendicular about seven feet. · A line or range of jet d'eaus in number thirteen are placed at the Head of the Cascade . . . There are also several jet d'eaus in the basin underneath. Also the knot of regularly figur'd Islets beset with Aquatick Plants."† Such Cascades were quite formal, all built of solid masonry, and are totally unlike the "Cascades" or miniature waterfalls of a later period. The gardens at Boughton were in the French style, but the head-gardener at this time was a Dutchman called Vandertmeulen.

The gardens described by Celia Fiennes have all alike gravel and grass walks, shady alleys of clipped trees, "some

---

* Loseley MSS.
† *Natural History of Northamptonshire.* By John Morton, 1712.

walks like arbours close, others shady, others open, some gravel, some grass." Standard cypress or yews "cut in severall forms were dotted about." Trim hedges of holly, laurel or box, divided the parts of the garden:—for instance, "the front garden wch has the largest fountaine," from "the garden of flower trees, and all sorts of herbage," or the one with "grass plotts" from the bowling-green. Occasionally, mention is made of "fine greens," and "dwarfs,"* or oranges and lemons; a shelter or greenhouse. Or, perhaps, the description of a broad terrace with stone steps; a wilderness planted with pines; a grove with alleys cut through; a pond, a canal, or a fine gateway, varies the recital of her travels and gives a reality to the scenes she recalls. At Mr. Thetwin's, near Stafford, she admires the "fine rows of trees" in the park. "ffirs Scots and Noroway, and ye picanther." She remarks, at Trygothy, in Cornwall, the drawing-room opened into the garden, "wch has gravell walks round and across, but ye squares are full of goosebery and shrub trees, and looks more like a kitchen-garden." Of Blith, near Worksop, she says, "I eate good fruite there," and she made her first acquaintance with orange trees at Lady Brook's house in Wiltshire. "Here was fine flowers and greens, Dwarfe trees and Oring and Lemon trees in rows wth fruite and flowers at once and some ripe, they are ye first oring trees I ever saw."

She evidently admires gardens in the new French or Dutch style, more than the gardens of the last generation. She passes over Haddon, merely observing, "it's a good old house, all built of stone on a hill, and behind it is a ffine grove of high trees and good gardens, but nothing very curious as ye mode now is." Again, of "Mr. Paul Folie's seate called Stoake," near Hereford, she writes, "it's a very good old house of timber worke but old ffashion'd, and good roome for gardens, but all in an old fform and mode and Mr. Folie intends to make both a new house and gardens. The latter I saw staked out . . . y^e ffine Bowling-green walled in and a Summer-house in it all new." At Barmstone, in Yorkshire,

* = *fruit trees cut small.*

... are capable of being ... remain in the old fashion ... ...Huntingdon, was having a new garden made. "The gardens and wilderness and greenhouse will be very fine when quite ffinished, with the dwarf trees and gravell walks. There is a large fountaine or bason which is to resemble that in the privy garden at Whitehall, which will front the house. The high terrass walks look out on the road."

At Sir John St. Barbe's house, near Rumsey, new gardens were also being made—"not finish'd but will be very ffine, wth Large Gates open to the Grounds beyond, some of wch are planted with trees." Such walls with "severall places with grates to Look through," was the latest development of the craving to look beyond the garden, which we have noticed in earlier times. Such arrangement of spaces, with gates or iron bars, in the walls, is constantly noticeable in the views of gardens early in the eighteenth century. This desire to extend the view, led to the planning of the park and avenues to correspond with the open spaces at the side or end of the garden walks. These attempts to harmonize the garden with its surroundings, gradually developed, until the walls were dispensed with, and the "landscape" style superseded the older forms. In studying the changes in design, it seems to me that there was no sudden "leaping the garden wall." We must look for the beginnings of the landscape style in the gradual change or decadence of the old formal school. The Dutch style, introduced by William III., was an exaggeration of the old manner of clipping trees. Topiary work in yew, box, and other "greens," was carried to such an excess; the gardens were so overcrowded with cut trees, as to become the laughing-stock of the succeeding generation, and so bring about their own destruction.

The word "knot" does not often occur in books of this date, and the word "parterre," which takes its place, requires some explanation. Meager, in the *English Gardener*, 1688, gives a list of herbs "fit to set knots with," of which "Dutch or French Box, it is the handsomest, the most durable, and the cheapest to keep." And in the same chapter, he refers his reader to the plates at the end of the book, where he has "presented to view divers forms or plots for gardens." In 1697, he speaks of parterres, and his designs are very similar. Sir Thomas Hanmer, in notes for his proposed work on gardening, also uses the two words:—" If the ground be spacious, the next adjacent quarters or parterres, as the French call them, are often of fine turf, but as low as any green to bowl on; cut out curiously into embroidery of flowers, and shapes of arabesques, animals, or birds, or feuillages, and the small alleys or intervals filled with several

PARTERRE. FROM LONDON AND WISE.

coloured sands and dust with much art, with but few flowers in such knots, and those only such as grow very low lest they spoil the beauty of the embroidery." Parterre is thus explained in Miller's Dictionary, 1724: "A level division of ground, which for the most part faces the South, and is best in front of a House, and is generally furnished with greens and flowers. There are several sorts of parterres, as bowling-green, or plain parterres, and parterres of embroidery . . . . Plain parterres most beautiful in England by reason of their turf, and that decency and unaffected simplicity it affords the eye:

others are cut into shell and scroll-work, with sand alleys between them, which are the finest parterre works esteemed in England."

In the *Retired Gardener*, translated from the French of Louis Liger, by London and Wise, no less than eleven sorts of parterres are described, but all are merely variations of design in grass, beds or cut-work, and patterns of scrolls and foliage or "embroidery, like we have on our cloaths." The two following are examples of his descriptions: No. VI. "The Form of a Parterre partly cut-work and partly

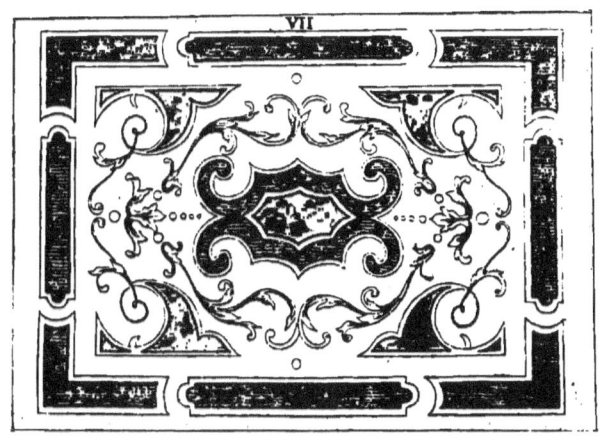

PARTERRE. FROM LONDON AND WISE.

green Turf with Borders. These Parterres are esteem'd according to their Design and their Symmetry. They look very well in great gardens as well as small, the verdure of the grass, and the Enamel of the Flowers with which the Compartments ought to be fill'd according to the different seasons of the year, present a charming object to the sight. These parterres may likewise be set off with such Pots as I mentioned before (*i.e.* Dutch jars) or surrounded with Boxes fill'd with Orange Trees or with other shrubs of like Nature."
VII. "The Form of a Parterre with cut-work of Grass and

Imbroidery in the middle and with Borders of Grass on the outsides. This sort of Design is very agreeable and serves for a great ornament to a garden, especially where the grass-work is well kept up, the Box well order'd, and the grass-work well cut ; and to give it yet a farther Beauty, you may fill the Flourishings and Branch-work with a black earth, provided the Paths or Alleys be cover'd with a yellow or white sand, different colours serving to set off the Parterre the better." In some cases the plot was filled with one design, in others it was divided into four, and the pattern repeated in each section.

Between the parterres were borders, formed either of a sanded path with a strip of grass or flowers, on either side, or shrubs placed at intervals, but the "most common" borders "are wrought with a sharp rising in the middle, like the back of an ass, and set with yews, shrubs, and flowers." Canons Ashby, as it is at the present day, is a good example of th's date of garden, and the parterres, as shown in the plan kindly made by the present owner, Sir Henry Dryden, are such as might have been seen in any garden of this date, though the design perhaps is more simple than in many of them. The garden, originally made in 1550, was altered in 1708, and has defied the changes of fashion for nearly two centuries. It is just such a garden as Celia Fiennes described as "neatly kept, with fine gravel walks, grass-plotts and beyond a garden of flower-trees and all sorts of herbage and store of fruits."

Incidental remarks in that lady's journal, throw light upon Town-gardening. Before such great difficulties in the way of smoke had to be contended with, town-gardens needed no more care than country ones, and many town-houses had fine ones attached to them. When gardens were simple, small, and enclosed, there was no reason why as pleasant and secluded ones should not be made in towns as in the open country. We still find old-fashioned gardens in the Cathedral towns, or in some few large market-towns, where smoke and overcrowding have not destroyed them. But long ago, when each good house had its garden, the aspect of the towns must indeed have been different. Public parks and gardens

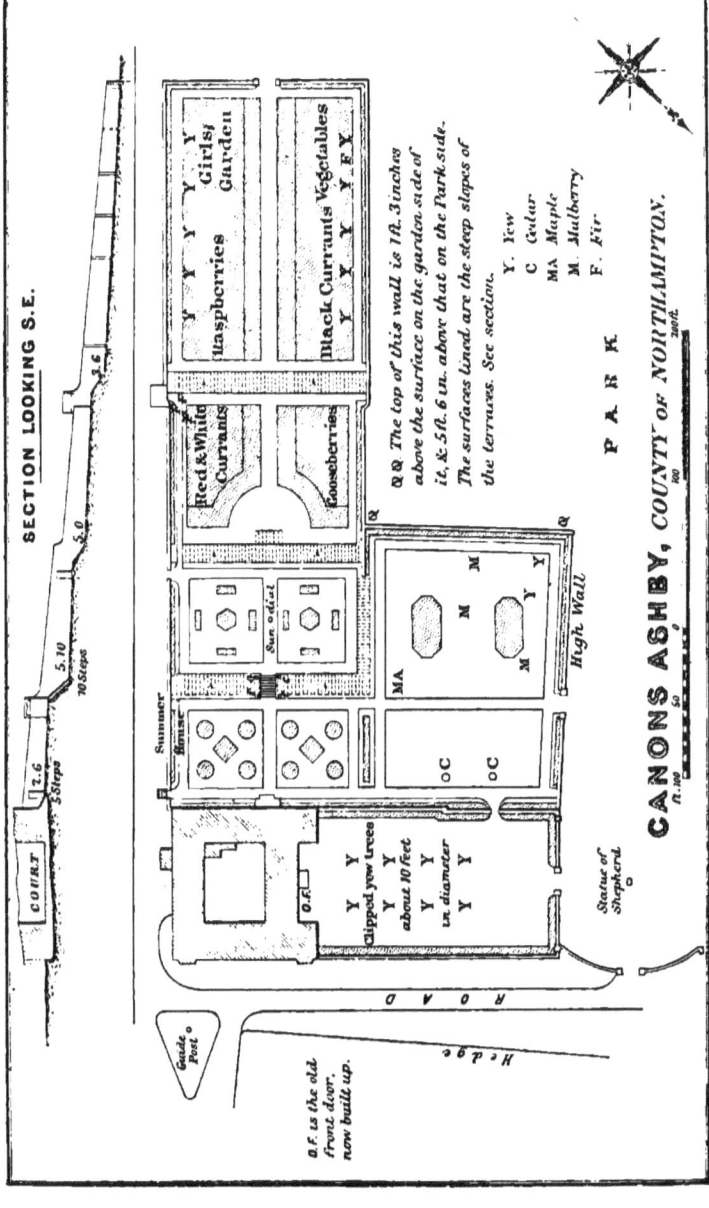

are no new invention, although so vastly improved even of
late years, in spite of all the disadvantages of fog, smoke, and
darkness. Certainly from Cowley's poem one would imagine
the smoke nuisance to have been as troublesome in the
middle of the seventeenth, as at the end of the nineteenth
century:—

> "Who that has reason and a smell
> Would not among Roses and Jesamine dwell
> Rather then all his spirits choak
> With exhalations of dust and smoak,
> And all the uncleanness which does drown
> In pestilential clouds a populous town."

Leeds, though then but a village in comparison with the
Leeds of to-day, is thus described by Celia Fiennes:—"A large
town, severall large streetes, cleane and well pitch'd, and good
houses all built of stone. Some have good gardens and steps up
to their houses, and walls before them." Of Bedford she
writes:—"It is an old building washed by the river Ouse ... its
stored with very good ffish, and those which have gardens on its
brinke keepes sort of ... Baskets which keeps the ffish by chaines
to the sides of the Banks in each man's garden. It (the river)
runs by a ground which is made into a fine bowling green ... well
kept with seates and summer houses in it." At Newcastle, she
finds—"This country all about is full of this Coale ye sulphur of it
taints ye aire and it smells strongly to strangers ... its a noble
town ... and most resembles London of any place in England.
... There is a pleasant bowling-green, a Little walk out of
the town wth a Large gravel walk round it, wth two Rows of trees
on each side. ... There is a pretty Garden, by ye side a shady
walk, its a sort of spring garden where the Gentlemen and
Ladyes walke in the evening;—there is a green house in the
garden."

Spring Gardens, which she here refers to, were the favourite
resort of fashion in London. They had been in existence since
the first quarter of the century, and originally were part of the
royal park of St. James', as appears from entries in the
Exchequer rolls:—

1617. "digging planting etc: of roses in the Spring garden in (St. James')
Park ... Gardeners, women weeders: in the spring garden ... Pheasants and
wild fowl in the spring garden."

By the middle of the century, however, it was a public garden, of which the street now bearing its name marks the site.

In London many old gardens were already disappearing, for Celia Fiennes writes thus in her Diary :—" There was formerly in ye Citty severall houses of ye Noblemens wth Large gardens and out houses and great attendances, but of Late are pulled down, and built into streetes and squares and called by ye names of ye noblemen :—and this practise by almost all even just to ye Court, excepting one or two. Northumberland and Bedford House, Lord Montagues, . . . and Whitehall with its privy garden and famous fountain." A description of the gardens near London in 1691, by Gibson, has been preserved.* He enumerates twenty-eight gardens, five of those being nursery-gardens—the Brompton Nursery, one "Clements" at Mile End, and Ricketts, Pearson and Darby, all three at Hoxton. Some of the gardens are more distant from London, as Hampton Court, Sir Henry Capel's at Kew, and Sir William Temple's at Sheen. At Beddington where the first orange trees in England had been planted by the Carew family, they had been so well taken care of that it still held the foremost place among the orangeries in the country. This orangery was two hundred feet long, and the trees were about thirteen feet high, and in one year yielded ten thousand oranges. Gibson also tells us that the Queen Dowager, at Hammersmith, had a good greenhouse, but was not " for curious plants or flowers" ; however, her gardener, Monsieur Hermon Van Guine, raised orange and lemon trees, which he had " to dispose of." Arlington garden was "a fair plat." Sir Thomas Cooke's, at Hackney, though very large was still being added to;† Lord Ranelagh's "elegantly-designed," though "but newly-made." The Archbishop, at Lambeth, was then improving the garden there, and putting up a greenhouse, "of three rooms, the middle having a stove under it ;—the foresides of the rooms are almost all glass, the roof covered with lead." Gibson only mentions those gardens which he

---

* Printed in the *Archæologia*, 1794, and lately reprinted in Hazlitt : *Gleanings in Old Garden Literature*.

† Rams Chapel was built in 1723 on part of the site of this garden. In a deed, dated July 20th, 1704, in the possession of the chapel authorities, two summer-houses are mentioned, one of which is used as the vestry.

visited in December, 1691 : others equally well known he passes over. He does not notice the large nursery between Spitalfields and Whitechapel, the owner of which Meager refers to as "my very Loving friend Captain Qarrle," and gives a long list of fruit trees, any one of which this friend can "furnish," besides "divers other rare and choice plants."* He omits, also, Essex House in the Strand, and Somerset House; also Southampton House, Bloomsbury, where the gardens were designed by Lord William Russell, who was beheaded in 1683. The garden at Fulham, which had been made famous by Bishop Grindal, who introduced the tamarisk in Elizabeth's reign, was further improved by Bishop Compton at this date, and there are splendid hickory and other trees of his planting still to be seen there:—"He had a thousand species of exotick plants in his stoves and gardens, in which last place he had endenizoned a great many that have been formerly thought too tender for this cold climate. There were few days in the year, till towards the latter part of his life, but he was actually in his garden, ordering and directing the Removal and Replacing of his Trees and plants."†

Besides the private gardens, there were the parks, which even then added beauty to the country round London, St. James's Park, and "another much Larger, Hide parke, wch is for Riding on horseback, but mostly for coaches, there being a ring railed in, round wch a gravel way, ... the rest of the park is green, and full of deer, there are Large ponds wth fish and fowle." ‡ Beyond Hyde Park was Kensington, a favourite palace of King William, and there, again, was a good garden, begun by him, and completed under Queen Anne. The gardeners employed there were the famous London and Wise, who owned the large nursery at Brompton, hard by. This was the finest nursery of the day, and they kept an immense collection of plants. The tender greens from the gardens at Kensington were housed during the winter at Brompton, where, although a fine collection in themselves,

---

* Leonard Meager, *The English Gardener*, 1688, p. 60.
† Switzer, *Ichnographia*, 1718. Bishop Compton, born 1632, died 1713.
‡ Celia Fiennes' Diary.

they took " but little room in comparison with "* those belonging to the nursery.

George London, who was the principal founder of the Brompton Nurseries, was a pupil of John Rose, and at one time gardener to Bishop Compton. He travelled abroad, both before and after he established the nursery, and visited Versailles after the peace of Ryswick, when he went to France with the Earl of Portland. He died 1713. The nursery " was started by him in the reign of James II. in conjunction with Cook, gardener to the Earl of Essex at Cassiobury, Lucre, gardener to the Queen Dowager at Somerset House, and Field, gardener to the Earl of Bedford, at Bedford House, in the Strand."† These partners designed the gardens at Longleat, " The four took it in turns to go down to lay out "‡ the grounds. Lucre and Field died, then Cook retired, and London took Henry Wise into partnership. Johnson § says this occurred in 1694, but Gibson in 1691, describes the nursery as " Brompton Park garden, belonging to Mr. London and Mr. Wise." So it does not seem as if the original four were many years together. These two gardeners became very famous, not only for their horticulture at Brompton, but for the gardens they designed all over the kingdom. London was made Superintendent of the Royal Gardens, and a Page of the Backstairs to Queen Mary. Besides the work they did for the King at Kensington, they carried out considerable alterations at Hampton Court. One rather strange piece of work undertaken there, was the transplanting of one of the rows of lime trees which formed the avenue by the semi-circular canal. The trees on the northern bank were taken up and replanted on the south of what had been the most southern row. " Four hundred and three large lime trees ye dimensions of them from 4^{ft} 6^{in} to 3^{ft}, the charge of taking up these trees, bringing them to the place, digging holes of 10 or 12 feet diameter, carting 5 loades of earth to each tree one with another, with all charges 10s. per tree,

---

* Gibson, 1691.
† Switzer, *Ichnographia Rustica*, 1718.   ‡ Ibid.
§ *History of English Gardening*, 1829, p. 123.

£201. 10." This removal took place some thirty years after the trees had been planted. Other changes were made in the "Mount Garden" and the "Privy Garden," "Queen Mary's Bower," of pleached elms, was planted, the old orchard turned into a wilderness, the terrace along the river was made, and probably the maze was laid out about the same time. Wise also planned the "Broadwalk" which runs all along the front of the palace between it and the fountain garden.* Blenheim Garden was another of their great undertakings, and they were three years in finishing it. A fine specimen of their style is still to be seen at Melbourne in Derbyshire. The gardens of Sir Richard Child, at Wanstead in Essex, of Bushey Park, of Cranborne, and of Castle Howard, were some of their other works; at the last-mentioned place Switzer says they reached "the highest pitch that Natural and Polite gardening can ever arrive to." On the accession of Queen Anne, Wise was given the care of the Royal Gardens, and London confined himself chiefly to work in the country. He passed his time going a round of great gardens, frequently, it is said, riding a distance of fifty to sixty miles a day, in the course of his business.

Moses Cook, one of the original partners, published a work on fruit trees, but London and Wise were the popular writers, as well as designers, of the firm. They translated two works from the French, the *Complete Gardener*, from Jean de la Quintinye (first ed. 1699), and the *Retired Gardener*, from Louis Liger, with the *Solitary Gardener*, from Le Gentil. They added copious notes from their own experience; the information is all conveyed in the form of question and answer between a gentleman about to purchase a seat in the country, and "taste the Sweets of Country Life," and a gardener. The gentleman asks such questions as, "Suppose I have some cases sent me from abroad . . . when I receive them my ground is lock'd up by a frost . . . what must I do with them?" Gardener:—"Upon Receipt of your trees, which I suppose sent in cases with moss laid round the roots . . . . you must

---

* In the estimate for the work, the walk was to cost £650. 13s., and the turfing of the sides, and planting and making the borders, £400. 10s., and £210 respectively. -*Treasury Papers* lxiii., 48, &c.

keep 'em in a cellar till your ground is capable of receiving 'em. . . . Take your roots out of the cases, and trim their roots. . . . After steep the roots in water for a Day, and then set them. . . . If you observe this rule you won't lose one of your Trees, tho' they have been out of the ground for three or four months together." London and Wise's experience follows, and is rather contradictory: "We had some peaches grafted on Almond's Stocks from France, in 1698 . . . . which were three months out of the ground, notwithstanding all requisite care . . . we could not save ten trees out of the whole hundred." In another chapter it is recommended in sending layers and slips from abroad, to rub them first with honey, and then cover in damp moss, or stick them into "a piece of Potter's Earth tempered with honey," and wrap round with moss. In this work the growing of mushrooms, artificially, is recommended. The process, a very lengthy one, of preparing the beds, is described, which took nearly a year to complete. Jean de la Quintinye's work is confined to fruit culture, and he is especially minute in describing the correct pruning of fruit trees, standards, and espaliers and wall-fruit. The " History and Origin of Flowers," which forms a large part of the *Retired Gardener*, is a disappointing title, as it is merely a collection of the most fantastical myths and legends, such as the origin of the foxglove. Juno, working one day, lost her thimble. Jove, to pacify her, said he had turned it into a flower, and accordingly up came a foxglove. Ornithogalum was a spoilt child, fed only on white of egg, till he grew feeble and was dying, so Venus, pitying him, turned him into the flower which bears his name—and many other such stories. London and Wise give a quaint list of how some plants are propagated, or are "vivacious and lasting, which are commonly grown in our flower gardens." Anemonies are vivacious by their fangs, Asphodils by their tubers, Auriculas, Columbine, Gillyflowers, Grenadil or Passion-flower, Lavender, Scabious, Sunflower, Thyme, and the like, by their roots; Crown Imperials by the suckers produced from their roots, Ranunculus by their claws, Day-lily by its bulb, Daisy and Sea-thrift by their tufts, Tuberose by its suckers, and so on.

The English Gardener, by Leonard Meager, was a popular book, and went through several editions. But little notice has been taken of the author, who was much more old-fashioned than his contemporaries. This book, in a quiet way, gives a great deal of practical information about fruit and kitchen-gardening, and his "Catalogue of Flowers," "such as are only for ornament in their places where they grow, or for nose-gays," reminds us more of Parkinson than of Evelyn or London and Wise. He calls the flowers all by their homely English names:—Such as Coventry Bell flowers (*Campanula Medium*), Melancholy Gentlemen (*Hesperis*

NETHERTON. FROM A SKETCH BY EDMOND PRIDEAUX, 1727.

tristis), Goat's Rue (*Galega officinalis*), None-such or flower of Bristol (*Lychnis chalcedonica*) and King's Spear, yellow and white (*Asphodelus*). Meager, on the title-page of the 1688 edition of his book, says he had been "Thirty years a Practitioner in the Art of Gardening." From the dedication, it appears that for many years he was gardener to Philip Hollman, of Warkworth, in the county of Northampton. The Hollmans were a good old family, and Philip, who died in 1669, seems to have encouraged Meager in his work, as indeed Meager adds he assisted all his "other servants that had any inclination or endeavour to the Practise of Good Husbandry." Meager probably shows us a type of the quiet old-fashioned "neatly-ordered" gardens, throughout England. The quaint view of Netherton, in Cornwall, is from a sketch made by Edmond Prideaux, about 1712, of a garden of this kind. Coryton Park,* in Devonshire, is a good example still existing. It was laid out about 1680, and when alterations were made in 1756, the old garden was left as a kitchen-garden, and is still untouched. The old wall, which divides the upper or new from the lower or older garden, is of a quaint zig-zag form; the simple lines of the rest of the garden might have been taken from Meager's book. A path all round, two large square parterres, two smaller ones, with two corners curved to allow room for a path round a pond and fountain, and across the centre of each plat, a clipped yew-hedge following the same curve, and terminating at the edge of the gravel path with a cypress-tree, two statues, a sundial, and opposite the fountain against the outer wall an old garden house or orangery, compose the design.

This kind of plan was already becoming old-fashioned, and the tendency was to make larger gardens than could be kept up in a formal style. Sir William Temple, in 1685, saw the danger when he wrote, "As to the size of a garden which will perhaps in time grow extravagant among us, I think from four or five to seven acres is as much as any gentleman need design." His own garden at Sheen was not large, but beautifully kept; of this wrote Evelyn, in 1688: "the wall fruit trees are most exquisitely nail'd and train'd, far better than I ever noted." His "Retreat" later in life in Surrey he called Moor Park after the favourite garden of his youth, Moor Park, in Hertfordshire, which he describes so delightfully, as it was, he says, "the perfectest figure of a garden I ever saw."† At the new Moor Park he laid out a garden in the Dutch style. It is not to be won'ered at, that the statesman who negotiated the Triple Alliance shou'd refer the taste of the Netherlands to that of France, and he was minded enough to get what was good from

---

* Belonging to Rev. Marwood Tucker.
† Sir Wm. Temple's *Miscellaneous Works*.

prided himself on having introduced four new sorts of grapes into England :—1. The "Arboyse from Franche Compté, which is a small white grape . . . it agrees well with our climate . . . it is the most delicious of all grapes that are not muscat. 2. The Burgundy, which is a grizelin or pale red, and of all others surest to ripen in our climate, so that I have never known them to fail one summer these 15 years, when all others have; and have had it very good upon an east wall. 3. A Black Muscat, which is called the Dowager, and ripens as well as the common white grape. 4. The Grizelin Frontignac, the noblest of all grapes I ever ate in England, but it requires the hottest wall and the sharpest gravel, and must be favoured by the summer too, to be very good." Unlike the proud possessor of the "Tulipe noire," or Alphonse Karr's enthusiastic old savants who fought over a Buddlea.* Temple was very generous in distributing the vines he introduced, for he writes : " I ever thought of all things of this kind the commoner they are made the better."

Temple turned his attention chiefly to fruit culture. Of flowers he says :—" I only pleased myself with seeing or smelling them, and not troubled myself with the care, which is more the ladies' part than the man's." Perhaps he left the floral part of his garden to his charming wife, Dorothy Osborne. In her delightfully fresh and witty love-letters to Temple during the long years of their engagement, we have one reference which is enough to show that she, too, took interest in gardening. She writes, in 1654, of Sir Samuel Luke, a neighbour of hers at Chick Sands, in Bedfordshire : " But of late I know not how Sir Sam has grown so kind as to send to me for some things he desired out of this garden, and withal made the offer of what was in his, which I had reason to take for a high favour, for he is a nice florist."

Another gardener who helped to encourage grape growing by distributing vines was Rose, gardener to Charles II. and author of *The English Vineyard Vindicated*. He offered to " all that desire it sets and plants of all the best vines sufficiently tried in our soil and climate at reasonable prices." † And John Beale,

---

* Buddlea globosa, introduced 1774.
† Letter concerning Orchards and Vineyards, John Beale, 1676.

following the example set by Rose, used to offer to give plants of vines to "cottagers," but they generally answered "churlishly that they would not be troubled with grapes"; but when he explained that in a few years their grapes would fetch a good price in the markets, "they were soon of a more thankful mind."

In his Diary on June 10th, 1658, Evelyn made the following entry:—"I went to see ye medical garden at Westminster well stored with plants under Morgan, a very skilful botanist." Hugh Morgan is twice mentioned by Johnson, in his edition of Gerard's *Herbal*, as "The Queen's Apothecary," and "a curious conserver of rare simples," and he notices a large specimen of the "Lote or Nettle" tree, growing in Morgan's garden, near "Coleman Street, in London." This Morgan was probably the same man whose garden at Westminster Evelyn visited, but how long he kept up this garden is uncertain. When a physic-garden in Westminster, presumably this one, was bought by the Apothecaries' Company, in June, 1676, it was in other hands, as the Company bought the lease from Mrs. Gape, with the liberty of moving the plants to Chelsea Garden.* The Physic Garden at Chelsea was founded in 1673,† and after a few years entirely superseded the one at Westminster. The lease of the land at Chelsea from Charles Cheyne (afterwards Lord Cheyne) was signed August 29th, 1673, for a term of sixty-one years, the rent £5 per annum, and the following year a wall was built round the garden. The first gardener was Piggott, who was succeeded in 1677 by Richard Pratt. These gardeners received £30 a year, and their successor, John Watts, 1679, got £50. The garden was managed by a committee of twenty-one assistants, thirty liverymen, and twenty yeomanry. They built a greenhouse which cost them £138 in 1680. Two years after, Dr. Herman, of Leyden, visited the garden and offered to exchange some plants. To effect this, Watts was sent over to Holland. In 1685 the expenses of the garden, besides Watts' salary, reached £130, so the Company, unable to carry on the garden at that rate, arranged to give Watts £100 a year, out of which he was to keep up the garden,

---

* Faulkner's *Chelsea*, Vol. II., pp. 174-176.
† *History of the Apothecary's Garden*. By Henry Field, 1820.

APOTHECARIES GARDEN, CHELSEA, IN 1840

and he was allowed to sell fruit and plants. The same sort of arrangement was afterwards made with his successor, Doody, a good botanist, and famous collector of native plants, chiefly cryptogams, who was given the post in 1693. In 1722, Sir Hans Sloane, having acquired land at Chelsea which included the garden, gave the site to the Apothecaries' Company, on condition that it was always to be a Physic Garden, and Philip Miller was made the curator. Another condition of Sir Hans Sloane's, was that the Company should present fifty new plants annually to the Royal Society (of which he was President) until they had given two thousand. They, however, continued the annual gift until 1773, and gave in all 2550 species.

Sir Hans Sloane had for many years taken a lively interest in the garden. In 1684 he wrote Ray an account of a visit which he paid to it.* "I was the other day at Chelsea, and find that the artifices used by Mr. Watts have been very effectual for the Preservation of his plants, insomuch that this severe winter has scarce killed any of his fine plants. One thing I much wonder to see Cedrus Montis Libani . . . should thrive so well, as without pot or green House, to be able to propagate itself by Layers this spring. Seeds sown last Autumn have as yet thriven very well." There were four cedars planted in 1683, and two were flourishing in 1820, and one remains in 1894. Before this visit to the garden, he must have paid many others, as he made most of his botanical studies there, and was encouraged and assisted by Ray. Sloane (born 1660) had been abroad and studied medicine at Montpelier, where a Botanical Garden had existed since 1598. Long years before he conveyed the land to the apothecaries, he was famous for his assiduous studies of Natural History. The first volume of his great work on Jamaica and the West Indies, was published in 1707. He was in Jamaica as Physician to the Duke of Albemarle, the Governor, who died there suddenly, and Sloane returned to England, having in fifteen months collected a large amount of curiosities, and no less than eight hundred species of plants. He lived at Chelsea all the latter part of his life, and died there in 1752. His fame as a Naturalist is scarcely less

---

* Ray's *Philosophical Letters*, 1718.

than as a Physician. The great Linnaeus as a young man came to England to see him in 1736. On every occasion he was the encourager and friend of gardeners, of which the following letter is an example:—

> Sir Henry Goodricke to Sir Hans Sloane. Ribstan, near Boroughbridge, in Yorkshire, 17$\frac{12}{13}$.
>
> SIR,
>
> The civilitys I have received from you do incourage me to give the trouble of a letter, and knowing you to be one who loves to incourage curiosity makes me hope that the subject of my letter won't be so disagreable to you as to another. It is to desire of you that if among your rarities you have any number of seeds, nuts or kernells of foreign and rare trees especially those that are hardy I shall verily thankfully pay for 'em, my pleasure being to raise such things in hot beds and preserve 'em with care; and I would not rob you of any but what you have so many as you may readily spare a part to one who will as readily supply you again when any accident happens to yours, which I believe y^{rs} are more subject to near London than we are, here where I myself take the chief care of my curious trees. I have not yet been able to procure a tree of the true lotus (*Zizyphus Lotus*), nor the larch tree, both which Mr. Evelyn says grow well in our climate, and may be raised from seed; these seeds and any other exotics I doubt not to raise, I mean trees, for smaller plants are too numerous for me to attend; if you could procure me a small tree of each of those kinds I w^d repay you with thanks, being S^r y^r obliged and humble servant,
>
> H. GOODRICKE.

There are three or four very fine larch trees in the grounds at Ribston now, which are probably the very ones sent in answer to this appeal. Sir Henry Goodricke was the introducer of the well-known Ribston pippin. He had three pippins sent him from Normandy about the year 1707, one of them grew up, and was the original Ribston Pippin tree; it was blown down in 1839, but a sucker from the root is now a fair-sized tree, and still bears occasional fruit.

CHAPTER XI.

> "Shade above shade, a woody theatre
> Of stateliest view . . . . ."
> MILTON, *Paradise Lost.*
>
> "Shower every beauty, every fragrance shower,
> Herbs, flowers and fruits; . . . "
> THOMSON, *Seasons.*

THE gardeners who followed London and Wise as designers, as well as cultivators and planters, were Stephen Switzer, and after him Bridgeman. These men were busy at a time when formal gardening was on the wane. It was in Queen Anne's time that Addison and Pope first ridiculed the old style, and sought to bring in the fashion of "copying Nature." But the reaction and destruction of old gardens did not take place till later; when the theories they advanced had had time to spread. There is no lack of views and designs of gardens during this period. They are to be found in County Histories such as Plot's Staffordshire, Atkyns' Gloucester, and Dugdale's Warwickshire; also Beeverell, "Les Délices de la Grande Bretagne et de l'Irlande," published at Leyden in 1707, in *Britannia Illustrata*, 1709, with a large series of views by Kip, and in other similar works. If the authors had foreseen the annihilation that was to befall so many gardens, they could hardly have more carefully preserved their designs. But these pictures are mostly taken from some imaginary point, and give a bird's-eye view of house, garden, and surrounding landscape, in a conventional plan, regardless of perspective. Faithful representations though they may be in many cases, the formal garden, as they show it, has lost all its poetry; the pale tints of the tender shoots of the beech hedge in spring, the soft green of the sheltering yews in winter, the secluded alley, or the woodbine-covered arbour, have no charm when set down in these stiff lines of black and white. The garden at Ingestre was described by a traveller, John Loveday, of Caversham, in 1732: The house, he says, is situated on the side of a hill, "the Gardens

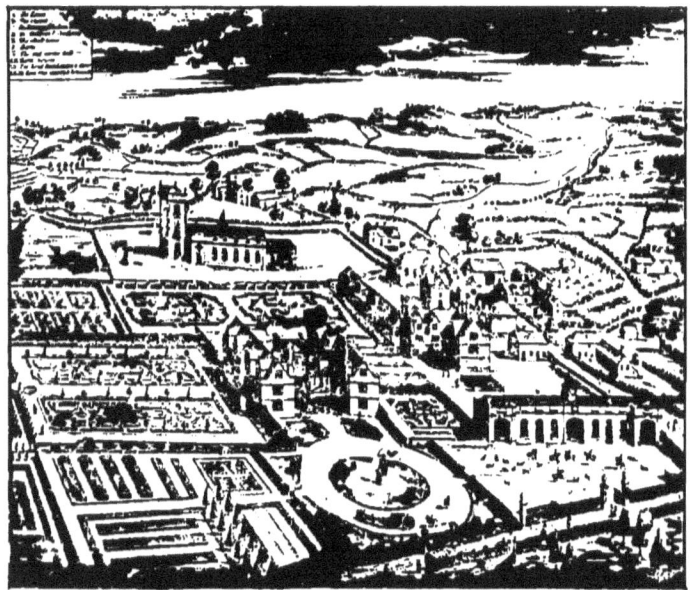

INGESTRE, THE SEAT OF LORD VISCOUNT CHETWYND, FROM PLOT'S "STAFFORDSHIRE." FIRST EDITION, 1686.

higher, They are large—laid out into the grandest walks between the stateliest Trees imaginable, Hares in abundance about the woody Garden, a Building erecting in the higher part for a Prospect . . . . which together with the Church is represented in Plot p. 299." The picture Loveday refers to is here reproduced and illustrates in a striking manner how inadequate these designs are to convey any idea of the beauties of the originals.

DAWN OF LANDSCAPE GARDENING.

It has been said that it was the decadence of art in the formal style which brought about its own fall, but it is difficult to imagine anything more charming than some of the gardens of the time of Queen Anne. Their chief characteristic was the prevalence of long walks between cut trees, not exactly hedges, but trees clipped up to a certain height, and allowed to feather naturally at the top. A most curious example of this is to be seen at Down, in Essex. The trees are cut to the

"CASHIOBURY, THE SEAT OF THE RT. HONBLE. THE EARLE OF ESSEX IN HARTFORDSHIRE."
FROM AN ENGRAVING BY KIP.*

height of sixty or seventy feet: the path between them is about fifteen feet wide, and seven hundred and eighty long, and closes with a view of Hatfield Broad Oak at the end. This garden was made when the place belonged conjointly to Prior, the poet, and Harley, Lord Oxford.† Prior wrote a humorous poem on the occasion of his first visit to "Derry

---

* See p. 184.
† Now the property of Lord Rookwood.

down down, hey derry down," as he called it. He expected to find there,

> " . . . . gardens so stately, and arbours so thick,
> A portal of stone, and a fabric of brick."

but on reaching his destination, the poet exclaimed to his friend,

> "O Morley, O Morley! if that be a hall,
> The fame with the building will suddenly fall."

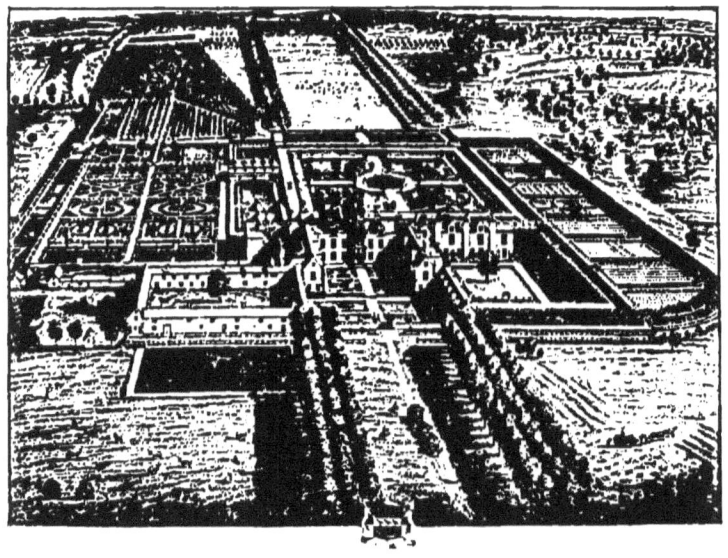

"BROME HALL, SUFFOLK, ONE OF THE SEATS OF THE RIGHT HON. CHARLES LORD CORNWALLIS."
FROM AN ENGRAVING BY KIP.

To which he received the answer,

> "I show'd you Down-Hall; did you look for Versailles?"

Prior lived here for many years, and designed new gardens, and these alterations which Lord Oxford carried out, included the present principal garden, with box hedges in the Dutch style, and the long wall of clipped hornbeams. Another charming example is at Bramham, in Yorkshire.* The ground-plan of the garden is like any figured in Switzer's

Belonging to Mr. Lane Fox.

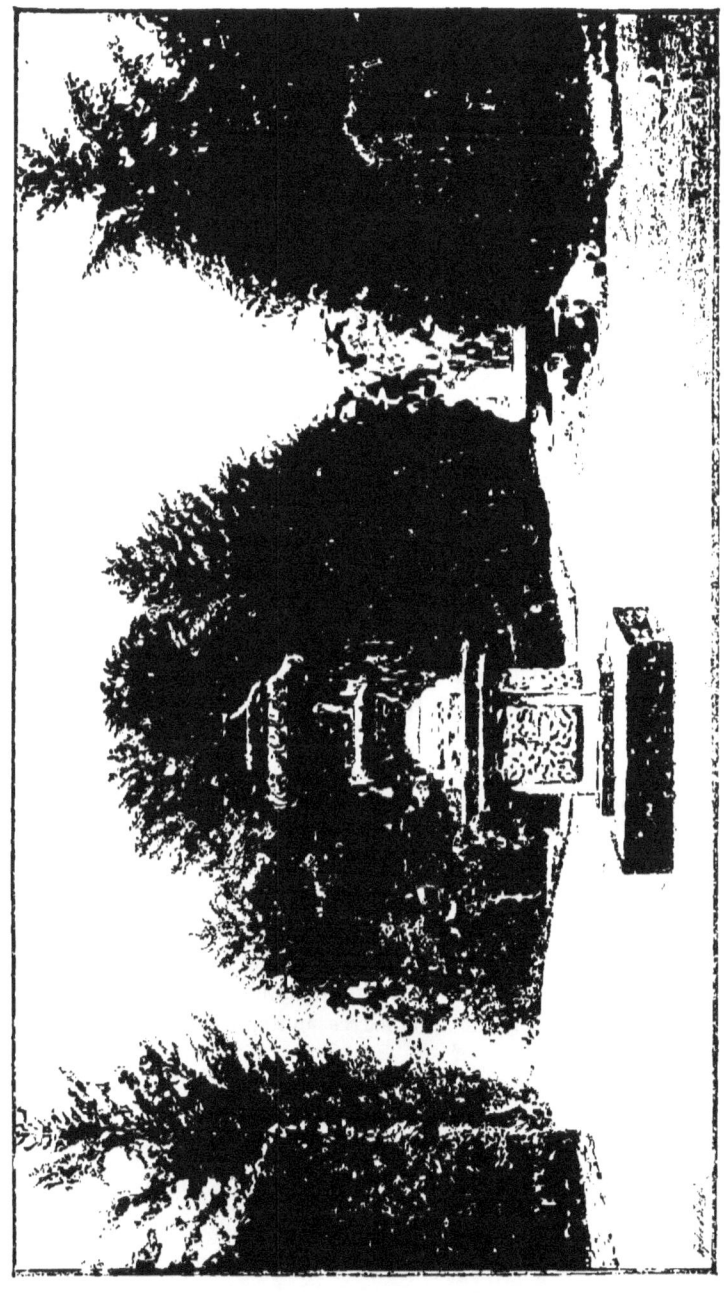

BRANTHAM

books. The house was burnt many years ago, and never restored, but the gardens have been kept up in their original state, as they were laid out by Mr. Benson. He was Ambassador to Spain, and Queen Anne gave him a grant of land on Bramham Moor; after he had built a house and made gardens round it, she paid him a visit there, and created him Lord Bingley. Along the house is a terrace, and in front of it a grass parterre. From thence are seen vistas through the beech and hornbeam woods beyond. From the northern end of the terrace, a straight walk between high cut hedges runs westward, and leads at once into the most entrancing maze of long walks diverging from each other at regular angles. At the end of some there is a small summer-house, a seat, or statue, or monument. From the ends of the walks furthest from the centre the view ranges over the open country beyond. The garden stands above the level of the park, therefore the terrace-wall which divides them has all the effect of a sunk fence. But the most delightful part, perhaps, is where the avenues are wider, where the walks skirt the edge of a canal, and the tall trees are reflected in its silent waters. There is an open space laid out as a "French garden." In this case it is an oval slope of grass, with large flower-beds in a regular pattern: a summer-house overlooks this garden, and to the back of the summer-house there is a large bowling-green, surrounded by trees, among which are the walks. At the opposite end of the oval garden there is a basin and "cascade," and a short distance from this point the path rejoins, at its southern end, the terrace which runs in front of the house. The effect of this garden at Bramham, on a fine autumn day, with the slanting beams of the evening sun, seen through the long vistas shining on the golden-brown foliage of the trees, is truly beautiful, and leaves an impression never to be forgotten.

There is a contemporary description of such a garden in a letter written by Lord Percival to his brother-in-law, Daniel Dering.* It is dated from Oxford, August 9th, 1724:—

"Friday morning left Becconsfield; we went half a mile out

---

* MS. belonging to Lord Egmont.

of our way to see Hall Barn, Mr. Waller's house—a London Box, if I may so call a house of 7 windows every way. He was gone a hunting, so we did not go into the house, which promised nothing extraordinary, but we spent a full hour and half in viewing the gardens, which you will think are fine, when I tell you they put us in mind of those at Versailles. He has 80 acres in garden

HALL BARN.

and wood, but the last is so managed as justly to be counted part of the former. From the parterre you have terraces and gravelled walks that lead up to and quite thro' the wood, in which several lesser ones cross the principal one, of different breadths, but all well gravelled and for the most part green sodded on the sides. The wood consists of tall beech trees and thick underwood, at least 30 foot high. The narrow winding walks and paths cut in it are innumerable; a woman in full health cannot walk them all, for which reason my wife was carry'd in a Windsor chair like those at Versailles, by which means she lost nothing worth seeing. The walks are terminated by Ha-hah's, over which you see a fine country and variety of prospects every time you come to the extremity of the close winding walks that shut out the sun. Versailles has indeed the advantage in fountains, for there is not one in all this garden; but there are two very noble pieces of water full of fish, and handsomely planted and teraced on the sides. In one part of the wood, and in a deep bottom, is a place to which one descends with horrour, for it seems the residence of some draggon; but there shines a gleam of light thro' the high wood that surrounds and shades it, which recovers the spirits, and makes you sensible a draggon would seek some place still more retired. This place may be call'd the Temple of Pan or Silvanus, consisting of several apartments, arches, corridores, &c., composed of high thriving ews cut very artfully. In the centre of the inner circle or court, if I may call it so, stands the figure of a guilt satyr on a stone pedestal. . . . I pass over the bowling-green, and large plantations about the house, which are but young, but I must not forget a bench or seat of the famous Edmond Waller's the Poet, which is so reverenced that, old as it is, it is never to be removed, but constantly repaired, like Sir Francis Drake's ship. The present Waller is his grandson. All this fine Improvement is made by himself or Aisleby, his father-in-law, who had this house and the lands about it, in right of his wife's joynture, but gave it up in the South Sea year to his Son-in-law. There is a great deal more still to be done, which will cost a prodigious sum, but this gentleman by marriage South Sea and his Paternal Estate [is able] to do what he pleases." After such a charming description it is pleasing to find that Hall Barn has been but little altered; and a seat bearing the poet's name remains to this day.

Lord Percival was a capital correspondent, and some other letters to Daniel Dering give his impressions of the gardens he saw on his tour about England in 1723, thus:—" To

Wickham, 7 miles to Lord Shelburn's [we thought] he would by this time have made some tolerable garden or cut fine walks in the woods that cover the hills about him, but we were entirely disappointed; the wood is neglected; the gardens which are but 4 acres, without tast and neglected too, and the house fourty times worse than Lady Bidulf's on blackheath." " Col. Tyrrel's called Shotover (near Oxford) about two miles [further on] . . . There is plenty of wood and water about the house, and both brought into the circuit of the garden, with regularity and bewty. A large octogon bason on the west, and two canals on the east; the walks, parterres, terraces, and avenues are agreably separated by groves of reverend oak, beech and elm trees; in a word, his garden is already compleated and yet he still goes on to gratify his good tast." Lord Percival was evidently a friend of Sir William Temple's nephew, as he refers to him frequently in other letters. It is interesting to follow the history of the garden at Moor Park. The following letter is dated August 25th, 1724 :—" Called on Jack Temple who lives a mile from Farnham. . . . It was purchased by the famous Sir William Temple, who took great delight in it, and made part of the garden, but this gentleman, his nephew, has greatly added to it, and rendered it indeed a very pleasant seat. He has the advantage of a branch of the River Wye, which is brought into the midst of his garden, and supply's two pretty cascades. In the Parterre are 4 Antique Statues a young Papyrius and his companion a Bacchus, and Diana."

The same year Lord Percival went into Norfolk and Suffolk. He visited Euston, which he thus describes :—" Neither are the gardens as yet considerable, being but young, and his trees not well grown. He has a very fine canal, that confines one side, and at the end of his gravel walk is a large bason with a lake beyond it." And Lord Oxford's place, " Chipman, 3 miles north of Newmarket. The gardens are 50 acres, and have a good deal of variety, a fine bowling-green, very high hedgerows cut into vistos, long tracts and walks, from which you see several miles into the country through well-grown avenues. There is a canal in the shape of a T 1000 foot long, and 70 broad." This again might be a description of the garden

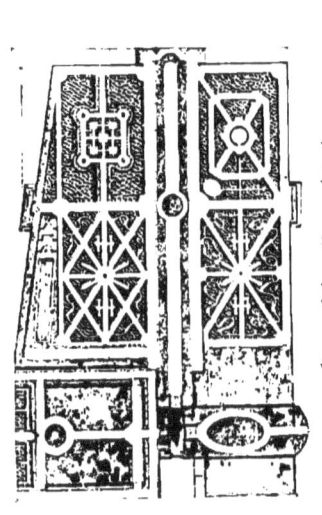

BELTON IN LINCOLNSHIRE. FROM AN ENGRAVING BY BADESLAD.

still existing at Bramham, or of one of Switzer's plans. Belton is another charming example of a garden of about this date which although somewhat altered still retains several features observable on these plans.

Switzer was a pupil of London and Wise, and avowed himself an admirer of Pope's ideas on gardens. He gives his views fully in *The Nobleman's, Gentleman's and Gardener's Recreation*, in 1715, published again with additions as *Ichnographia Rustica*, in 1718, " by which title is meant the general Designing and Distributing of County Seats into gardens woods Parks Paddocks &c.: which I therefore call forest, or in more easie stile Rural gardening." Here is a beginning of the end of Formal Gardening. This " Le grand Manier," he goes on to say, is " oppos'd to those crimping, diminutive and wretched Performances we every where meet with. . . . The top of these designs being in clipt plants, Flowers, and other trifling Decorations . . . fit only for little Town gardens, and not for the expansive Tracts of the Country." In another place,* he goes still further, and says his work is for the " Embellishment of the whole Estate." The grounds to be " handsomely divided by Avenues and Hedges . . . little walks and purling streams . . and why is not a level easy walk of gravel or sand shaded over with Trees and running thro' a corn field or Pasture ground as pleasing as the largest walk in the most magnificent garden one can think of? and why are not little gardens and Basons of water as useful and surprising (and indeed why not more so) at some considerable Distance from the Mansion House as they are near it." The gardens I have quoted above, and his own plans, however, do not go as far as admitting cornfields, but the garden had ceased to be an enclosure, and was already encroaching on the park and surrounding country. The movement in its beginning was doubtless a good one, this casting off some of the unnatural formality and stiffness that gardens of the Dutch type had reached. The French gardens that were copied gave a larger space to work upon, and involved much more expense, thus the natural surroundings were made

* Ed. of 1718.

use of, to help out a large design, and so if possible to cut down the expense.

I do not think that the pioneers of the landscape style can be blamed for the abuse of it a few years later; when the real flower-garden, "the terrestrial Paradise" of flowers was gradually banished, and instead of a garden encroaching on a park, the park came up to the house, and the flower-garden nearly disappeared. People were tiring of "Topiary" work, which had so long been popular. Instead of cut hedges, alleys, arbours, and a few standard trees, gardens were overcrowded with a confusion of cut bushes, and it is not surprising that any one with a love of the beauties of Nature, as she appears in woods and fields, should long to see, at any rate, an occasional tree left to grow in its own wild and graceful way. "Our British gardeners," wrote Addison,[*] "instead of humouring Nature, love to deviate from it as much as possible. Our Trees rise in Cones, Globes, and Pyramids. We see the marks of the scissars upon every Plant and Bush. I do not know whether I am singular in my Opinion, but, for my own part, I would rather look upon a tree in all its Luxuriancy and Diffusion of Boughs and Branches, than when it is thus cut and trimmed into a Mathematical Figure; and cannot but fancy that an Orchard in Flower looks infinitely more delightful than all the little Labyrinths of the most finished Parterre."

The next year (1713) Pope followed up this appeal for natural gardens in the *Guardian*, with some more cutting remarks on the fashion of "verdant sculpture." He supposes that "an eminent town gardiner" . . . who has "arrived to such perfection, that he cuts family pieces of men, women, or children in trees," has sent him his catalogue of greens for sale. A most witty list of trees follows; among them are "Adam and Eve in yew, Adam a little shattered by the fall of the tree of knowledge in the great storm; Eve and the Serpent, very flourishing. St. George in box, his arm scarce long enough, but will be in condition to stick the dragon by next April; A green dragon of the same, with a tail of ground-ivy for the present. (N.B.—These two not to be sold

[*] *Spectator*, 414, June 25th, 1712.

separately.) Divers eminent modern poets in bays, somewhat blighted, to be disposed of a pennyworth. A quickset hog, shot up into a porcupine, by its being forgot a week in rainy weather." In the beginning of the Essay from which the above is taken, Pope quotes Homer's description of the garden of Alcinous, in the *Odyssey*, and gives his own translation of the passage:—

> "Close to the gates a spacious garden lies,
> From storms defended and inclement skies;
> Four acres was the allotted space of ground,
> Fenc'd with a green inclosure all around.
> Tall thriving trees confess the fruitful mold,
> The red'ning apple ripens here to gold.
> * * * *
> Beds of all various herbs, for ever green,
> In beauteous order terminate the scene."

If such was Pope's ideal garden, it had little in common with the landscape style he helped so much to bring in. "How contrary to this simplicity is the modern practice of gardening!" he continues. "We seem to make it our study to recede from Nature, not only in the various tonsure of greens into the most regular and formal shapes, but even in monstrous attempts beyond the reach of the art itself. We run into sculpture, and are yet better pleased to have our trees in the most awkward figures of men and animals, than in the most regular of their own." No one, even the most ardent advocate of the formal garden, can deny that Pope and Addison had much right on their side. But there was no reason to rush to the other extreme, and have no arrangement, or no straight lines of any sort, in a garden. Two years later Pope settled at Twickenham, and his Villa there, far from being in the simple style he admired, became a complicated piece of mimicry of rural scenery of all sorts. He took infinite pains in planning and planting. "I thank God," he wrote in a letter to a friend, "for every wet day and for every fog, that gives me a headache, but prospers my work." His famous grotto, "composed of marbles, spars, gems, ores, and minerals," was the amusement of his declining years. It would hardly lay claim to being "natural," for nothing more fantastical can be imagined, although in Pope's own lines to his grotto, he invites the stranger thus:—"Approach! Great Nature studiously behold."

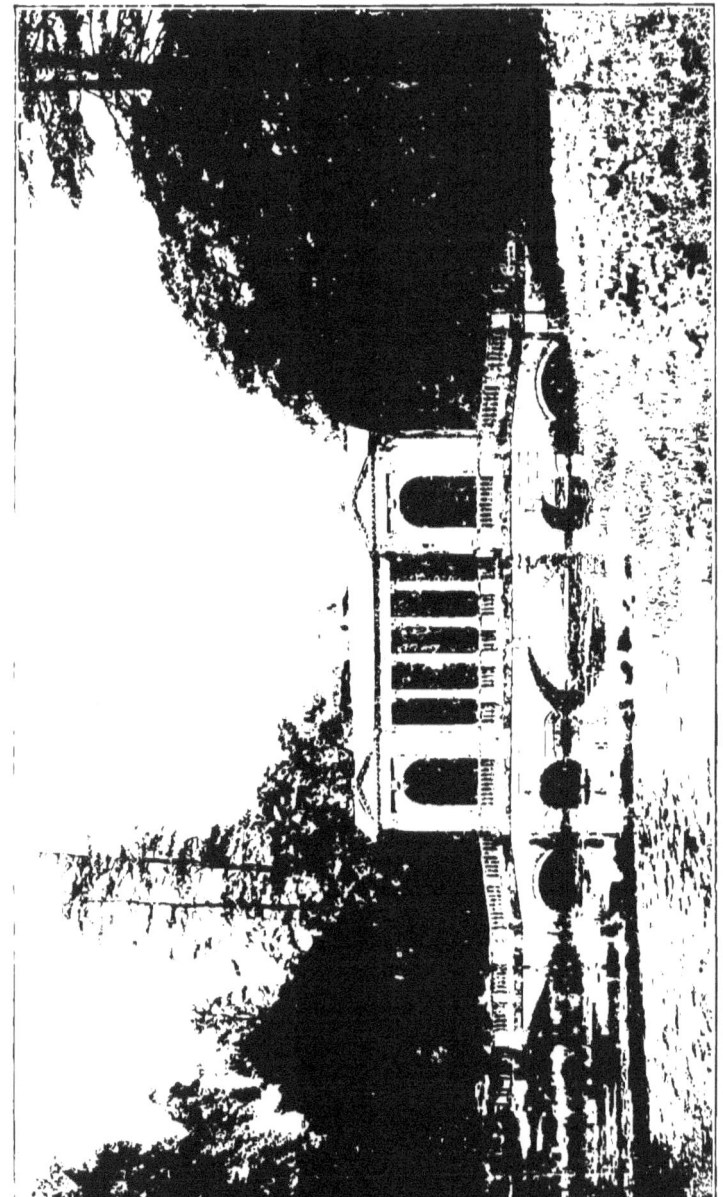

PAUAPDAN BRIDGE AT STOW

Addison lived at one time at Bilton, in Warwickshire, and his garden there is not in a "natural style" either. Part of the garden dates from 1623; some of it was altered in the early part of the century, but the arbour used by Addison is still there. It is of classical "Queen Anne" style of architecture, with a straight bench, facing a view of the garden, with nothing rustic about it. There are still, however, in the garden, two old cut yew arbours, also good yew and holly hedges.

Bridgeman, the other designer of this date, who followed up the ideas of these two writers, was not himself an author like Switzer, so one must look at his work to judge of his ideas. Walpole, writing some years later, praises Bridgeman very highly. He was the successor to London and Wise in the charge of the Royal Gardens, and was, writes Walpole, "far more chaste" than his predecessors. "He enlarged his plans, disdained to make every division tally to its opposite, and though he still adhered much to strait walks with high clipt hedges, they were only his great lines; the rest he diversified by wilderness, and with loose groves of oak, though still within surrounding hedges. I have observed in the gardens at Gubbins, in Hertfordshire, the seat of the late Sir Jeremy Sambrooke, many detached thoughts, that strongly indicate the dawn of modern taste. As his reformation gained footing, he ventured farther, and in the Royal Garden at Richmond, dared to introduce cultivated fields, and even morsels of a forest appearance. But this was not till other innovators had broke loose too, from rigid symmetry."

The names of several landscape-gardeners are known in connection with Stow, in Buckinghamshire, each in turn having added something to the place. The garden was looked upon as quite the acme of perfection, by this school of garden-designers. Pope's lines on the principles of landscape gardening are summed up in the one word, Stow :—

> " Still follow Sense, of ev'ry art the soul,
> Parts answ'ring parts shall slide into a whole ;
> Spontaneous beauties all around advance,
> Start ev'n from difficulty, strike from chance.
> Nature shall join you ; time shall make it grow,
> A work to wonder at—perhaps a STOW."

Sir Richard Temple, who died in 1697, commenced rebuilding the house at Stow, and his son, Lord Cobham, continued it, and began the gardens, which were constantly being added to until 1755. By that time they covered a space of five hundred acres. Bridgeman was the first designer, and after him, Kent, while Vanburgh constructed several of the temples and monuments. In one of the numerous descriptions of Stow, a pyramid is specially mentioned as being the last design Vanburgh executed *:—

> ". . . . . . ascends
> The pointed pyramid; this, too, is thine,
> Lamented Vanburgh! this thy last design,
> Among the various structures, that around,
> Formed by thy hand adorn this happy ground."

As this was the ideal garden of the period, there are several contemporary guides and descriptions to it published. As smaller places copied it, and were composed of the same sort of collection of temples, gardens, and vistas; it will be necessary to go through its varied features in detail, so I have transcribed in full a letter from that same delightful correspondent, Lord Percival, to his brother-in-law, Dering, giving his own impressions of the gardens, to which he paid a visit in 1724:†

"Brackley. 14 Aug: 1724 Friday night,
7 a clock.

"DEAR DANIEL,

"Yesterday we saw Lord Cobham's house, which within these five years, has gained the reputation of being the finest seat in England. . . . The gardens by reason of the good contrivance of the walks, seem to be three times as large as they are. They contain but 28 acres, yet took us up two hours. It is entirely new, and tho' begun but eleven years ago, is now almost finished. From the lower end you ascend a multitude of steps (but at several distances) to the parterre, and from thence several more to the house, which, standing high, commands a fine prospect. One way they can see 26 miles. It is impossible to give you an exact Idea of this garden, but

---

* *Stow. The Gardens of the Rt. Hon. Richard, Lord Viscount Cobham*, 1732. Anon.

† MS. belonging to the Earl of Egmont. In St. James's Place.

we shall shortly have a graving of it. It consists of a great
number of walks, terminated by summer houses, and heathen
Temples of different structure, and adorned with statues cast
from the Anticks. Here you see the Temple of Apollo, there
a Triumphal Arch. The garden of Venus is delightful; you
see her standing in her Temple, at the head of a noble bason of
water, and opposite to her an Amphitheater, with statues of
Gods and Goddesses; this bason is sorounded with walks and
groves, and overlook'd from a considerable heigth by a tall
Column of a Composite order on which stands a statue of
Pr: George in his Robes. At the end of the gravel walk leading
from the house, are two heathen Temples with a circle of water,
2 acres and a quarter large. In the midst whereof is a Gulio
or pyramid, at least 50 foot high, from the top of which it
is designed that water shall fall, being by pipes convey'd
thro' the heart of it. Half way up this walk is another
fine bason, with pyramid in it 30 foot high, and nearer the house
you meet a fountain that plays 40 foot. The cross walks end
in vistos, arches and statues, and the private ones cut thro'
groves are delightful. You think twenty times you have no
more to see, and of a sudden find yourself in some new garden
or walk, as finish'd and adorn'd as that you left. Nothing is
more irregular in the whole, nothing more regular in the parts,
which totally differ the one from the other. This shows my
Lord's good tast, and his fondness to the place appears by the
great expense he has been at. We all know how chargeable
it is to make a garden with tast; to make one of a sudden more
so; but to erect so many Summer houses, Temples, Pillars,
Piramids, and Statues, most of fine hewn stone, the rest of
guilded lead, would drain the richest purse, and I doubt not but
much of his wife's great fortune has been sunk in it. The
Pyramid at the end of one of the walks is a copy in mignature
of the most famous one in Egypt, and the only thing of the
kind, I think, in England. Bridgman* laid out the ground
and plan'd the whole, which cannot fail of recommending him
to business. What adds to the bewty of this garden is, that

* Note in the margin:—"Mr. Bridgman was afterwards made the Kings
ch: Gardiner."

it is not bounded by walls, but by a Ha-hah, which leaves you the sight of a bewtifull woody country, and makes you ignorant how far the high planted walks extend."

The garden thus by means of the ha-ha was becoming merged in the park. In many cases the actual garden was neglected to carry out larger designs in the parks. The changes at Boughton, in the reign of George I., were typical of the times; the extensive waterworks were done away with, the wilderness was enlarged, and many miles of avenues were planted.

"Who plants like Bathurst?" wrote Pope, and as Pope was a leader of fashion in planting, we may be sure that Bathurst's method was characteristic of this period. It was not a garden he planted at Cirencester, but a park, with miles of avenues skilfully planned, yet all distant from the house, and with but little of them visible from the small garden. The summer-house, where Pope used to sit, and enjoy the beauty of the planting, is where seven avenues diverge more than a mile from the house. A still finer point is two miles further off where ten avenues meet. The same idea was carried out at Badminton, where the avenues extended for miles into the country, and met at a distant point.* This is all quite beyond the scope of a garden, and therefore beyond my subject, but as we have reached the time when, according to Walpole, "Kent leapt the fence and saw all nature was a garden," we were bound to take a glance beyond.

To the lovers of flowers, a garden was always a garden; under their protection, horticulture and botany were making steady progress, in spite of the new rage for merging the garden in the park. The workers in the practical branches of gardening were many. Richard Bradley, Philip Miller, Thomas Fairchild, and John Lawrence, were among the most famous. Bradley was a very voluminous writer on Natural History, Gardening, and Botany. He entered into various questions concerning the growth of plants, the movements of the sap, and fertilization. "The sap of plants," he wrote, "circulates much after the same manner as the Fluids do in Animal Bodies." On fertilization

* See Kip's Views, reproduced in Blomfield and Thomas's *Formal Garden*.

he says he received "many hints from a gentleman of Paris and
Mr. Samuel Moreland . . . how the pollen powder (or male
dust) fertilizes the embryo seeds in the ovary." Probably the
"gentleman of Paris" was Sebastien Vaillant (1669-1722), who
wrote on the subject, and agreed with the theories first
propounded by Dr. Grew and Thomas Middleton, Ray, and
others, regarding the sexes of plants. Samuel Moreland
wrote a paper for the Royal Society in 1703; his theory
varying only slightly from the others, as to the process
by which the pollen reached the ovary. Scientists made
experiments on plants to prove their theories, and practical
gardeners were not slow in giving their help. The natural
result was that, before long, they succeeded in improving
and increasing the varieties of well-known species. Bradley
instances examples of cross fertilization, as shown by the
changes of colours in auriculas and tulips, and by a plant in
Fairchild's garden grown from carnation seed fertilized by
the pollen of the Sweet William.

Fairchild's garden at Hoxton was the scene of many
experiments. Bradley frequently refers to him as one of
the most skilful gardeners of his acquaintance. Fairchild
was the author of *The City Gardener*. In this work he gives
a list of evergreens, trees and flowers "which will thrive
best in the London gardens," as "everything will not
prosper . . . because of the smoke of the sea-coal . . .
but," he continues, " I find that most persons whose business
requires them to be constantly in town, will have something
of a garden at any rate. One may guess the general love my
fellow citizens have of gardening, in furnishing their rooms
and chambers with basons of flowers and Bough pots, rather
than not have something of a garden before them." In the
course of the work he mentions several trees which were
then (1722) to be seen flourishing in different parts of London:
the ilex, Spanish broom, guelder rose, syringa and lilac in
Soho Square; pears, in several "confined alleys" about Barbican,
Aldersgate, and Bishopsgate. A vine bearing good grapes in
Leicester fields; figs in Roll's Garden in Chancery Lane, and
in Dr. Bennet's in Cripplegate; lily of the valley in a close

place at the back of Guild Hall; plane trees by St. Dunstan-
in-the-East, above forty feet high, besides all the numerous
plants seen growing to perfection at Westminster, and "the
parts of London near the river." So many curious plants
were raised by this enthusiastic gardener in his own garden
at Hoxton, that he thought with proper care almost anything
would grow in the town. He completes a list by saying, "I
am almost persuaded that the olive would do well in London."

The name of Fairchild is still remembered in the part
of London in which he lived. "The Fairchild Lecture" is
delivered annually in St. Leonard's, Shoreditch, in accordance
with the bequest left by him. The subject of the sermon,
which is preached on Whit-Tuesday, is either on "The
Wonderful Works of God in the Creation," or "On the
certainty of the resurrection of the dead, proved by the certain
changes of the animal and vegetable parts of the creation."
The preacher, appointed yearly by the Bishop of London,
still expresses from the pulpit the founder's views.

Fairchild was a member of a Society of Gardeners, and seems
to have taken a leading part, as his name stands first upon the
list of the members given at the end of the preface of the
work published by them, the year after Fairchild's death.* This
book is one of great interest. Only one part was published,
others were to follow if the first met sufficient encouragement,
and that this was not so is much to be regretted. The following
gardeners were the joint authors :—

Thomas Fairchild.	George Singleton.
Robert Furber.	Thomas Bickerstaff.
John Alston.	William Hood.
Obadiah Lowe.	Richard Cole.
Philip Miller.	William Welstead.
John Thompson.	Benjamin Whitmill.
Christopher Gray.	Samuel Hunt.
Francis Hunt.	John James.
Samuel Driver.	Stephen Bacon.
Moses James.	William Spencer.

* *Catalogus Plantarum. A catalogue of Trees, Shrubs, &c; for sale in
the Gardens near London, by a Society of Gardeners*, 1730. The British
Museum copy is under Fairchild's name, 452, h. 2.

TITLE-PAGE OF CATALOGUE OF THE SOCIETY OF GARDENERS 1730

Most of these men were nursery-gardeners, and all lived in London or the suburbs:—Furber at Kensington; Alston, Miller, and Thompson at Chelsea; Lowe and Cole in Battersea; Fairchild, Whitmill, and Bacon at Hoxton; Francis and Samuel Hunt at Putney; Gray at Fulham,* James in Lambeth, George Singleton at the Neat Houses; and Wm. Hood at the Wheatsheaf near Hyde Park Corner. Every month, for five or six years, this Society met at Newhall's Coffee-house in Chelsea, or some other convenient place. Each member brought some plants of his own growing, which were discussed by the assembled gardeners. The names and descriptions were then carefully registered. At the end of five or six years, they decided to have all the plants they had catalogued, "drawn and painted by an able hand." For this purpose they engaged the services of Jacob van Huysum; a good artist, and brother of the famous Dutch flower painter. They got together a large collection of drawings, and finally agreed to publish them. The first part only, containing hardy shrubs, appeared. It was to have been followed by other volumes, for more tender exotics, then "flowers for the pleasure-gardens," and also a part devoted to fruits. The great value of the part we have, is that it mentions all the synonyms and refers to many previous writers to identify each plant, and gives the history of the introduction of some of the new varieties; their monograph on the honeysuckle, which occupies several pages, is of great worth. They also refer to good specimens of trees in some well-known London gardens. The following is an instance, the service tree (= *Pyrus Sorbus*): "In the garden which was formerly in the possession of John Tradescant at South Lambeth, as also at Mr. Marsh's at Hammersmith, a curious collector of rare and uncommon trees, in both which places, these Trees annually produce large quantities of Fruits which ripen perfectly well." Again, there is a note added to the description of the "Three Thorned Acacia or Locust Tree" (= *Gleditschia tricanthos*), "that

---

* The magnolia grandiflora was first planted in Gray's garden. See Johnson's *Hist. Eng. Gar.*, p. 202.

it hath produced pods in the gardens of the Bishop of London at Fulham this year 1729." The Naturalist Catesby is often referred to in these pages, as the introducer of several plants. The following are among the number:—"Bignonia Americana," the Catalpa, which had not flowered in England in 1730; the yellow-berried hawthorn (= *Cratacgus flava*), sent from Carolina in 1724; the Carolina ash (= *Fraxinus caroliniana*) "raised from seeds sent over from South Carolina by Mr. Catesby, anno 1724; Tilia Caroliniana (= *T. americana*) in 1726; the Carolina kidney bean tree (= *Wistaria fructescens*), 1724, which had only flowered (in 1730) in Robert Furber's garden at Kensington; the scarlet flowering acacia, and the "Water Acacia" (= *Gleditschia tricanthos inermis*), both sent home in 1723.

Mark Catesby was an eminent naturalist. He first collected in Virginia, and being induced by Sir Hans Sloane and others to return to America to work still further in the cause of science, he went out again for some six or seven years, and during his stay sent home seeds from time to time. On his return in 1726, he began his great work, *Natural History of Carolina, Florida, and the Bahama Islands*, the first part of which was published 1731. The genus Catesbaea or lily-thorn, was named after him by his contemporary, Gronovius, the Dutch naturalist.

The most celebrated member of this Society of Gardeners was Philip Miller, keeper of the Chelsea Physic-garden, and author of a well-known Gardener's Dictionary. This work first appeared in 1731, and was so popular, that a seventh edition was brought out in 1759, and it was translated into Dutch, German, and French. Each successive edition shows some progress in the science of botany, and an immense increase in the number of foreign plants. In the seventh edition, Miller adopted the Linnaean system of classification. Miller had become acquainted with the great Swede during his visit to England in 1736. It was the year following that Linnaeus' first great work, which revolutionized classification, *Genera Plantarum*, appeared. Miller was a man well suited to the work he undertook; he was both practical and scientific; he first followed the system of Tournefort, then that of Ray, but was sufficiently learned and clear-sighted to go with the times, and adopt the improved nomenclature of Linnaeus. The quantities of new plants coming in not only required skilful growing, but careful arrangement and classification, and Philip Miller did much good work in both ways.

Not only were plants coming in from America, but new treasures found their way to England from other parts of the Old World also. William Sherard, a learned botanist and friend of Ray and Sloane, and patron of Catesby, was, in 1702, appointed Consul at Smyrna, and during his stay there, until 1718, employed much of his time in making a collection of the plants of Greece and Asia Minor. His younger brother, James, at Eltham in Kent, had a famous garden, and cultivated many of the new exotics sent home by William. Besides foreign importations, gardeners at home added to the number of cultivated plants by trying experiments of hybridising, producing double varieties, and more especially variegation. Such things as variegated "silver-striped," or "gold-blotched," lilacs, syringa, privet, phillyrea or maple, were great favourites.

Improved methods of heating and building conservatories and hot-houses made it possible not only to shelter " tender exotics " and grow fruit, but to force vegetables. Attempts were made to force grapes, and the experiment was tried by the Duke of Rutland at Belvoir. Bradley and Switzer describe the process, which was to " build ovens at certain distances at the back of walls, and keeping them continually warm from January till the Sun's Power is sufficient of itself to maintain the growth of the plants growing against such walls . . . . whereby the latest kinds of grapes are commonly ripen'd about July or August." Bradley adds a caution which takes one a step further towards a modern vinery, " Take notice, that during the cold season, when these Fruits are forced to shoot unseasonably, the Plants must be cover'd with glasses to prevent the injuries they might receive from frosts."* At Lord Derby's, at Knowsley Hall, near Liverpool, there was another method of heating a

---

* Bradley, *Works of Nature*, 1721.

wall to produce early grapes, thus described by a traveller in
1732 :—" An hot wall here for Vines, ye wall is built hollow,
or you may say two walls are run up just together at each
end are Stoves where you put in the coal & there is a chimney
in ye halfway of ye wall; ye fires are lighted every night."*
Philip Miller had a method of forcing apricots and cherries
by nailing the trees on to a screen of boards, facing south,
covering the front with glass, and piling up the back of the
boards with a hot bed.

Rose is said to have raised a pine-apple in England, and
presented it to Charles II., but for many years that remained a
unique specimen and an unrivalled feat. The culture was not
understood until this period. Henry Tellende, gardener to Sir
Matthew Decker, at Richmond, was the first who brought the
" Ananas or Pine Apple to rejoice in our climate." † Before
long, several growers gave their attention to Pines, and within
fifty years books entirely devoted to their culture, found ready
sale.‡

Fairchild, at Hoxton, and Green at Brentford, had two of
the best fruit gardens, the latter being exceptionally good for
figs. But it was more especially in vegetable culture that
great advances were made. There had for long been a fair
supply of vegetables in England; but when anything special,
anything early, or out of season, was wanted on great
festive occasions, it was procured from abroad, chiefly
from Holland. But enterprising gardeners, early in the
eighteenth century, began to make attempts at forcing greens
and salads, asparagus, and cucumbers. The first to raise the
latter in the autumn for fruiting in winter was Fowler, gardener
to Sir N. Gould, at Stoke Newington. He presented George I.
with two fine cucumbers on New Year's Day, 1721. Samuel
Collins, in 1717, wrote a Treatise on the culture of melons and

---

* *Diary of a Tour in 1732 made by John Loveday of Caversham*, ed. by
his Grandson. Roxburghe Club, 1890.

† Bradley, *Dictionarium Botanicum*, 1728.

‡ *Ananas, a Treatise on the Pine Apple*, by John Giles, 1767. *A Treatise
on the Anana*, by Adam Taylor, Devizes, 1769. *Treatise on the Pine Apple*,
by W. Speechley, 1779.

cucumbers, suggesting various glasses and frames, for their protection. The following is quoted from Bradley, and gives the names of some of the pioneers in early forcing:—"The first which are Kitchen Gardens and exceed all the other gardens in Europe for wholesome Produce and variety of Herbs are those at the Neat-Houses near Tuttle-fields, Westminster, which abound in Salads, early Cucumbers, Colliflowers, Melons, Winter Asparagus, and almost every Herb fitting the Table; and I think there is no where so good a school for a Kitchen gardener as this Place; tho' Battersea affords the largest natural Asparagus and the earliest Cabbages. Again, the Gardens about Hammersmith are as famous for Strawberries, Rasberries, Currants, Gooseberries, and such like; and if early Fruit is our Desire Mr. Millet's, at North End, near the same Place, affords us Cherries, Apricocks, and Curiosities of those kinds, some months before the Natural Season." Another good nurseryman near London was Nicholas Parker at Chiswick. He is highly recommended by Lawrence as known to all men for his "honesty, skill and integrity," which seems more than could be said of all in the same trade. They were inclined to cheat and send out inferior varieties of fruit, in the place of those ordered by the purchaser, "a dry insipid Nectorine" instead of "an old Newington Peach, or instead of a rich French Pear a gritty choak-pear or Warden."*

Kalm, the great Swedish horticulturalist, after whom the genus Kalmia was named, who passed through England on his way to America, in 1748, was struck by the market-gardens and early vegetables which he saw. He describes some gardens where the beds were raised, sloping a little towards the sun, and "most of them were at this time (February) covered with glass frames, which could be taken off at will. . . . Russian matting over these, and straw over that four inches thick. These contained cauliflowers some four inches high. In the rest of the field were 'bell-glasses,' under which also cauliflower-plants were set 3 or 4 under each bell-glass. Besides the

* *The Clergyman's Recreation.* John Lawrence, Rector of Yelvertoft, Northamptonshire.

afore-named beds, there were here long asparagus-beds. Their height above the ground was two feet: on the top they were similarly covered with glass, matting, and straw, which had just been all taken off at midday. The Asparagus under them was one inch high and considerably thick."* Radishes were also grown in the same garden, and the beds covered with mats. In the month of May, he says, the vegetables, which were most numerous round London, were beans, peas, cabbages of different sorts, leeks, chives, radishes, lettuce (salad), asparagus, and spinach. He writes of Chelsea, "There is scarcely anything else than either orchards or vegetable market gardens, and large fields all planted full of all kinds of small trees for sale."

Thus it will be seen that great strides had been made in vegetable-culture. In some things, however, gardeners still had very primitive ideas. When, in 1729, an aloe (*Agave*) flowered in "Mr. Cowell's garden at Hoxton," there was great excitement as to how it should be kept through the winter.† The plant was then twenty feet high, and an erection of wood and glass was built over it, and stoves placed outside with pipes to "convey a due proportion of heat," and it was so arranged that the structure could be heightened, if necessary, to suit the "unexpected growth of this famous plant." They must have been much distressed to find all this care and expense of little use, as not only the flower, but most of the plant itself would soon perish.

A great many of the vegetables, grown in these market gardens, would be sold in the streets of London. The various cries of the hawkers were a notable feature of London life. One among the many refrains of this perpetual chorus is recalled by Addison,‡ when he writes:—"I am always pleased with that particular time of the year which is proper for the pickling of dill and cucumbers, but alas! this cry, like the song of the

---

* *Kalm's Visit to England.* Translated by Joseph Lucas, 1892.
† *A True Account of the Aloe Americana or Africana now in flower in Mr. Cowell's Garden at Hoxton. . . . The like whereof has never been seen in England before.* 1729.
‡ *Spectator,* 251.

nightingale, is not heard above two months." Some of the best-known cries are preserved in an old ballad of early, but uncertain date, from which the following is an extract * :—

> " Here's fine rosemary, sage and thyme
> Come buy my ground ivy,
> Here fatherfew, gilliflowers and rue
> Come buy my knotted marjorum ho!
> Come buy my mint my fine green mint
> Here's fine lavender for your cloaths
> Here's parseley and winter savory
> And heart's-ease which all do choose
> Here's balm and hissop and cinque foil
> All fine herbs it is well known
> Let none despise the merry merry cries
> Of Famous London Town.
>
> " Here's penny royal and marygolds
> Come buy my nettle-tops
> Here's water-cresses and scurvy-grass
> Come buy my sage of virtue ho!
> Come buy my wormwood and mugwort
> Here's all fine herbs of every sort
> Here's southern wood that's very good
> Dandelion and houseleek
> Here's dragon's-tongue and wood sorrel
> With bear's-foot and horehound
> Let none despise the merry merry cries
> Of Famous London Town.
>
> " Here's green coleworts and brocoli
> Come buy my radishes
> Here's fine savorys and ripe hautboys
> Come buy my young green hastings ho!
> Come buy my beans right Windsor beans
> Two pence a bunch young carrots ho!
> Here's fine nosegays ripe strawberries
> With ready pickled salad also
> Here's collyflowers and asparagus
> New prunes twopence a pound
> Let none despise the merry merry cries
> Of Famous London Town.

---

* " Roxburghe Ballads," 1560-1700. *History of the Cries of London.* Charles Hindley. 2nd ed., 1884.

"Here's cucumbers spinage and frinch beans
Come buy my nice sallery
Here's parsnips and fine leeks
Come buy my potatoes ho!
Come buy my plumbs and fine ripe plumbs
A groat a pound ripe filberts ho!
Here's corn-poppies and mulberries
Goose berries and currants also
Fine nectarines peaches and apricots
New rice two pence a pound
Let none despise the merry merry cries
Of Famous London Town."

CHAPTER XII.

> ". . . So will I rest in hope
> To see wide plains, fair trees, and lawney slope;
> The morn, the eve, the light, the shade, the flowers;
> Clear streams, smooth lakes, and overlooking towers."
>
> KEATS.

"Is there anything more shocking than a stiff regular garden?"* What a revolution of the taste in gardening these words reveal! Yet such a complete change in fashion had taken place, that this was the opinion held by all the garden designers of the latter half of the eighteenth century. Nor were they content to lay out new gardens to suit the prevailing style, but they freely destroyed, and abused, where they could not obliterate, the work of former generations. The leader of this new departure in garden design was Kent. He was the successor of Bridgeman, and at first made gardens on the same plan. Soon, however, he went so far beyond him as to entirely leave the formal garden, and substitute for it the landscape style. Walpole considers the first step towards this revolution to have been the introduction of the sunk fence. And certainly he there touched the key-note, for as soon as walls and enclosures were dispensed with, any piece of natural and rural scenery could be included in the garden. "The capital stroke,"† he wrote, "the leading step to all that has followed, was (I believe the first thought was Bridgeman's) the destruction of

---

* Batty Langley, *New Principles of Gardening*, 1728.
† *Essay on Modern Gardening*. By Horace Walpole, 1785.

walls for boundaries, and the invention of fosses . . . an attempt then deemed so astonishing, that the common people called them Ha! Ha's! to express their surprise at finding a sudden and unperceived check to their walk." " No sooner was this simple enchantment made, than levelling, mowing, and rolling, followed. The contiguous ground of the park without the sunk fence, was to be harmonized with the lawn within; and the garden in its turn was to be set free from its prim regularity, that it might assort with the wilder country without. . . . At that moment appeared Kent, painter enough to taste the charms of landscape, bold and opinionative enough to dare and to dictate, and born with a genius to strike out a great system from the twilight of imperfect essays. He leaped the fence, and saw that all Nature was a garden. He felt the delicious contrast of hill and valley changing imperceptibly into each other, tasted the beauty of the gentle swell, or concave scoop, and remarked how loose groves crowned an easy eminence with happy ornament, and while they called in the distant view between their graceful stems, removed and extended the perspective by delusive comparison."

This shows the ideal at which Kent was aiming. To copy Nature was the aim of the new school:—" Nature abhors a straight line," was one of Kent's ruling principles, so avenues and straight walks and hedges were an eyesore to him, and this feeling of dislike was shared by other landscape gardeners. Batty Langley wrote, " To be condemned to pass along the famous vista from Moscow to Petersburg, or that other from Agra to Lahor in India, must be as disagreeable a sentence, as to be condemned to labour at the gallies. I conceiv'd some idea of the sensation . . . from walking but a few minutes, immured, betwixt Lord D——'s high shorn yew hedges." This is but a specimen of the exaggerated language in which the new school of gardeners expressed their contempt for the work of their predecessors.

This passion for the imitation of Nature, was part of the general reaction which was taking place, not only in gardening, but in the world of letters and of fashion. The extremely artificial French taste had for long taken the lead in civilized Europe, and now there was an attempt to shake off the shackles of its exaggerated formalism. The poets of the age were also pioneers of this school of Nature. Dyer, in his poem of "Grongar Hill," and Thomson, in his *Seasons*, called up pictures which the gardeners and architects of the day strove to imitate in the scenery they planned. The idea was to create a landscape such as poets celebrated or as Claude immortalized on canvass. But the lovers of the beauties of Nature soon became as hopelessly fettered by rules and theories as had been the designers of the more formal schools. The gardens they laid out were planned to produce a set impression on the beholder :—"Garden scenes," wrote the poet Shenstone, "may perhaps be divided into the sublime, the beautiful, and the melancholy or pensive." * "Art," says this same writer, "should never be allowed to set foot in the province of Nature," and yet these gardeners advocated every sort of artifice to impose on the spectator, and to make the landscape appear different from what it really was. Shenstone himself suggests a means by which an avenue may be made to appear longer than its true length. "An avenue that is widened in front and planted there with yew trees, then firs, then with trees more and more fady, till they end in the almond willow or silver osier; will produce a very remarkable deception." His own garden at Leasowes was held by all who practised this "art of gardening" to be a most perfect specimen of this style. There was a lake, and small streams, and cascades, which George Mason describes as "living fountains," and says they were here "carried to the pitch of perfection." A seat overlooking one of these streams was inscribed with a poem in its praise, which ends thus :—

> "Flow, gentle stream, nor let the vain
> Thy small unsullied stores disdain :—
> Nor let the pensive sage repine
> Whose latent course resembles thine."

All through the garden, in the dingle, or by the side of the serpentine walks, seats, grottoes, ruins or urns, appeared at

---

* *Unconnected Thoughts on Gardening.* By Wm. Shenstone.

unexpected places, and were inscribed with lines addressed to some friend, or singing the praises of some natural beauty.

Most conspicuous among the innovations was the change in the form of the ornamental sheets of water. "Stone basons," marble fountains, and straight canals, were swept away, or converted into miniature waterfalls, winding streams, or artificial lakes. Lord Bathurst, at Ryskins, near Colebrook,* was the first to make a winding stream through a garden, and so unusual was the effect that his friend, Lord Stafford, could not believe it had been done on purpose, and supposing it to have been for economy, asked him "to own fairly how little more it would have cost to have made the course of the brook in a strait direction." About this time Queen Caroline " threw a string of ponds in Hyde Park into one to form what is called the Serpentine River." This is only one among many instances which show that these so-called reforms, undertaken with the aim of increased simplicity, resulted in greater stiffness and formality. This is not to be wondered at, when we take into account the influence of Chinese gardening on this school of design. Sir William Chambers, one of this new class of gardeners, had, in his youth, made a voyage to China and brought back from that country ideas which he set forth in his work entitled, *Dissertations on Oriental Gardening*. The Pagoda at Kew, designed by him, is a well-known monument of this passing fashion. A Chinese writer, Lien-tschen, himself lays down the principles which ruled their gardening†:—" The Art of laying out gardens consists in an endeavour to combine cheerfulness of aspect, luxuriance of growth, shade, solitude, and repose, in such a manner that the senses may be deluded by an imitation of rural Nature." Alluding to this supposed resemblance of English gardens to those of China, Oliver Goldsmith wrote, " The English have not yet brought the art of gardening to the same perfection with the Chinese, but have lately begun to imitate them. Nature is now followed

---

* *Progress of Gardening.* By Barrington. *Archæologia*, Vol. V.
† *Praise of Gardens.* By Siveking, p. 17.

with greater assiduity than formerly: the trees are suffered to shoot out into the utmost luxuriance;—the streams, no longer forced from their native beds, are permitted to wind along the valleys: spontaneous flowers take the place of the finished parterre, and the enamelled meadow of the shaven green."

Batty Langley was one of the exponents of the principles which guided some of these Landscape-Gardeners. The chief of them he lays down in twenty-eight rules, among which are the following:—" The grand front of the building lies open upon an elegant lawn, adorned with statues, terminated on its sides with open groves." " Such walks whose views cannot be extended terminate in Woods, Forests, misshapen Rocks, strange Precipices, Mountains, old Ruins, grand buildings, &c." " No regular evergreens in any part of an open plain or parterre." " No borders or scroll work cut in any lawn or parterre." " That all gardens be grand, beautiful and natural." " That all the trees in your shady walks and groves be planted with sweet Briar, white Jessemine, and Honeysuckle, environed at the Bottom with a small circle of Dwarf stock, Candy tuft and Pinks." " Hills and Dales be made by art where Nature has not performed the act before." " That the intersections of walks be adorned with statues," and many like rules for the correct way of making " rivulets, aviaries, grottoes, cascades, rocks, ruins, niches, canals, and fishponds." He also gives a long list of what statues were most suitable for each place:—Pomona in the Orchard, Harpocrates, the God of Silence, for a grove, and so on. This subject of statues much perturbed some of the designers. "The use of statues," wrote George Mason, "is a dangerous attempt in gardening, not impossible, however, to be practised with success: how peculiarly happy is the position of the river God at Stourhead (Sir Richard Hoare's) in Wiltshire! . . . I remember a figure at Hagley,* which one could fancy darting across the Alley of a grove . . . and only wished the pedestal had been concealed." These statues, urns and

---

* Laid out by Lord Lyttleton.

monuments, were arranged to impart to the beholder a particular impression, on first discovering them. Shenstone discusses the various sensations produced by an urn, and comes to the conclusion that "Solemnity is perhaps their point, and the situation of them should still co-operate with it." "They are more solemn, if large and plain." A clump of trees, a lake, or wilderness, had to be "sublime," "beautiful," "picturesque," "solemn," "grand," "dignified," or "elegant." A wood was planted for "rudeness or grandeur," a "grove for beauty," a cave or grotto was to strike "horror or terror." "A feigned steeple of a distant church or an unreal bridge, to disguise the termination of water,"* were brought in to "improve the landscape." These designers were careful not only of form but of colour, the "solemn grove" had to be planted with trees of dark foliage, and some touch of bright colour was introduced to give effect to the landscape. "An object of a sober tint unexpectedly gilded by the sun, is like a serious countenance suddenly lighted up by a smile, a whitened object, like the eternal grin of a fool,"† wrote one authority on the subject. Such were the high-flown ideas which inspired these designers, but in their efforts to reproduce the beauties of Nature they fell into the most artificial system that one can possibly imagine. William Mason's poem, "The English Garden," addressed to "Divine Simplicity," is characteristic of the spirit which guided these "reformers," of which Sir Walter Scott said it "is not simplicity but affectation labouring to seem simple."

Many places were laid out on this new plan by Kent. The gardens at Esher,

"Where Kent and Nature vied for Pelham's love,"

and at Claremont, were considered some of his best productions; also Carlton House, which he designed for the Prince of Wales. Walpole thought "the most engaging of all Kent's works," and most "elegant and antique," was Rousham, in Oxfordshire. Kent began life as an apprentice to a

---

* Horace Walpole, *Essay on Modern Gardening*.
† Sir Uvedale Price, *On the Picturesque*.

coachbuilder; with the assistance of friends he went to Italy, and studied painting. He, however, never attained any good results in that art, but succeeded better as an architect, and designed temples and ruins for gardens. By the help of his patron, Lord Burlington, he was noticed by the Queen, and made Architect and then Painter to the Crown. He was looked up to by all the designers who followed, as the originator of the idea, and founder of the School of Landscape-Gardening. At one time, his wish to follow Nature, carried him so far that he planted dead trees in Kensington Garden "to give a greater air of truth to the scene." But Walpole says "he was soon laughed out of this excess." Philip Southcote appears to have been one of the first of those in whom Kent's "Elysian scenes excited the idea of improving their own domains," and "the elegance of Wooburn Farm (designed by him) was so conspicuous, that even its faults were imposing."* Pain's-hill, in Surrey, begun about the same time by Charles Hamilton, was "a perfect example of this mode." † Hagley, laid out by Lord Lyttleton, was another garden, or "ferme ornée" in the same style, frequently referred to by contemporary writers, who praised "the new modelling of the shades and unfettering of the the rills." ‡ In spite of the admiration lavished by many on this place, Gilpin § remarks that although "there are certainly many beautiful views in these extensive gardens, yet we can easily conceive the same variety of ground . . . . so combined as to produce a much nobler whole." Hagley, in Worcestershire, was only a short distance from the Leasowes, already referred to, which was perhaps the most admired garden of this type. Goldsmith and others who had seen the place during the life-time of its poet-possessor,

---

* George Mason, *Essay on Design in Gardening.* Wooburn Farm, near Chertsey, no longer existed in 1829.— G. W. JOHNSON, *Hist. Eng. Gardening.*

† Walpole.

‡ George Mason.

§ *Observations on Picturesque Beauty made in 1772, Particularly the Mountains and Lakes.* By Wm. Gilpin.

lamented the changes and decay which marred it, only a few years after Shenstone's death. Wright was another designer of this landscape-school, who succeeded Kent. He planned and sketched designs, but did not himself superintend the carrying out of the works.

The name which stands out most conspicuously in connexion with landscape-gardening is that of Brown. From his habit of saying of any place he was asked to improve, or lay out afresh, that it "had great capabilities," he became known by the name of "Capability Brown." For a time he was the most popular of all designers. He was born in Northumberland in 1715, and began as a kitchen-gardener, first at a small place near Woodstock, and then at Stow. He remained with Lord Cobham, in that capacity, until 1750, and it was not until, as head-gardener to the Duke of Grafton, he planned and executed a lake at Wakefield Lodge, that he attempted any designing. This brought him into notice, and through the influence of Lord Cobham, he was appointed Royal Gardener at Hampton Court, "and it was he who planted the celebrated vine[*] there, in 1769."[†] He was next employed at Blenheim, and the way in which he made the lake there established his reputation, and soon every one who wished to alter their grounds, or lay out new ones, employed Brown. He laid out Croome, Luton, Trentham, Nuneham, Burghley, and many other places, and altered in some way or the other half the gardens in the country. He became the fashion, and was consulted by nearly every one in England who had a garden of any consideration. Had Brown confined himself to creating new landscapes and gardens, posterity could not have borne such a grudge against him. As it is, in studying the designs he carried out, it is difficult to look with an unprejudiced eye at his work, for before the results he produced can be admired, one is filled with regret for the beauties he swept away.

[*] London, *Encyclopædia of Gardening*.
[†] Phillips, *Pomarium*. 1820. The parent of the Hampton Court Vine was a Black Hamburg planted by Mr. Eden at Valentine House, Essex, 1758.

> " Improvement too, the idol of the age,
> Is fed with many a victim. Lo he comes !
> The omnipotent magician, Brown, appears !
> Down falls the venerable pile, the abode
> Of our forefathers. . . . .
> He speaks—The lake in front becomes a lawn ;
> Woods vanish, hills subside, and valleys rise,
> And streams, as if created for his use
> Pursue the track of his directing wand."
>
> COWPER, *The Garden.*

Old gardens in every part of England disappeared before the transforming influence of Brown, but luckily before many years had passed a reaction set in, or it is doubtful whether a single garden would have survived. Sir Uvedale Price* described his pleasure on approaching "a venerable castle-like mansion built in the beginning of the fifteenth century," through an avenue of fine old trees. "I was much hurt," he continues, "to learn from the master of the place, that I might take my leave of the avenue and its romantic effects, for that its death-warrant was signed. The destruction of so many of these venerable approaches, is a fatal consequence of the present excessive horror of straight lines. . . . As to saving a few of the trees, I own I never saw it done with a good effect ; they always pointed out the old line, and the spot was haunted by the ghost of the departed avenue. . . . At a gentleman's place in Cheshire, there is an avenue of oaks. Mr. Brown absolutely condemned it, but it now stands a noble monument of the triumph of the natural feelings of the owner over the narrow and systematic ideas of a professed improver." One is thankful that a few people had strength of mind enough to resist the all-powerful Brown.

The management of water was considered Brown's strong point. A pleasing example of a sheet of water laid out by him is that at Castle Ashby.† As it is now "improved by time" it could not fail to please even the most determined detractors of Brown. But here, too, Brown's hand worked

---

* Sir U. Price, *On the Picturesque.*
† Belonging to the Marquess of Northampton.

destruction as well as improvement, for two rows of trees, forming part of one of the avenues planted about 1699, were felled by his orders. He was lost in admiration of the rivers and lakes he created. Having completed one of these, he thought he had achieved such a success as to surpass the Thames, and is said to have exclaimed:—" Thames! Thames! thou wilt never forgive me!" At Hackwood Park,* in Hampshire, Brown effected various changes, which were thus spoken of a few years later :—" Alterations on a con-

CASTLE ASHBY.

siderable scale" were carried out, particularly on the south of the house, where there had been a garden, " in the old style, with terraces, ascended by flights of steps, and adorned with statues on pedestals, a great reservoir of water, angular ramparts, &c.; the view from the house was also interrupted by high yew hedges skirting long and formal

* The seat of Lord Bolton.

avenues. Nature has now regained her rights; the avenues have been broken into walks and glades, and several distant views admitted." It never seems to have occurred to these landscape-gardeners that an avenue and a yew hedge were in themselves beautiful objects. It is almost like a Norfolk girl who visited Switzerland, and complained that the mountains shut out the view! Another scheme of wholesale devastation he suggested, was luckily not acted upon. He proposed to blast away that part of the rock on which Powis Castle stands, which forms the first or "Sundial terrace," and make it into a flat lawn. This change would have been completely out of all keeping with the rest of the lovely garden, which had been made in the time of William and Mary, by Lord Rochfort, a Dutchman, who for a few years held the estates. The alterations he carried out at Burghley were also typical of his method. He took away the walls and hedges, entirely swept away a terraced kitchen-garden on a slope near the house, and in its place planted trees; beyond this wooded-eminence of his own creating, and in front of the site of the old formal garden, he made a lake. "How far the fashionable array, in which Mr. Brown has dressed the grounds, about this venerable building, agrees with its formality, and antique appendages, I dare not take upon me to say," wrote Gilpin, a few years after Brown's work was completed. "A doubt arises," he continues, "whether the old decoration of avenues and parterres was not in a more suitable style of ornament. It is, however, a nice question, that would admit of many plausible arguments on both sides."

Gilpin also doubts the expedience of the alterations Brown was carrying out at Roche Abbey, when he visited that place. Brown, it is said, was himself unable to draw a line, and had had no artistic training, or education sufficient to understand the historical interest, or natural beauties of the scenes he tried to improve. It is therefore not to be wondered at that he signally failed, on many occasions, in his endeavours to create a more suitable landscape. "Many modern places," wrote Gilpin, "he has adorned and beautified, but a ruin presented a new idea, which

BURGHLEY THE TEMPLE DESIGNED BY BROWN.

I doubt whether he has sufficiently considered. He has finished one of the valleys which look towards Laughton spire: he has floated it with a lake, and formed it into a very beautiful scene. But I fear it is too-magnificent and too-artificial an appendage to be in unison with the ruins of an Abbey."* He levelled all the ground round the old Abbey, leaving the walls and pillars standing in "a neat bowling-green," and he removed all the overgrown pieces of ruin and mounds, which showed the old lines of the building, and even took stones from the Abbey to make the dam in the river, and get the effect of a water-fall. Gilpin most sarcastically remarks, "If Mr. Brown should proceed a step further, pull down the ruin, and build an elegant mansion, everything would then be right." Some of Brown's handiwork about the ruins has of late been removed, and their former conditions, as much as possible, restored.

The following is the Agreement between Brown and Lord Scarbrough made at the time of these alterations †:—

The Agreement between Lord Scarbrough and "Capability Brown," 1774:—

September the 12th, 1774.

Then an Agreement made between the Earl of Scarbrough on the one Part, and Lancelot Brown on the other, for the underwritten Articles of Work to be Performed at Sandbeck in the County of York (To Wit):

Article the 1st.—To compleat the sunk Fence which separates the Park from the Farm, and to Build a Wall in it, as also to make a proper Drain at the Bottom of the Sunk Fence to keep it Dry.

Article the 2nd.—To demolish all the old Ponds which are in the Lawn, and to level and Drain all the ground where they are.

Article the 3rd.—To Drain and level all the ground which is between the above mentioned Sunk Fence and the old

---

* Gilpin, *Obs. on Picturesque Beauty*, 1776, *Particularly the Highlands of Scotland.*

† Copied from the original MS. at Sandbeck, by the kind permission of the Earl of Scarbrough.

Canals mentioned in the Second Article. To plant whatever Trees may be thought necessary for Ornament in that Space discribed in this Article, and to sow with grass seeds and Dutch Clover the whole of the ground wherever the Turf has been broke up or disturbed by Drains, Leveling or by making the Sunk Fence.

ROCHE ABBEY.

Article the 4th.—To make good and keep up a Pond for the use of the stables.

Article the 5th.—To finish all the Valley of Roach Abbey in all its Parts, according to the Ideas fixed on with Lord

Scarbrough (with Poet's feeling and with Painter's eye) beginning at the Heads of the Hammer Pond and continuing up the Valley towards Loton, als: Loughton in the Morn, as far as Lord Scarbroughs ground goes, and to continue the Water and Dress the Valley up by the Present Farm House untill it comes to the seperation fixed for the Boundary of the New Farm. N : B: The Paths in the Wood are included in this Discription and every thing but the Buildings. The said Lancelot Brown does Promise for himself His Heirs Executors and Administrators to perform or cause to be Performed in the Best manner in His or Their Power between the Date hereof and December one Thousand Seven Hundred and Seventy Seven, the above written five Articles. For the Due Performance of the above written five Articles The Earl of Scarbrovgh does Promise for himself His Heirs Administrators and Executors to Pay or cause to be Paid at the underwritten Times of Payment Two Thousand Seven Hundred Pounds of Lawfull money of England, and three Hundred Pounds in consideration of and for the Plans and trouble Brown has had for his Lordship at Sandbeck, previous to this Agreement. Lord Scarbrough to find Rough Timber, four able Horses, carts, and Harness for them, wheelbarrows and Planks, as also Trees and Shrubbs.

The Times of Payment in

           June, 1775   £800
           Feb. 1776   400
           June, D⁰.   400
           Feb. 1777   600
On Finishing the work   800

                     £3000

      (Signed)   Scarbrough.
                    Lancelot Brown.

The melancholy spectacle presented by some of the stately houses surrounded by the stiff and unreal "natural landscape" substituted by Brown for the carefully designed and well-kept old gardens, is thus described by Knight*:—

> "Oft when I've seen some lonely mansion stand
> Fresh from the improver's desolating hand,
> 'Midst shaven lawns that far around it creep
> In one eternal undulating sweep;
> And scatter'd clumps, that nod at one another,
> Each stiffly waving to its formal brother:
> Tired with the extensive scene, so dull and bare,
> To Heaven devoutly I've address'd my prayer;
> Again the moss-grown terraces to raise,
> And spread the labyrinth's perplexing maze;
> Replace in even lines the ductile yew,
> And plant again the ancient avenue.
> Some features then, at least, we should obtain
> To mark this flat, insipid, waving plain:
> Some vary'd tints and forms would intervene
> To break this uniform, eternal green."

Although Brown was assailed by Gilpin, Price, Knight, and Mason, he had many adherents and imitators. Repton is the best known of these. He was an admirer of Brown's works, and carried out designs in the same style. As, however, men had now begun to find out Brown's mistakes, and reflect on his destruction of old places and historical relics, Repton could scarcely venture to suggest such sweeping alterations as Brown had made. Repton was openly an opponent of those who wrote against Brown, yet their ideas evidently influenced his judgment. He did not always alter all he found at a place, before commencing additions; and he did not entirely confine himself to the "landscape" style. He maintained that a "Flower garden should be an object detached and distinct from the general scenery of the place; and whether large or small, whether varied or formal, it ought to be protected from hares and smaller animals by an inner fence; within this enclosure rare plants of every description should be encouraged, and provision made of soil, and aspect for every different class. Beds of bog earth should be prepared for the American plants: the aquatic plants, some

---

* "The Landscape," A didactic poem in III Books, addressed to Sir Uvedale Price by R. P. Knight, 2nd Ed., 1795.

of which are peculiarly beautiful, should grow on the surface or near the edges of water. The numerous class of rockplants should have beds of rugged stone provided, without the affectation of such stones being the natural production of the soil; but, above all, there should be poles or hoops for those kind of creeping plants which spontaneously form themselves into graceful festoons when encouraged and supported by art."* Such was Repton's idea of a flowergarden, but that was to form but a small portion of the design, and its very existence seemed to him to require an

WOODFORD. NO. 1. FROM A DRAWING BY H. REPTON.

apology. He boasts that he had "frequently been the means of restoring acres of useless garden to the deer or sheep, to which they more properly belong," yet he sometimes designed a small formal garden for flowers. The "Dutch garden" at Hewell Grange was made according to his suggestions.† It is a semicircle, surrounded by a cut

---

* Repton, *Obs. on Landscape Gardening*, 1803.
† MS. "Red Book" of Repton, belonging to Lord Windsor.

Thuja hedge, and a high brick wall across the straight side.
The beds within are edged with box, between which are
small gravel paths tiled in the middle, and a sundial in the
centre of the garden. He also designed the lawn and rock-
garden while an older French garden, approached by cut yew
hedges, he did not interfere with. Much as he disliked avenues
as being "utterly inconsistent with Natural scenery,"* he
occasionally respected "such marks of ancient dignity." At
Finedon, although he thought the view "encumbered" by the
vicarage and church, and said the garden wall, malt-house,

WOODFORD. NO. 2. FROM THE SAME DRAWING BY H. REPTON, SHOWING THE
SUGGESTED IMPROVEMENTS.

pigeon-house, and even part of the village "must be removed," he
spared the avenue called the "Holly Walk."

When asked to make suggestions for the improvement of
a place, Repton prepared what he called his "Red Book,"
with plans and views of the garden as it was, and as he
proposed to make it. He published a collection of these
"Red Books," amplifying it with expositions of his own views

* MS. "Red Book" by Repton, 1793, belonging to Miss Mackworth
Dolben, Finedon, Northamptonshire.

on landscape gardens. The best way to understand what these views were, is by a study of these "Red Books," many of which are still unpublished. The above illustrations are taken from his MS. "Red Book," of Woodford, in Essex.* The first sketch in the book represents the house as it was, viewed from the grounds, "with the kitchen-garden on the one side, and the naked village on the other. That the former ought to be removed, and the latter planted out, are such obvious improvements that I do not take upon myself the merit of suggesting them." The second view shows the place as it would be when these designs were carried out. The further alterations were chiefly made to gain a more pleasing view from the house, the planting and turfing of a ploughed field, and the "floating the bottom of the lawn with water." Repton was not the last of this school to admire and extol Brown, some few still spoke of him in glowing terms:—

> "Born to grace Nature, and her works complete
> With all that's beautiful, sublime and great,
> For him each Muse enwreathes the laurel crown,
> And consecrates to Fame immortal Brown." †

As late as 1835 Dennis refers to him as a great "improver of English taste." ‡ This author also bestows praise on some changes that Brown himself might have been proud of, if his achievements were measured by the amount he swept away. He speaks of the alterations in St. James's Park as "the best obliteration of avenues" that "has been effected . . . but it has involved a tremendous destruction of fine elms. Certainly considerable credit redounds to the projector of these improvements for astounding ingenuity in converting a Dutch Canal into a fine flowing river, with incurvated banks, terminated at one end by a planted island, and at the other by a peninsula." This was "planned and executed" by Eyton. The grounds of Buckingham Palace were about this time laid out by William Aiton the younger, son of the author of *Hortus Kewensis*, the royal gardener at Kensington

---

* Reproduced from the original MS. belonging to Courtenay Warner, Esq.
† "The Rise and Progress of the Present Taste of Planting," an epistle to Charles Lord Viscount Irwin—1767. MS. in Guildhall Library.
‡ *The Landscape Gardener*. By J. Dennis, 1835.

and Kew. Davis was another landscape gardener of this school, said by his contemporaries to have "displayed considerable taste," especially in the alterations he carried out at Longleat. The two views of Narford show how complete the change from a formal to a landscape garden can be. The cascade pond was sketched between 1716 and 1724 by Edmond Prideaux, of Prideaux in Cornwall, when on a tour in Norfolk. The second view is from a photograph taken in 1894 from as nearly as possible the same point of view. The lake which covers seventy acres was made about 1842, and all traces of the stiff pond have vanished.

Landscape gardening had by this time become the recognized National style of England, and it was copied on the Continent, in France, Italy and Germany. "English gardens" became the fashion, and books were written abroad to extol the English taste, and invite other nations to copy it,* and old gardens were destroyed to give place to the new style. But on the Continent one thing was lacking, which was the redeeming point in all these landscapes, and that was the green turf. Nowhere is the grass so fair and green as in England, and landscape-gardeners appreciated this great advantage.

It is strange the way in which the writers of this school pointed to Milton and Bacon as the founders of their taste. They claimed Bacon because he devotes a part of his ideal garden to a "natural wildness," and also praises "green grass kept finely shorn," and Milton, because he says that in Paradise there were :—

> "Flowers worthy of Paradise, which not nice art
> In beds and curious knots, but nature boon
> Poured forth profuse on hill, and dale, and plain." †

Yet how opposed to all ideas of landscape gardeners would these two men have been. Bacon, who loved the green grass, and yet would have his garden full of flowers in bloom in every month of the year, would have been shocked by the idea of "a garden . . . disgracing by discordant character the contiguous lawn," or by being told that "the

---

* *Del' Arte dei Giardini Inglesi*, Milan, 1801, &c.   † *Paradise Lost*—Book IV.

flower-garden ought never to be visible from the windows of the house." Sir Walter Scott,* in one of his charming articles on landscape gardening, points out that Milton never intended to censure the "trim gardens" of his own day, although he pictured the natural beauties in the newly-created Paradise. Scott well understood the great mistake that had been made in destroying such a large number of old gardens. He saw how perfectly an Elizabethan garden harmonized with the house, and while he could not vindicate the "paltry imitations

NARFORD. NO. I. FROM A SKETCH BY EDMOND PRIDEAUX ABOUT 1761.

of the Dutch, who clipped yews into monsters," he acknowledged that there existed gardens. "The work of London and Wise, and such persons as laid out ground in the Dutch taste, which would be much better subjects for modification than for absolute destruction." He admired fine terraces, flights of steps, balustrades, and vases of gardens in the Italian style, and the fountains and waterworks of the French.

* *Quarterly*, Vol. 37, 1828, and *Criticism*, Vol. V.

Sir Uvedale Price, although he was the champion of rational landscape gardening, could only justify a "jet d'eau," because such things were to be seen in the form of Geysers. Sir Walter Scott, still more large-minded, felt sure that the captivating beauty "of a magnificent fountain . . . flinging up its waters into the air, and returning down in showers of mist," was in itself sufficient justification. These men who pointed out that some beauties were to be found in

NARFORD. NO. 2. 1894.

the formal garden, and the great folly of ruthlessly destroying everything in that style, gradually arrested the progress of destruction. The taste became modified, and further attempts to improve were not accompanied by such disastrous results. Great thanks are due to those who first saw the mistake that was being made, and who then had the courage to try and arrest the onward progress of the fashion. The writings of some of those who first appealed against the "Natural

School," were as strong in their language as that used but a few years before by the abusers of the formal style. The following lines from Knight, the opponent of Repton, are a fair example:—

> "Hence, hence! that haggard fiend however call'd,
> Thin meagre genius of the bare and bald;
> Thy spade and mattock here at length lay down
> And follow to the tomb thy favourite Brown:
> Thy favourite Brown, whose innovating hand,
> First dealt thy curses o'er this fertile land."*

The absurdity of trying to make small villa gardens in the landscape style, with miniature lawns, "clumps and strips of trees," was pointed out by Loudon.† He recommended instead, designs in a more formal style, and gives plans of villa grounds of six acres laid out in "the geometric style," and others combining that with the newer fashions. Regent's Park was made in the early years of this century, and Loudon speaks of it to illustrate his theories. "The magnificent design of the late Mr. Fordyce, Surveyor-General, now executing (1812) in Marylebone Farm, will in a few years afford a noble example of the unison of the ancient and modern styles of planting."

The flower-garden began once more to hold a more conspicuous position, and to be considered as separate from the shrubbery, or less formal part, while that again was kept more distinct from the park beyond. The planting of the grounds outside the flower-garden was also much improved: the stiff clumps and belts broken into, and trees arranged more ornamentally. Sir Henry Steuart, of Allanton, whose work, *The Planter's Guide*, occasioned the review by Sir Walter Scott in the *Quarterly*, already referred to, was a good authority on the subject of planting, and by his own plantations, as well as in his works, gave useful hints as to the management of trees, and the choice of suitable ones for different situations.

Thus the garden and its surroundings were again being

---

* *Landscape.* By R. P. Knight, 1795.
† *Hints on the Formation of Gardens and Pleasure Grounds.* By J. C. Loudon, 1812.

treated with more skill and taste. Although other styles are now practised as well, the landscape, in its reformed character, still finds admirers, and skilful designers.* Architects have made garden design more of a study, and artists and gardeners also have, in many instances, shown that, with careful handling, the landscape style can be reconciled to the house, and most pleasing effects of scenery produced, well suited to this country and its climate.

* *The Art and Practice of Landscape Gardening.* By Henry Ernest Milner, 1890.

CHAPTER XIII.

> "Hence through the garden I was drawn,
> A realm of pleasance, many a mound,
> And many a shadow-chequered lawn,
> Full of the city's stilly sound;
> And deep myrrh thickets blowing round
> The stately cedar, tamarisks,
> Thick rosaries of scented thorn,
> Tall orient shrubs and obelisks,
> Graven with emblems of the time."
> LORD TENNYSON.

THE progress of gardening during the last hundred years has been so great and so rapid, that it would be a well-nigh endless task to take even a very cursory review of it in all its branches. The immense advance in Botany and classification, the improved methods of cultivation, the vast hot-houses and stoves, and the countless treasures from tropical climes with which to stock them, the numberless plants collected from all parts of the world to beautify the flower-garden, and the endless florists' varieties improved and added to year by year, all these have combined to make the garden of the nineteenth century what it is. Much as we may praise the gardens of our forefathers, and we have seen how much there was to admire or imitate in them, it is difficult to imagine our gardens deprived of the many floral treasures which have been added to them of late years. Many flowers have become so familiar, that it is hard to picture a garden without them, yet numbers of plants now to be seen almost everywhere had not been brought to our shores one hundred years ago. To bring about such changes many men have been at work, in every department, each contributing something towards the progress of gardening. There have been practical gardeners and nurserymen, great botanists and men of knowledge and daring, whose lives have been risked in the cause of science, and to whose courage and perseverance we owe so many of the treasures of a modern garden.

While the rage for landscape-gardening was at its height, there were many practical gardeners busy in a quiet way

ARLEY, A GARDEN LAID OUT FIFTY YEARS AGO IN THE OLD FORMAL STYLE.

carrying on the work of horticulture. One of these was Abercrombie, whose writings were popular for many years. He was the son of a market gardener near Edinburgh, and was born in the year 1726. The Battle of Preston Pans was fought close to his father's garden wall, and he was present at the time. His first place as gardener was with Sir James Douglas, and later on he married a rich girl, a relative of his former employer. In 1770 he settled with his family, consisting of two sons and sixteen daughters, between Mile End and

Hackney, and there started a nursery garden. His first book, *Every Man his own Gardener*, came out in 1767, and he was so afraid of failure that he paid Mawe, gardener to the Duke of Leeds, the sum of £20 to allow his name also to appear on the title-page. Hence the book has become known as the work of Mawe and Abercrombie, although the latter wrote it entirely. His other writings, *Amateur Gardening*, *The Gardener's Daily Assistant*, and such like, were equally popular. Another book of this date, by Wm. Hanbury, also gives full directions for the cultivation of a great number of trees, shrubs, perennial and annual hardy flowers, and greenhouse and stove plants.* Among those mentioned in these books we find many things which had just been introduced, such as the Pontic Rhododendron, Azalea nudiflora, or "American upright honeysuckle," as Hanbury calls it; Andromeda polifolia, varieties of Allspice (*Calycanthus*) of Sumach (*Rhus*) and of Magnolia (*grandiflora* and others), the snowdrop tree (*Halesia*), Hydrangias, and Spireas, and other hardy plants. There were also many additions to the half hardy and stove plants. Crinum capense or "lily Asphodel," and the more tender Belladonna lily (*Amaryllis Belladonna*). The Scarborough lily (*Vallota purpurea*) appeared about this time; the same kind of story being told of its origin as of that of the Guernsey lily (*Nerine sarniensis*), which was said to have grown in Guernsey from bulbs washed ashore from a wreck of a ship from Japan about 1659. The camellia or "Japanese rose" (*Camellia japonica*), was grown by the middle of the eighteenth century. The "gardenia, or the Cape Jasmine" (*Gardenia florida*), Plumbago (*rosea*) and other "tender sorts of leadwort," the Gloriosa superba, and Allamanda (*Allamanda cathartica*) were among the climbing plants which adorned the stove.

Some families of plants were becoming so conspicuous as to have a special literature of their own. The geraniums and heaths were treated of by Andrews, the Mesembryanthemums

---

* *Complete Body of Planting and Gardening*. By Wm. Hanbury, 1770. 2 vols. folio.

by Haworth, and the Proteæ by Knight. The literature of the orchard was also carried on by able hands. Speechly, gardener to the Duke of Portland, was the author of treatises on the pine and the vine. He describes fifty of the varieties of grapes grown at Welbeck, and mentions many of the fine vines to be seen then in England.* The Black Hamburgh at Valentine, in Essex, the parent of the Hampton Court one, yielded so much fruit that the gardener frequently made £100 a year by selling the bunches. A vine growing at Northallerton outside a house in 1789 covered 137 square yards of wall. He notices the vineyards near Bath, also those of Sir William Basset, in Somerset, who made some hogsheads of wine annually, and the Hon. Charles Hamilton, at Pain's Hill (the famous landscape garden), made wine from "Burgundy" and "black cluster" grapes, which sold for 7s 6d to 10s the bottle. Speechly himself grew a famous bunch of grapes at Welbeck, in 1781, which weighed 19½ lbs., and measured 20 in. in diameter. It was sent by the Duke of Portland to the Marquess of Rockingham, carried by men, like the spies returning from the promised land. Early in this century a vine was brought from abroad and planted at Cannon Hall, Yorkshire, which has since produced the well-known variety bearing that name. Haynes wrote on the strawberry, gooseberry, and raspberry. The strawberry was being much improved, and new and large varieties produced by crossing the Virginian with the Chilian, a species introduced early in the eighteenth century. Old-fashioned gardens still retained the hautboy (*F. elatior*) now so rarely to be seen, having been entirely superseded by the finer American species.

A fine work on fruit trees, with well drawn and coloured plates, by Brookshaw, *Pomona Britannica*, 1817, is principally taken from the fruit grown in the royal gardens at Hampton Court. In this book, besides some varieties which were then quite new, there are drawings of many of the old favourites. The "Catherine Pear" is figured and described as ripening in August, "sweet and juicy, with a degree of

---

* *Culture of the Vine.* By Wm. Speechly. York, 1790.

musky flavour; but at best is considered as a common pear." "The old Newington Peach," "Duke Cherry," "Norfolk Beefin Apple," "Red Streak Pippin," and many others are still favourites, and of Tradescant's Cherry Brookshaw writes: "I am doubtful whether we have a better black cherry than this, and yet it is so very scarce, and so little known, that it would be the most difficult task to find it. It is a cherry that was raised by Sir John Tradescant, gardener to King Charles I., different in shape from any other black cherry; and its flavour is unlike that of any other cherry; it ripens about 20th June." The history such as this of many fruits and vegetables has been handed down by Phillips,* who was the author of several valuable works on the subject. Another gardener who turned his attention chiefly to fruit trees was William Forsyth (1737-1804), who succeeded Miller as Curator of the Chelsea garden, and was afterwards appointed Royal gardener at Kensington. His works on fruit trees and the best methods of training and pruning, went through many editions. He is said to have done more for the improvement of fruit culture than any other gardener, although Knight disagreed with him on some of his methods of treating trees. Thomas Andrew Knight, President of the Horticultural Society, was himself an improver of fruit, especially of apples. He produced the Grange Apple in 1802, a cross between the golden and the orange pippin. George Johnson, the historian of Gardening, dedicated his work to Knight, and speaks of him in glowing terms as one "who unites to a knowledge of the Practices of Gardening, the most perfect knowledge of the sciences that assist it."† To "this distinguished vegetable physiologist" the Horticultural Society owed its origin. Being born in Herefordshire, in 1759, and brought up in the midst of orchards, he began early in life to watch the growth of trees, and try experiments. He felt the want of some stimulus to horticulture, and thought the formation of a Society "whose object should be the improvement of

* *Pomarium Britannicum*, 1820. *History of Cultivated Vegetables*, 1822. *Sylva Florifera*, 1823. *Flora Historica*, 1824, etc. All by Henry Phillips.
† *History of English Gardening*. By Geo. W. Johnson, 1829.

Horticulture in all its branches," * would have that effect. Accordingly, with the co-operation of Sir Joseph Banks, he organized the Horticultural Society, and a meeting to inaugurate it was held on March 7th, 1804. The first President was the Earl of Dartmouth, John Wedgewood the first Treasurer, and Cleeve the first Secretary, who was soon superseded by R. A. Salisbury. Price, the Clerk of the Linnæan Society, was also engaged as Clerk to the New Horticultural. In 1809, on April 17th, the charter of incorporation was signed by King George the Third. The next year, the first number of the *Transactions* was brought out. These quarto volumes were elaborately got up, and were so costly that the sum spent on them by 1830 amounted to £25,250.† In 1811, on the death of the Earl of Dartmouth, Thomas Andrew Knight was elected President. Under his energetic presidency, the affairs of the Society prospered. In 1818 their first experimental gardens were started at Kensington and at Ealing, but these were discontinued when the Society obtained a long lease of the Chiswick gardens four years later, and carried on their experiments there.

About the same time the Society began its greatest work, which was not only the receiving of plants from abroad, but the sending out of collectors also. The first plant of Wistaria (*Wistaria sinensis*) was sent by Mr. Reeves, from China, in 1818, and the original specimen is still at Chiswick, and other Chinese plants—Peonies, Roses and Chrysanthemums— were also received from abroad. The first collector sent out was George Don, who went to West Africa, and on to South America, in 1822-3. John Forbes was sent to East Africa the same year; he died while going up the Zambesi, but not before he had despatched home many new species. John Potts, who went in search of plants in China and the East Indies, also died from the effects of the climate. John Dampier Parks followed him to China, and also found a number of plants, and James Roe searched successfully 'in

---

* *The Book of the Royal Horticultural Society.* By Andrew Murray, 1863.
† From Notes kindly furnished by the present Assistant Secretary to the Society, Mr. John Weathers.

America and the Sandwich Islands. The well-known collector, David Douglas, was also employed by the Horticultural Society. He was born at Scone in 1799, and as a lad came under the notice of Sir William Hooker, then Professor at Glasgow. Hooker recommended him to Joseph Sabine, the Secretary of the Society, and Douglas was sent out to North America and California. The wealth of plants there discovered by him was unprecedented, flowers as well as trees. The number of conifers he sent home was so astonishing he wrote on one occasion to Hooker, "You will begin to think that I manufacture Pines at my pleasure." Besides the well-known Douglas pine (*Abies Douglasii*) he enriched this country with many others, Pinus Lambertiana, Pinus insignis, Pinus ponderosa, Pinus Sabiniana, Picea nobilis, Pinus grandis, the beautiful Taxodium sempervirens, and many more, which now adorn Pinetums and woods in all parts of England. At Dropmore there is a Douglas pine grown from seed given by the Horticultural Society to Lord Grenville in 1827. The tree was planted out in 1830, and in 1886 was 124 feet high, with a girth of fifteen feet. Besides these wonderful conifers we owe many other plants to Douglas.* The red-flowering Ribies now so common he sent home; also Calochorti, Clarkias, Gaillardias Godetias, Collinsias, Lupines, Eschscholtzias, Mimuli, and Penstemons. After many years of search in America, he went to seek more treasures in the Sandwich Islands, and met his death in a very sad way soon after his arrival there in 1834. He fell into a deep hole cut by natives for catching wild cattle, and was killed by one of the animals in it. Such a tragic end to one who had done so much, did not deter others from risking their lives in search of plants in strange countries. More pines were collected in California by Theodor Hartweg. Pinus Benthamiana, Pinus Devoniana and others; also Lupines, Berberries, and Fuchsias, and several Achimenes were discovered by him.

Perhaps the most successful of all adventurous collectors was Robert Fortune. He was born in 1813 and died in 1880.

* The plants are described by Hooker, *Flora Boreali Americana*, and in the *Botanical Magazine*.

He first entered the Edinburgh Botanical Gardens, and was subsequently superintendent of the hot-houses at Chiswick. In 1842 he started for China, and during the years which followed he was constantly sending home fresh treasures. Some of the best known garden flowers we owe to him :—Anemone japonica, Dielytra (or Dicentra) spectabilis, Kerria japonica, varieties of Prunus, Viburnum, Spirea and many Azaleas and Chrysanthemums, Gardenia Fortuniana, Daphne Fortuni, Berberis Fortuni, Forsythia viridissima, Weigela rosea, Jasminum nudiflorum, the white variety of Wistaria, and many other valuable plants. His greatest feat was to go to Loo Chow, disguised as a Chinaman, and there he obtained the double yellow rose, and the fan-leafed or Chusan palm, which bear his name. Since then this work of discovery has been carried on by able hands. To Sir Joseph Hooker we owe the Sikkim Rhododendrons and a large number of Himalayan plants. Lobb collected for Veitch and introduced many new things. Mr. F. C. Burbidge, especially in Borneo, has brought to light many treasures; Mr. Edward Whittall, at Smyrna, has sent many charming hardy bulbs from Asia Minor, and there are still numerous other active workers in this branch of science.

The number of roses in our gardens now is infinite, and a very large proportion has only been known in this country during this century. In addition to the old-fashioned species, the Gallica, the Damask, Sulphurea, Scotch, Austrian, Moss, Sempervirens and Musk, there are now many more species, besides endless hybrids. Most of the new species have come to us from Eastern Asia. The little Banksian Rose came from China in 1807, and smaller Fairy Rose in 1810; the Tea-scented Rose about the same time, Monthly Roses in 1789, and multiflora in 1822. Since then numerous varieties have been added, Boursault's, Noisette, Polyantha, Bourbon, and so on. In the Catalogue of the great nurseryman, Loddiges, in Hackney, in 1826, there are "no less than 1393 species and varieties of Roses," numbered as existing in their nurseries, and Lee, of Hammersmith, also had great quantities. Ever since then roses have been multiplying yearly. In 1861-2 Paul* brought out as

* *The Rose Garden.* By Wm. Paul. 9th Ed., 1888.

many as sixty-two new varieties, and during the next ten years he added many more, including such favourites as Marechal Niel, Louis Van Houtte, and Paul Neyron. This profusion of new roses which is still being added to year by year, has banished many of the old ones, such as the sweet Moss Rose and Damask, which deserve a place, as well as the hybrid perpetuals and teas.

The Dahlia,* a native of Mexico, was first introduced in 1789 from Spain by Lady Bute, but was lost and re-introduced in 1804 by Lady Holland, and twenty years later the craze for these flowers reached its height. The Fuchsia appeared in this country within the first five-and-twenty years of this century, although named by Plumier after Fuchs about a hundred years earlier. The story is told of how Lee saw a Fuchsia plant in a window of a small house in Wapping. He was so struck with the flower that he went in and asked the old woman to whom it belonged whether she would sell it to him. She, however, at first refused to part with it, as it had been sent to her by her husband who was a sailor, but was persuaded to let him have it when he offered her eight guineas and promised to give her two of the first plants he reared. He succeeded in getting some three hundred cuttings to strike, and presented the old woman with her share, while the rest, with their graceful hanging flowers, astonished the visitors to his Nursery, and brought him in a profit of about £300.†

That which perhaps would most astonish a gardener of the fifteenth century, could he but for one moment see it, would be an orchid house. Numerous as orchids are to-day, they nearly all have been imported during the last fifty years. There are still tracts of country which have not been searched, but most of the orchid-growing portions of the globe have been ransacked, and these glorious plants packed off by thousands to this country, leaving in some cases their native habitats bare. One reads accounts of whole districts being denuded of these treasures; for instance, a certain locality, once the home of Miltonia vexillaria, was so pillaged that the woods in the vicinity

---

* Named after Dahl the Swedish botanist, and quite distinct from the Dalea called after Dr. Samuel Dale (1659-1739).

† N. and Q., Sept., 1894.

"have become pretty well cleared." During one search for Odontoglossum crispum, when ten thousand plants were collected, four thousand trees were cut down to obtain them, the camp of the explorers was moved on week by week as they exhausted the plants in their neighbourhood.* The sight of this glorious wealth of flowers, which has gladdened many orchid hunters, will be denied to future generations, if the searchers are not more moderate in their demands on the virgin forests of the Old and New World.

The first tropical orchid which flowered in this country was a specimen of Bletia verecunda, which was sent from Providence Island, one of the Bahamas, in 1731, to Peter Collinson.† In Miller's Dictionary, two or three tropical orchids are mentioned, and some were grown by him at Chelsea. He says of the Vanilla which was sent to him "from Carthagena in New Spain," that "this plant flowered in the Chelsea Garden, but wanting its proper support it lived but one year." In 1778 Dr. John Fothergill brought home two species from China, one of which, Phaius grandifolius, flowered soon after in the stove of his niece, Mrs. Hird, at Appesby Bridge in Yorkshire. In 1787 Epidendrum cochleatum flowered at the Royal Gardens, Kew,‡ and Epidendrum fragrans the following year. Soon after the beginning of this century, several species were cultivated for sale by the Loddiges at Hackney, and this firm held for many years a most conspicuous place among orchid growers. As early as 1812 they grew a plant of Oncidium bifolium, which was brought from Monte Video, and about the same year the first of the Vandas, Aërides, and Dendrobiums were sent from India by Dr. Roxburgh. Although plants of many orchids were coming to this country during the first thirty years of this century, so little was known of their native places, and their conditions of life, that their cultivation was extremely difficult, and orchid growers met with constant failures. A house was set apart for them at Kew, and Lindley at the Horticultural Society also, by

---

* *Travels and Adventures of an Orchid Hunter.* By Albert Millican, 1891.
† W. B. Hemsley, *Gardener's Chronicle*, 1887.
‡ *A Manual of Orchidaceous Plants*, part x. By James Veitch and Sons, 1894.

careful study of their habits, tried to discover the right treatment. One of the earliest private orchid-houses was that of the Earl FitzWilliam, at Wentworth Woodhouse, the genus Miltonia being named in his honour. His gardener, Joseph Cooper, was one of the first successful growers. In 1833 the orchid collection at Chatsworth was begun. The Duke of Devonshire procured plants from the East, and Paxton, who was his gardener at the time, was enabled to cultivate many successfully, and publish the interesting records in the *Magazine of Botany*, which he edited. The orchid growers since then that have been successful, are too numerous to mention. Such collections as that of Sir Trevor Lawrence are one of the wonders of the nineteenth century.

The history of the introduction of many of these orchids reads like an exciting adventure or fairy tale. The story of the lost orchid Cattleya labiata vera is known to all orchid lovers. The plant was originally sent home from Brazil to Dr. Lindley by Mr. W. Swainson, as a packing round some lichens, in 1818,[*] and Lindley described and named it in memory of Mr. Cattley, a great horticulturalist. For years after other species were sent home, which passed for the true labiata, until it was discovered that the "vera" no longer existed in cultivation, and that its native home was forgotten. For fifty years it was the aim of all collectors to find this treasure again. By chance at last in 1889 some plants were sent home to M. Moreau, of Paris, from whom Messrs. Sanders learnt its habitat, and sent off in search of it, and soon all orchid growers were able to add the long-lost treasure to their collections. Many fruitless voyages have been made to procure these floral wonders, and frequently the collector has at last met with them when least expected. One plant of Cyprepedium Curtisi was sent home by Mr. Curtis from Penang in 1882, and no more were forthcoming until collectors despaired of ever finding it. At last, Ericsson, climbing a mountain in Sumatra, took shelter in a little hut. On the walls he saw among the names of the travellers who had rested there, a drawing of the very flower he was in search of, and underneath was written

[*] *About Orchids*. By Frederick Boyle, 1893.

"C. C.'s contribution to the adornment of the house." He at once set to work to look for it in the neighbourhood, and at length he found it in a most unlikely place, just as he was about to return home in despair. Such stories could be multiplied *ad infinitum*, as every year collectors are going through toilsome expeditions in order to procure these plants. One firm alone, Messrs. Sanders, at St. Albans, have often as many as twenty collectors working at one time. In the Spring of 1894 they had two in Brazil, two in Columbia, two in Peru and Ecuador, one in Mexico, one in Madagascar, one in New Guinea, three in India, Burmah, and Straits Settlements. Besides those species sent home from all tropical lands, the numerous hybrids brought out each year by large firms, as Veitch, Bull, or Low, or from private collections, must be taken into account to form an estimate of the numbers of orchids now in cultivation in England.

In every branch of gardening the changes have been rapid. The florists' varieties of Begonia, Gloxinia, Geranium, Cyclamen, Cineraria, Primula, Streptocarpus, Carnations, Achimenes, Chrysanthemum, Violas, Dahlias, Asters, Verbenas, and many such-like things, were unknown during the early part of this century. Donald Beaton, writing his recollections in 1854, of his early life as a gardener, tells how he remembers seeing the first Petunia that ever flowered in this country, at Lower Boughton, near Manchester, and the first Calceolaria in the Epsom Nursery. The institution of Shows and Awards of Merit has doubtless done much to stimulate the energy of florists and promote the production of new varieties. In Thomas Hogg's Treatise on the culture of the carnation and other flowers in 1820, he submits the Rules of two "Societies of Florists," in Islington and Chelsea, which had been started some years previously, for encouraging the cultivation of "Auriculas, Pinks, and Carnations." There were, he says, "several other societies of the same description in the neighbourhood of London, but these two are not only the most numerous in point of numbers, but likewise the most respectable in regard to the members composing them." The Rules of this Society are given at length. The subscription was £1. 11s 6d a year, and the value of the prizes, six in number, was presented to the successful candidates on Show Days. On the appointed days a dinner was held, and each member had to buy a dinner-ticket for the Auricula, the Carnation, and Pink shows. The flowers were judged by three members selected from among those present, and the flowers passed round the table while all were sitting at dinner, " beginning on the President's right hand, and returning on his left, in order that each person may distinctly view them." Many such societies have been started since then to encourage the florist varieties of different classes of flowers. Perhaps the most conspicuous have been those in connection with the rose, and more recently the chrysanthemum, which now boast of National Societies. The National Chrysanthemum Society originated in the one at Stoke Newington. That locality of London, which has for centuries been the haunt of gardens, from the times of L'Obel and Fairchild, and on to that of Loddiges, has not forgotten its old traditions; even in the midst of fog and smoke the dwellers in the East of London try to cultivate flowers. The chrysanthemum occupies much of their attention, and that they can cultivate them with success can be seen by the local Exhibitions.* The Horticultural Society held their first *fête* in 1831, and soon after the regular Exhibitions began. Since then their shows and those of the Botanical Society and of local societies in every town and county of England, have become events of yearly, or almost weekly occurrence, and the stimulus to Floriculture promoted by these institutions must be apparent to all. The Botanical Society of London was incorporated in 1839. That part of the grounds which were devoted to the illustration of the Natural Orders, were arranged by Sowerby, then the Secretary, and his father, Dr. F. J. Fane and Dr. Sigmund; and everything was done to facilitate the labours of students of Scientific Botany.†

In the hasty review that has been taken of the progress of Horticulture, the prominent position of the Royal Gardens at

---

* The Shows of the Dalston and De Beauvoir Town Amateur Chrysanthemum Society, held annually, are an example of what care and attention can achieve.

† *Memoirs of Dr. Frederic J. Fane.* 1886.

Kew has not been properly pointed out. They were begun by the Princess of Wales, Mother of George III., about 1760. In the extremely quaint and original Poem, "The Botanic Garden," in 1791, Erasmus Darwin alludes to the wonders of Kew in his usual stilted verse :—

> "So sits enthroned, in vegetable pride,
> Imperial Kew by Thames' glittering side;
> Obedient sails from realms unfurrow'd bring
> For her the unnam'd progeny of Spring;
> Attendant Nymphs her dulcet mandates hear,
> And nurse in fostering arms the tender year;
> Plant the young bulb, inhume the living seed,
> Prop the weak stem, the erring tendril lead;
> Or fan in glass-built fanes the stranger flowers,
> With milder gales, and steep with warmer showers.
> Delighted Thames through tropic umbrage glides,
> And flowers antarctic, bending o'er his tides;
> Drinks the new tints, the sweets unknown inhales,
> And calls the sons of Science to his vales."

The importance of Kew gradually increased under the management of William Aiton. This able gardener was born in 1731, and obtained the appointment of Botanical Superintendent at Kew through the influence of Philip Miller. He brought out a catalogue of the plants grown at Kew in 1789. To each plant Aiton added the native habitat, and the date of introduction, and records, from his own recollection, those that were grown by Philip Miller at Chelsea. He identified those introduced by Peter Collinson with the help of his son Michael; James Lee, of Hammersmith, and Knowlton, who had been gardener to James Sherard, also gave him what information they could. The plants are arranged on the Linnæan System, and include between five and six thousand species, this number being raised to eleven thousand in the second edition to which Dryander and R. Brown largely contributed, published by the younger Aiton in 1810-1813. William Aiton died in 1793, and was succeeded by his son, William Townsend Aiton. Since then, under the many able botanists connected with it, Kew has assumed more and more the first place among the Botanical Institutions of the world. Of the work of the great botanists of this century, Lindley, Hooker, Brown, Smith,

Loudon, Henslow, Sowerby, and the great Darwin himself, and many others, it is impossible to speak, but it is to these great men that the wonderful progress of this century is due, to say nothing of the men still living who are looked up to with respect and admiration by the practical gardeners in this close of the nineteenth century, not only in England itself but throughout her vast dominions.

In England a garden appears to have been attached even to the humblest home. As early as in Tudor times the peasant tried to grow a few plants around his cottage door; and many an old cottage is still covered with a vine that has stood there for centuries, and many an apple tree has born its ruddy crops year by year undisturbed, while the gardens of the more imposing mansions hard by have passed away. Of late years the desire to cultivate again some of the old-fashioned plants which had been discarded, has led many to search for them in cottage gardens, and thus numerous treasures have been found which had for long remained hidden in some retired spot. The fruit and vegetables now grown by cottagers are often an example to their more wealthy but less skilful neighbours. In coldest winter, it is wonderful to see the bright masses of flowers in their cottage windows. Even in the towns the poor man tries to have some plants to "serve him with a hint that nature lives."

> "Mark the dim windows ye shall pass
> And see the petted myrtle here;
> While there upraised in tinted glass,
> The curling hyacinths appear.
>
> The gay geranium in its pride,
> Looks out to kiss the scanty gleam;
> And rosebud nurslings by its side,
> Are gently brought to share the beam.
>
> Hands with their daily bread to gain
> May oft be seen at twilight hour,
> Decking their dingy garret pane
> With wreathing stem or sickly flower."
>
> ELIZA COOK.

The Italian style of design, which was the prevailing one by the middle of this century, was easily adapted to suit the new florist flowers which were then rapidly increasing. The fashion of what is known as "bedding out," came in, and old-fashioned plants, which had been the pride of our gardens for centuries, were banished to make room for these newcomers. In an Essay on Landscape Gardening by Morris, in 1825,* he advocates this plan, which was then quite a new one. "The beauty of the flower-garden, in the summer season," he writes, "may be heightened by planting in beds some of the most freely-flowering young and healthy greenhouse plants. Where there is an extent of greenhouse, a sufficient quantity of plants should be grown annually for this purpose, and should be sunk in the beds about the middle or end of May. The following are among the most beautiful of this species: Anagallis grandiflora, Anagallis Monelli, Heliotropium grandiflorum, Fuchsia coccinea, Lobelia Erinus and unidentata, Hemimeris urticifolia, Alstroemeria peregrina, Bouvardia triphylla, Geraniums of sorts, Lychnis coronaria, Linum trigynum." These are what Morris suggests, but other plants, Petunias, Zinnias, Begonias, Ageratum, Calceolarias, and many more, might now be added to the list, besides the numerous foliage plants, such as Coleus, Echiverias, Cerastiums, Dracænas, also Alternanthera, and other low growing things which are used for carpet bedding. More skill is now used in the selection of colours and arrangements of plants, some fine effects being thus produced with these combinations. Graceful and more feathering plants are planted among the old-fashioned bedding plants, such as a groundwork of some self-coloured viola, relieved by tall standards of ivy-leafed Geranium, Dracænas, Cannas, or Grevillea robusta. At first the bedding-out consisted in merely filling the beds with flowers to produce as great a blaze of colour as possible. Trentham garden is described in 1859 as a "startling mass of Geraniums and Calceolarias," and this alone was the aim of the gardeners in many places.

There is a very large folio volume by A. E. Brooke, in which are depicted what were then considered the finest gardens in England.† Most of them are Italian in

* *Essay on Landscape Gardening.* By Richard Morris, 1825.
† *Gardens of England.* By A. E. Brooke, 1858.

HAREWOOD.

design, and the beds are filled with these gaudy but perishable flowers. Among the number he illustrates may be mentioned Woburn, Worsley, Eaton, Trentham, Castle Howard, and Teddesley, designed by Nesfield, all laid out between 1845 and 1858. Sir Joseph Paxton, the well-known Editor of the *Magazine of Botany*, and gardener to the Duke of Devonshire at Chatsworth, was also the designer of the Crystal Palace Gardens, in a pseudo-Italian style, for which he was Knighted. The taste must not be judged from this crude and uninteresting example, as many charming gardens of a stiff Italian design exist. Besides those already quoted, Harwood is a fine example. It was planned by Lady Harwood, and the designs for the fountains and stone balustrades were made by Sir Charles Barry. Shrublands* was begun to be laid out by Sir William Middleton about 1830. There is in front of the house at Shrublands a wide terrace with flower-beds like that at Harewood, but without fountains, from it a long flight of steps leads to a semi-circular terrace garden below (*see illustration*). This, like all gardens in this style, was formerly "bedded-out" each summer. It is easy to see what an immense expense this involves, and how difficult it was to keep up a garden under those conditions. For unless the beds are to be empty except for four months in the year, there must be spring bedding of hyacinths, crocuses, tulips, &c., as well as geraniums, and such like, for the summer.

There are now in cultivation such an immense variety of hardy perennial plants, which, with a little care, will thrive well in this country, that if a judicious selection is made from these, the beds can be made as bright in summer as with the more delicate plants which perish with the first touch of frost; and the beauty of the garden can be considerably prolonged. By planting such things as violas or "tufted pansies," a mass of colour from early spring until late autumn can be obtained. The garden at Shrublands has been thus arranged, according to the suggestions of Mr. W. Robinson, with great success. The beds are filled with roses, pinks, and carnations, and many hardy plants, the masses of colour

* In Suffolk, belonging to Lord de Saumarez.

being skilfully arranged. One bed, shown in the illustration, is composed of Lobelia cardinalis in the centre with a border of Centaurea ragusina, which makes a striking effect until late in the year. This bringing back to our gardens the numerous hardy plants which were banished, and in many cases ruthlessly torn up and thrown away when the rage for " bedding-out " came in, is the greatest improvement of the end of the nineteenth century. They are once again holding their proper place, and with all the new species which every year come to swell the list

SHRUBLANDS.

of things which will endure our cold climate, more lovely effects could be produced than ever were possible with the stiff bedding plants of forty years ago. But no one would wish to discard altogether these half hardy things ; our green-houses, a blaze of bright colours with tuberous Begonias, or some such flowers, are a wonderful sight, and even from a practical point of view it is a good plan to make room in the houses, by planting out some of these things in the summer. Very different is this arrangement from devoting all the glass to nurture up geraniums to fill the whole garden. Bacon's aim was to have flowers in the garden during every month of the year, and in his essay he mentions some for each successive season. Surely after a lapse of three centuries we ought to be able to attain that object and arrange that no month should be without its brightening flowers.

> "The daughters of the year
> One after one through that still garden passed,
> Danced into light and died into the shade."

Among the many plants which have been introduced of late years the class of Alpines has been very largely represented. We now possess an immense variety of plants whose natural place of growth is on rocks, or between the crevices of stones. It is only reasonable to try and give these plants, as nearly as possible, the same conditions of life here in England as on their native hills. The result of this has been the formation of several rock gardens, very different from the old-fashioned pile of stones which went by the name of a rockery. These new rock gardens are in every way successful, as rare Alpines, which it was thought almost impossible to grow in this country, are now made to thrive. The illustration of a typical rock garden is a part of the very large one at Batsford (Gloucestershire), made by Mr. Mitford within the last few years. The one at Kew is a well-known example. Every year there are new things of interest there. It is wonderful to see plants from nearly all the mountain ranges of the world perfectly at home within a few miles of the City of London.

Another development of gardening during the last few years has been sub-tropical gardening. Mr. Robinson has kindly pointed out to me that this kind of gardening, which came to us from Paris some twenty years ago, did something to relieve the formality of "bedding-out," although not nearly as important an improvement as the newer movement towards hardy flowers. Groups of Cannas, Caladiums, and such like, in beds, help to render them less stiff. There can also be obtained fine results from planting out the hardier kinds of tree ferns and palms during the summer months. But the best kind of sub-tropical garden is the

ROCK GARDEN, BATSFORD

permanent one. Even in the coldest districts of England, numerous plants will grow which give a tropical appearance.* In Norfolk and Suffolk, where the late frosts are most trying to gardeners, various bamboos will flourish: Bambusa meteke, Simonii, viridiglaucescens, and edulis are perfectly hardy, and besides these many things such as Berberis, Aralias, Gunnera scabra, Aristolochias, giant Heracleums, Arundo Donax, several species of Rhus and Spirea, Polygonum cuspidatum, Tamarix, Yuccas, Polygonatum multiflorum, Solomons seal, Bocconia cordata, and several sorts of Acanthus, besides taller trees, such as the Ailanthus glandulosa, and Japanese maples; these when grouped on grass with smaller ferns and grasses, produce a very tropical effect. Green gardens composed of such things would be a pleasant variation from the brighter flowering plants. In the warmer districts of England this could be more easily accomplished. Some of the hardier palms do well and appear almost at home among the familiar English trees.

> "But fair the exiled palm tree grew
> Midst foliage of no kindred hue;
> Through the laburnum's dropping gold
> Rose the light shaft of orient mould,
> And Europe's violets faintly sweet
> Purpled the moss-beds at its feet.
>
> Strange looked it there! the willow streamed
> Where silvery waters near it gleamed;
> The lime-bough lured the honey bee
> To murmur by the desert tree,
> And showers of snowy roses made
> A lustre in its fan-like shade."
>
> MRS. HEMANS.

Parts of Cornwall are so mild that many things will do well there which are considered as greenhouse plants in other parts of England. There are in that county some gardens that would astonish gardeners from less-favoured districts. Pengerrick, Menabilly, Heligan, Tregothuan, and Carclew are among the finest of these Cornish gardens. Camelias grow into fine trees,† and Sikkim Rhododendrons flower

---

* *The Sub-tropical Garden.* By W. Robinson. 2nd Edition, 1870.
† Also in Hampshire and some other Southern and Western Counties.

in the open-air, and Lapargerias will grow like ivy on sheltered walls. At Carclew, Rhododendrons Thomsoni, Hodgsoni, campylocarpum, argenteum, and many other tender varieties were covered with bloom last spring. In that garden there are many interesting plants thriving well, which are usually kept in green-houses in England. Choisya ternata, Embothrium coccineum, Phyllocladus rhomboidalis, Azara microphylla, are among the number, and Benthamias, the seeds of which were

NARCISSUS IN THE SCILLY ISLES.

first sent home to England from Ceylon by Sir Anthony Buller, flourish; some of the original ones still grow in the garden at Heligan, where they were first planted. Still more favourable is the climate of the Scilly Isles, and lately this has been taken advantage of, for growing narcissi. Mr. Dorrien Smith started the industry, and within the last ten years it has been steadily increasing, and thousands of cut flowers are sent to the London markets. In the Islands in February there are acres of narcissi in bloom, which are picked and sent off to London. The illustration shows a field of Poets Narcissus, and there are also quantities of the polyanthus varieties grown. The daffodil is a flower which has come prominently into notice of late years. Each type has been enormously developed, some of the new Trumpet varities being of special beauty.*

The spring garden now is no longer only a few tulips and hyacinths bedded out, but these narcissi and many other bulbs such as Scilla siberica, Chionodoxa Luciliæ or Tulipa silvestris, can be naturalized, and if planted in masses on grass, in glades, or on the edges of lawns, they will give a brilliant effect before the summer flowers have made their appearance, and can be mown over with the rest of the grass if necessary when their flowers are over. Bulb culture is a favourite pursuit in the manufacturing districts of north-west England. It is thought that the taste was carried thither by the Flemish weavers, who in earlier times had brought the love of these plants with them from the Low Countries, when they first settled in East Anglia, Essex and Kent. There is also the kind of spring garden which has been most successfully carried out at Belvoir. Not only are the beds filled with such things as "Forget-me-nots," Iris reticulata and Iris siberica, Silenes, Violas, Wall-flowers or Heuchera sanguinea, Aubretias, Cerastium tomentosum, but many Primulas, Anemones, Gentians, Cyclamens, and various alpines, are naturalized on a vast rock garden.

The idea of naturalizing plants in shrubberies, grassy banks and wild places, is also a new departure of the late nineteenth century. Mr. W. Robinson, by his works, the *Wild Garden*, and the *English Flower Garden*, has done more than any one to bring in the taste. By grouping flowers naturally in this way, fine picturesque effects can be obtained. It is the reverse of the "Landscape Gardening," which brought green undulations of park-like appearance up to the house, and banished the flower-garden; it extends the flower-garden into

* *Ye Narcissus, a Daffodil Flower.* By Barre, 1884.

the surrounding country. In practising the art of wild gardening, that is, naturalising of plants, which are not natives,

LILIES IN WILD GARDEN.

but which are hardy in our climate, and if once planted will take care of themselves, there is no need to banish the " formal garden." The formal garden certainly seems to be the most suitable to place near a house, and its design should harmonize with the architecture. This kind of garden is necessary, if any tender plants, or those that require special care and treatment are to be reared, but beyond this formal garden, and separated from it by some suitable enclosure, the wild garden, judiciously planted, proves a continuing source of interest and pleasure. The accompanying illustration of some giant lilies about ten feet high, shows one of the many effects such a "wild garden" can produce. They are planted in a wood, and are spreading and thriving, and look quite in keeping with their surroundings. The low bushes in the background are varieties of cistus, all quite at home in the Surrey copse.* There is great scope for wild gardening on the banks of streams and lakes, and even in the water itself. The new hybrid water lilies raised by Marliac in France, and coming to us from that country, are one of the latest additions to gardens, and in a few years their worth will be recognized.† The numbers of lilies imported from Japan have added yet another feature to nineteenth-century gardens, and the varieties of hardy Rhododendrons and Azaleas are further precious contributions from that country.

There has been a movement of late years in favour of the formal garden,‡ and the study of old works on gardens naturally has a tendency to increase this. Some formal gardens have been laid out in England within this century which are equal in beauty to any older ones. Those of Penshurst in Kent, Arley in Cheshire,§ Blickling in Norfolk, and Montacute in Somerset are well-known instances, all differing in style, and by their beauty bear better testimony to the many advantages of a formal garden than any written arguments could do. The garden has always been considered as, and always must be, an adjunct of the house, and therefore must accord with it, if it is to look well. No one would put

* Miss Jekyll's garden, Munstead, Godalming.
† Nymphea rosea, N. sulphurea, N. odorata, N. Marliacia, and its varieties, rosea, rubra, carnea, &c.
‡ *The Formal Garden.* By Blomfield and Thomas. *Garden Craft Old and New.* By John Sedding.
§ Belonging to P. Egerton Warburton, Esq. See illustration on page 281.

an Elizabethan garden in front of an Italian house, or *vice versâ*, and an old-fashioned formal garden would not look well in front of a new looking suburban villa, but no hard and fast rules of style can be laid down, as the selection depends on the architecture, scenery, climate and many other things.

With the many beautiful gardens which exist throughout England there need be no plea of ignorance. Anyone laying out a garden can see examples of every style. While such places as Knole, Ham, Bromwich, Wrest, Melbourne, Haddon, and Levens, exist, there can be no lack of inspiration. This is an age of progress in Gardening, as in other arts, and if garden design is carefully studied, and the wealth of hardy as well as tender plants made proper use of, the newest gardens of the nineteenth century might easily surpass anything that has yet been seen in England.

PARLIAMENTARY SURVEYS.

SURREY, No. 72.

SURVEY OF WIMBLEDON.*

A SURVEY of the Manor of Wymbledon, alias Wimbleton, with the Rights, Members, and Appurtenances thereof, lying and being in the county of Surrey, late parcel of the possessions of Henrietta Maria, the relict and late Queen of Charles Stuart, late King of England, made and taken by us whose names are hereunto subscribed in the month of November, 1649; by virtue of a commission grounded upon an Act of the Commons assembled in Parliament for sale of the Honors, Manors, and Lands heretofore belonging to the late King, Queen, or Prince, under the hands and seals of five or more of the Trustees in the said Act named and appointed.

[The Hall, the rooms, the stone gallery, the grotto, the marble parlour, the organs, the Chapel, the King's chamber, the West stairs, the East stairs, a room called the Den of Lions, the great gallery, the Queen's chamber, the second, third, and fourth floors, the Clock stairs, the Wardrobe stairs, the back stairs, the leads, &c., are described.]

AND also consisting of one garden called the Oringe garden, adjoining to the East end of the said Manor or Mansion House, severed from the Pheasant Garden with a high brick wall upon the East and North sides thereof, and from the upper or greater garden with an open pale on the South side thereof, containing upon admeasurement one rood and twenty perches of ground, worth per annum £1.

MEMORANDUM, that in the said Oringe Garden there are four knotts, fitted for the growth of choice flowers, bordered with box in the points, angles, squares, and roundlets, and handsomely turfed in the intervals or little walks thereof; which knotts, and the flower roots therein growing, we estimate to be worth £24. 10s.

In the middle part of which four knotts is one large round, Marble paved with small pebble stones; in the middle whereof stands Fountain.

---

* Transcribed from the original MS. in the Record Office. It has been printed before with the complete Survey in Vol. X. of the *Archæologia*.

The Oringe Garden.

one handsome Fountain of white marble, which, with the pipes of lead and cocks thereunto belonging, we value to be worth £20.

Unto which Fountain one pavement of Flanders brick, six foot broad, extends itself from the East of the said Manor or Mansion House, up the middle of the said Oringe Garden, which we value to be worth ——.

The other three alleys or little walks betwixt the said four knotts are paved with pebble stone, worth in both £2.

The middle of which said three allies leadeth from the said Fountain unto a garden or Shadow house, paved with Flanders brick, and handsomely benched, standing in the middle of the East wall of the said Oringe Garden; the materials of which house are worth £5. 10s.

There are four large and handsome gravelled walks inclosing the said four knotts; the value whereof we include in the foresaid yearly value of the said Oringe garden.

In the side of which said Oringe Garden there stands one large Garden House; the out walls of brick; fitted for the keeping of Oringe trees; neatly covered with blue slate, and ridged and guttered with lead; the materials of which house, with the great doors and the iron thereof, with a certain stone pavement lying before those doors, in nature of a little walk 4 foot broad and seventy-nine foot long, we value to be worth £66. 13s. 4d.

Oringe trees.

In which said Garden House there are now standing in square boxes, fitted for that purpose, forty-two Oringe trees bearing fair and large oringes; which trees, with the boxes and the earth and materials therein feeding the same, we value at ten pounds a tree, one tree with another, in toto amounting unto £420.

Lemon tree.

In the said Garden House there now also is one Lemon tree bearing great and very large lemons, which, together with the box that it grows in, and the earth and materials therein feeding the same, we value at £20.

Pomecitron tree.

In the said Garden House there now also is one Pomecitron tree, which, together with the box that it grows in, and the earth and materials feeding the same, we value at £10.

Pomegranet tree(s).

There are also belonging to the said Oringe Garden 6 Pomegranet trees, bearing fair and large fruits, which, together with the square boxes they grow in, and the earth and materials therein feeding the same, we value at three pounds a tree, one with another, in toto £18.

There are also belonging to the said Oringe Garden 18 Oringe
trees, that have not yet borne fruit, which, with their boxes, earth,
and materials therein feeding the same, we value at five pounds
a tree, one with another, in toto amounting unto the sum of £90.

MEMORANDUM, that the foresaid six Pomegranet trees and the
said eighteen Oringe trees now stand and are placed with their
boxes in one little room of the said Mansion House called the
lower Spanish Room, and opening to the said Oringe Garden.

In the head of every of the said four knotts there is one
Cypress tree growing, which 4 together we value at £1.

There are two Apricot trees growing to the wall on the
North Side of the said Oringe Garden, worth £1.

There are also 14 Laurel trees planted in several places of
the said Oringe Garden, which we value in the gross at £1. 8s.

In the South East corner of the said Oringe Garden, there is
one fair Bay tree, which we value at £1.

MEMORANDUM, that the said Oringe Garden extends no farther
in breadth than the East end of the said Manor or Mansion
House doth extend itself; but is exceedingly graced with the said
two long galleries or walks adjoining to the East end of the said
Manor or Mansion House; the one leaded, standing four yards
above the said Garden, and the other floored with free stones
lying level with the said Oringe Garden, and extending to the
whole breadth thereof; the value of the materials of which said
galleries are contained in the valuation of the said Manor or
Mansion House, as in the particulars thereof may appear.

AND also of one other garden called the Upper or Great
Garden, adjoining to the South side of the said Manor or Mansion
House; severed from the said Oringe Garden with the said raised
pale on the South side of the said Oringe Garden, and lying
between the said Manor or Mansion House and the Vineyard
Garden, from which it is severed with a long brick wall ten foot
high on the South side thereof; and from Wymbledon Park with
a brick wall of ten foot high on the East side thereof; and from
the Churchyard with another brick wall of ten foot high on the
West side thereof; and from the Wood yard with a brick wall of ten
foot high on the South side thereof; containing upon admeasurement
6 acres and 26 perches of land, worth per annum £12.

MEMORANDUM, that the said Upper or Great Garden is divided
into two several levels or parts by an ascent of ten steps; the

lower level or part whereof adjoins to the South side of the said Manor or Mansion House, and lies level with the floor of the Hall of that Mansion House, containing in itself 4 several squares, having one fair and spacious gravelled walk, neatly ordered, running from East to West all along the said South side of the said Manor or Mansion House, being twenty-five foot broad and one hundred three score and ten yards long; at either end of which Lower Level is one other gravelled walk running up in a regular form to the Upper or Higher Level. These three walks include within them the whole extent of the said lower level, and are comprised in the yearly value of the whole Garden.

The said Lower Level is divided and cut out into 4 great squares, the two middlemost whereof contain within them eight several squares, and well ordered knotts, stored with the roots of very many and choice flowers; bordered with box, well planted and ordered, in the points, angles, squares, and roundlets; the four innermost quarters thereof being paved with Flanders bricks in the intervals, spaces, or little walks thereof: which knotts, borders, and roots of flowers, and the said Flanders bricks, we estimate to be worth £60.

Up the middle of which eight knotts, runs one walk or alley of paved stone from the hall door of the said Manor or Mansion House to the foot of the ascent of the said Higher or Upper Level; containing in breadth 16 foot and in length 127 foot; the stones whereof we value to be worth £20.

The said eight knotts are compassed about on three sides thereof with very handsome rails, piked with spired posts in every corner and angle, all of wood, varnished with white, [which] very much adorns and sets forth the Garden; all along the insides of which rails grow divers Cypress trees in a very decent order, having the outsides bordered with choice and pleasant flowers; in the two angles of which rails inwards stand two stone statues of good ornament: which rails, spired posts, and statues we estimate to be worth £29. 8s.

In the middle of the 4 of the foresaid eight knotts which lie on the West side of the said pavement, there stands one Fountain of white marble, having a statue of Diana upon it, and a fair lead cistern belonging to it, from whence runs a channelled pavement of stone into the Birdcage, being shadowed round with twelve Cherry trees, which stand in the points and angles of those four knotts; which fountain, statue, cistern, and channelled pavement we estimate to be worth £7.

In the middle of the 4 knotts which lie on the East side of the said pavement, there is one other Fountain of white marble, having a statue of a mermaid upon it, and a cistern of lead, being also shadowed round with twelve Cherry trees, which stand in the points and angles of those 4 knotts; which Fountain, statue, and cistern we value to be worth £10. *Mermaid Fountain*

The other two great squares of the said Lower Level, each of them contains within its own square four square grass plots, with one handsome round grass plot in the middle thereof, and lie at the East and West ends of the said eight knotts: in the middle of each of which four grass plots stands one fair Cypress tree. The four grass plots are bordered on all sides and angles with neat and well ordered thorn hedges, and well planted with many Cherry trees; but the value of the said two squares is not otherwise valuable than as comprised within the yearly estimate of the whole Garden. *The Lower Level.*

At the west end of the gravelled alley which adjoins to the South side of the said Manor House, there stands one Garden House, part of boards, part of rails, covered with blue slate, and ridged and guttered with lead, paved with square stone, having one door going into the said gravelled alley, one other door going into the end alley leading to the said Upper Level, and one other door opening into the Hartichoke Garden; the materials of which house we value to be worth £9. *Garden House.*

In the middle of the East wall of the said Lower Level there stands one garden, summer, or shadow house, covered with blue slate, handsomely benched and wainscoted in part, and paved with bricks, the materials whereof we value to be worth £5. *Shadow House.*

In the North side of the said alley, next adjoining to the said Manor House, and in the very end of the pale which divides the said Lower Level from the Oringe Garden, there stands one Banqueting House, covered with blue slate, ridged and guttered with lead, having one room above, floored with boards, the door whereof opens into the said alley; and one other room below, paved with tile, the door whereof opens into the Oringe Garden; having also in the sides thereof several lights of glass; the materials of which house we value to be worth £30. *Banqueting House.*

The North side of the said alley, very near as far as the said

Lower Level

Manor House doth extend itself in length, to wit, from the East end thereof to the end of the Birdcage westward, is railed with turned ballusters of free stone, well battled with stone, and cemented with lead and iron; betwixt which rails and the said Manor House are several little grass plot courts, which lie level with the lowest rooms of the said Manor House; over the middle of which courts lies the said pavement that leads from the said Hall door to the ascent of the said Upper Level, railed with the said stone rails on each side thereof, in a very graceful manner; in two of which courts there grow three great and fair Figtrees, the branches whereof by the spreading and dilating of themselves in a very large proportion, but yet in a most decent manner, cover a very great part of the walls of the South side of the said Manor House, being a very great and munificent ornament thereunto; into which little courts there are several descents of 16 steps from the said alley; in one of which courts there is an oval cistern of lead, set about with stone, having a pipe of lead in it; the outward walls of which little courts are planted with young Figtrees: the profits and contents of which little courts are comprised in the foresaid yearly value and admeasurement of the said Upper or Higher Garden; but we value the said oval cistern at two pounds, and the said 3 great Fig-trees and other young Fig-trees at twelve pounds ten shillings, and the said free-stone rails at, in all, £34. 10s.

Fig trees

Birdcage Fountain.

One other of the said little courts is fitted with a birdcage, having three open turrets, very well wrought for the sitting and perching of birds; and also having standing in it one very fair and handsome fountain, with three cisterns of lead belonging to it, and many several small pipes of lead, gilded, which, when they flow and fall into the cisterns, make a pleasant noise. The turrets, fountain, and little court are all covered with strong iron wire, and lie directly under the windows of the two rooms of the said Manor House called the Balcony Room and the Lord's Chamber; from which Balcony Room, one pavement of black and white marble containing 104 foot, railed with rails of wood on each side thereof, extends itself into the said alley over the middle of the said birdcage. This birdcage is a great ornament both to the House and Garden: the materials whereof and the said fountains and cistern, and the said marble pavement and rails, we value to be worth in the whole at £25. 4s.

In the height of the said Higher Level there is one fair green tarras or walk, very well turfed, extending itself two hundred and thirty yards from East to West, and containing twenty-five foot in the breadth thereof; the North side whereof is planted with lime trees of very good bulks, and of a very high growth, growing, both tops, bodies, and branches, in a most uniform and regular manner; the height whereof, being perspicuous to the country round about, renders them a very special ornament to the whole house. The south side of the said turfed tarras is planted with Elms, betwixt every one whereof grows a Cypress tree, well planted and ordered, much adorning and setting forth the completeness of the tarras; besides which there are on either side of the said tarras, betwixt every tree, borders of box, very well ordered, adding also a further ornament thereunto; which tarras and borders we value to be worth £17. 2s. 6d. *Tarras. (Terrace.)*

At the east end of the said turfed tarras there stands one fair banqueting house, most of wood; the model thereof containing a fair round in the middle of four angles, covered with blue slate, and ridged and guttered with lead, wainscoted round from the bottom to the roof, varnished with green within and without, benched in the angles, having sixteen windows or covers of the same wainscot, to open or shut at pleasure, and having also sixteen half rounds of glass to enlighten the room when those covers are shut up; the floor paved with painted tile in the angles, and with squared stone in the middle; in one of which angles stands a table of artificial stone very well polished; and in every of the said angles, besides the said benches, there stands one wainscot chair. There are to the said banqueting house, two double leaved doors, the one pair of which doors opens in the very middle of the said tarras, the outside thereof being gilt, with several coats of arms; the other of the said leaved doors opens into a fair walk within the Park, planted with Elms and Lime trees, extending itself from the said banqueting house in a direct line eastward, to the very Park pale. The round of the said banqueting house is handsomely arched; within which thirteen heads or statues, gilded, stand in a circular form, adding very much to the beauty of the whole room. The materials of this house, the said table and chairs, we value to be worth £66. 13s. 4d. *Banqueting House. The Upper Garden. The Higher Level.*

At the west end of the said turfed tarras there stands one other Garden or Summer house, covered with blue slate, and *Garden House.*

ridged, and guttered with lead, wainscoted and benched round, paved with square tile; in which stands one table of Rance stone, set in a frame of wood. There are two doors belonging to this garden house, the one opening into the said tarras, and the other opening into the Churchyard, into an alley or walk therein, leading to the Church door, planted on either side thereof with Sicamore trees. The materials of this house, and the said table, we value to be worth £13. 6s. 8d.

Betwixt the ascent from the said Lower Level and the said turfed tarras, there are on each side of the gravelled alley that leads from that ascent to the said tarras, three grass plot walks planted with fruit trees of divers sorts and kinds, both pleasant for taste and profitable for use; the borders of which grass plots are Coran* trees; the value of which trees and borders doth herein and hereafter appear in the several particulars thereof: the value of the grass plots being comprised in the foresaid yearly value of the whole Upper Garden.

Maze.

In the South of the said turfed tarras there are planted one great Maze, and one Wilderness, which being severed with one gravelled alley in or near the middle of the said turfed tarras, sets forth the Maze to lie towards the east, and the Wilderness towards the west. The Maze consists of young trees, wood[s], and sprays of a good growth and height, cut out into several meanders, circles, semicircles, windings, and intricate turnings, the walks or intervals whereof are all grass plots. This Maze, as it is now ordered, adds very much to the worth of the Upper Level. The Wilderness (a work of a vast expense to the maker thereof) consists of many young trees, woods, and sprays of a good growth and height, cut and formed into several ovals, squares, and angles, very well ordered; in most of the angular points whereof, as also in the centre of every oval, stands one Lime tree or Elm. All the alleys of this wilderness, being in number eighteen, are of a gravelled earth, very well ordered and maintained: the whole work being compiled with such order and decency, as that it is not one of the least of the ornaments of the said Manor or Mansion House. The foresaid alley dividing the said Maze and Wilderness is planted on each side thereof with Lime trees and Elms, betwixt every tree whereof grows a Cypress tree; at the south end of

The Wilderness.

The Higher Level.

* Currant.

which alley, and in the wall that parts the said Upper Garden from the Vineyard Garden, betwixt two fair pillars of brick, there are set fair and large pair of railed gates, of good ornament to both the said gardens. On the South side of the said Maze and Wilderness there is one close or private gravelled walk, inclosed on each side thereof with a very high and well grown hedge of thorn, extending itself from the East wall to the West wall of the said Upper Garden; at each end of which close walk there stands one little shadow or summer house, covered with blue slate and ridged with lead, and fitted for resting places. Which Maze and Wilderness, over and besides the trees thereof, which are herein hereafter valued amongst the other trees of the Upper Garden, and the materials of the said two shadow or summer houses, we value to be worth £90. [Private Walk. Shadow houses.]

There are in the said Upper Garden one hundred thirty one Lime trees and sixty eight elms, of good growths, worth in the gross at £44. 13s. [Lime trees and Elms.]

There are in the said higher and lower level of the said Upper Garden one hundred twenty three Cypress trees of divers growths, which, though they are not of any great profit, yet, as they are now planted, they exceedingly adorn and set forth the said upper garden, which trees, one with another, we value to be worth in the whole £30. 15s. [Cypress trees.]

There are also in the said higher and lower level an hundred and nineteen Cherry trees, well planted and ordered, and of a great growth in themselves, the fruit whereof cannot but be of a great yearly value; which trees we value to be worth £29. 15s. [Cherry trees.]

There are also in the said higher and lower level one hundred and fifty fruit trees, of divers kinds of apples and pears, pleasant and profitable; these trees we value to be worth £37. 10s. [Fruit trees.]

There are growing to the walls of the said Upper Garden, fifty three wall fruit trees of divers sorts of fruit, as apricots, may cherries, duke cherries, pear, plums, boone crityans,* french pears, and many other sorts of most rare and choice fruits; which trees, one with another, in the whole we value at £13. 5s. [Wall Fruits.]

In and about the said upper garden there are thirteen muskadine Vines, well ordered and planted, bearing very sweet grapes, and those in abundance at the season of the year; which we value to be worth £3. 5s. [Vines.]

* = bon chrétiens, pears.

Fig trees.	There also are in the said upper garden two other fair Fig-trees, well planted and ordered, which we value to be worth 10s.
Box borders.	The borders of box, rosemary, corants, and the roots of flowers and herbs belonging to the said upper garden, and not herein before valued, we estimate to be worth £27. 17s. 6d.
The Hartichoke garden.	There is one parcel of land belonging to the said upper garden, containing forty four perches of land, called the Hartichoke Garden, lying on the west end of the said lower level; unto which there are 12 steps of descent; the ground whereof is ordered for the growth of hartichokes, the value and contents whereof are comprised in the foresaid yearly value and admeasurement of the said upper garden; but the roots and plants of hartichokes therein now growing and planted we value at £1. 10s.

There are in the said Hartichoke Garden five very handsome Bay trees, which we value to be worth £1.

The Phesant Garden.	And also of one parcel of ground adjoining to the North and East wall of the Oringe Garden, commonly called the Phesant Garden, severed from the Park with a pale of deal boards of 10 foot high; within which is one phesant house, boarded within and without, containing 6 rooms, tiled overhead, and also one shed, tiled, containing 4 rooms, wherein the phesant keeper used to live and lodge; one great partition of deal boards, ten foot high and fifty yards long; twenty partitions of lattices, sixty three young sicamore trees, two oaks, two ash trees, three birch trees, ten fruit trees, and a descent of twenty three steps of stone; all which we value to be worth £26. 13s. The Phesant garden contains upon admeasurement one acre, — roods, and 5 perches, [and] is worth per annum £1.
The Vineyard Garden.	And also of one other garden called the Vineyard [Garden], adjoining to the foresaid upper or great girden upon the East side thereof, and severed from it with a brick wall of ten foot high, and also severed from Wymbledon Park with a brick wall ten foot high upon the east side thereof, and severed from the highway or lane leading from Wymbledon town to the Iron Plate Mills with a brick wall of nine foot high upon the South side thereof, and from the Kitchen garden with another wall of bricks of ten foot high on the West side thereof, containing upon admeasurement ten acres, one rood, twenty-three perches; worth per annum £10. 5s.

MEMORANDUM, that the said Vineyard Garden is divided into twelve several triangles, inclosed within four fair walks or allies, twenty three foot broad, lying round the said garden, two whereof are gravelled walks, and the other two grass plots. Eight of the foresaid twelve triangles make in themselves one square, in the middle whereof, is one fair round or circle of gravelled earth, in the centre whereof stands one Lime tree, having eight several walks or allies, 23 foot broad, running cross and angular ways, answerable to the foresaid eight triangles; the insides of which eight walks or allies are planted with Lime trees, and other young and well planted trees and borders of Currant trees and Respass* trees. The other four triangles, having angular and cross walks within them, though not so fully completed as the other eight triangles, make one square, and, being reduced to a regular form with the other eight triangles, make a very complete garden plot. Within which said twelve several triangles there are growing five hundred and seven fruit trees of divers sorts and kinds of fruits, pleasant and profitable, which we value, one tree with another, in the whole at £83. 11s.

There are also one hundred forty four Lime trees, very well planted and ordered, which, growing in a regular form in the insides of the said triangles, are a great grace and special ornament to the whole garden; which Lime trees we value, one tree with another, in the whole at £28. 16s. Lime trees.

The insides of three of the outward walks or allies are of latticed rails, upon which lattices there are growing one hundred and six trees of divers kinds of wall fruit, which one with another we value to be worth £10. 12s. Wall Fruit.

In the inside of the fourth outward walk or alley are sixteen quince trees, well planted and ordered, worth £2. 13s. Quince trees.

And also upon the out borders there are growing thirty eight fruit trees of pears and cherries, worth £3. 16s. [Fruit trees.]

There are growing upon three of the walls of the said Vineyard Garden two hundred fifty and four trees, of divers special sorts and kinds of wall fruits, as apricots, pears, pear plums, may cherries, boone critians, and divers other kinds of fruits, both curious for taste and variety, and very profitable for use; the trees, being very well planted and ordered, we estimate to be worth, one tree with another, in the whole, at £84. 13s. 4d. Special Wall Fruit.

There are also forty six Sicamore trees, growing along the fourth wall of the said Vineyard Garden in a regular form; which wall standing to the highway or lane, the said trees are a great Sicamore trees.

* Raspberry.

ornament to that part of the Vineyard Garden; which we value to be worth £7. 13s. 4d.

Dutch Elms. There also are seven Dutch Elms growing in some of the borders of the said eight triangles in a regular form, which we value to be worth £1. 15s.

Coran trees, &c. There are in the said Vineyard Garden, divers neat and handsome borders of coran trees, respasses, strawberry beds, roots, flowers, and herbs, all very well ordered, which we value to be worth £5.

Garden Houses. There are also in the said Vineyard Garden, two little garden, summer, or shadow houses, covered with blue slate, ceiled and benched and floored with brick; the one standing in the wall at the end of the walk that leads in a line diametrically opposite to the hall door of the said Manor or Mansion House, and very much graces that walk; the other, standing in the East wall of the said Vineyard garden, at the end of the walk or alley that leads up the middle of the Vineyard, from East to West: the materials of which two garden houses we value to be worth £14.

Rollers. There are in and belonging to the said Vineyard Garden, two rollers of stone with very large and handsome frames of Iron; and also there are belonging to that said Oringe and Upper Garden 6 rollers of stone, fitted as aforesaid, worth in all £16.

The Kitchen Garden. And also of one other garden called the Kitchen Garden, lying and being between the said Vineyard Garden and the highway or lane leading from the town of Wymbledon unto the Iron Plate Mills, and fenced with a pale upon the North west and South west side thereof, and with the South west wall of the said Vineyard garden on the North east side thereof, containing upon admeasurement two roods and twenty six perches of ground, worth per annum £1. 10s.

MEMORANDUM, that in the said Kitchen Garden there are forty trees of very good growth, and pleasant wall fruits, well planted and ordered, which we value (one tree with another) in the whole at £10.

[Laurels] There are also ten Laurel trees, well planted and ordered, which we estimate to be well worth in the gross £1. 10s.

Arbutis tree. There is also one fair tree, called the Irish Arbutis, standing in the middle of the said Kitchen Garden, very lovely to look upon, worth £1. 10s.

There are also thirty eight Cherry trees well planted and

ordered, in the said Kitchen Garden, which we value one with [Cherry another to be worth in the whole the sum of £4. 15s. trees.]

There are also in the said Kitchen Garden very great and [Borders large borders of Rosemary, Rue, White Lavender, and great variety of Rosemary, of excellent herbs, and some choice flowers, and in the South east &c.] end of the said Kitchen Garden there is a Muskmilion* ground, trenched, manured, and very well ordered for the growth of Mus[k]milions; which borders, herbs, flowers, and Mus[k]milion ground we value to be worth £3.

MEMORANDUM, that there is one door belonging to the said Kitchen Garden, opening into the Vineyard Garden, and one other door which opens into the highway or lane that leads from Wymbledon town to Wymbledon Churchyard.

The brick walls of all the gardens aforesaid and of the courts Walls. hereafter mentioned do contain one hundred and seventy pole or square rod of wall, at 16 foot and $\frac{1}{2}$ to the pole, which we value to be worth £3. per rod, in toto, £510.

The rest of the Survey relates to the Courts, ascents, woodyard, dairy house, slaughterhouse yard, the site, the paddock, the Brewer's close, barns, Wymbledon Park, a Dutch barn, deer, timber trees in the Park, paddock, &c. (valued at £2174. 0s. 6d.); springs and coppices of wood (£2020. 3s. 10d.); fishponds, Harpham's farm, a dovecote, meadow called the Great Bitterns, Wymbledon Common, Putney Common, Moreclack Common, pollard trees growing on the said Commons (£500.), &c.

It is signed by Hu: Hindley, John Inwood, John Wale, and John Webb, and examined by William Webb, Surveyor General.

PARLIAMENTARY SURVEYS,

HERTFORD, No. 26.

SURVEY OF THEOBALDS.†

EXTRACTS from the Survey of the Manor of Theobalds, April, 1650.
House, rooms, galleries, &c.

THE Pheasant Garden.—A long description of a house called the Coale Courte or Scaldinge House, &c.: "which said house adjoyneth unto an orchard or garden called by yᵉ name of the

---

* Melons.   † Transcribed from the original MS. in the Record Office.

Pheasant garden, wch conteynes one Roode ten poles and 4 primes, and is excellentlie well planted with wall trees, (vizt) 7 Figg trees, 4 Cherrie trees, and one Roase marie tree, 2 Vines, fower Peach trees, 5 Apricock trees, one Peare tree, six Damas and Damson Plumme trees, one Currant trée, and one Bay tree, planted in ye middle with Gooseberrie trees, and other younge fruite trees, and a bricke wall aboute ye same, abuttinge North on ye passage leadinge east from ye Laundrie," (&c.) Value of the house, court, and garden, £9. 10s.

The Laundrie house,—rooms, barn, stables, &c. Alsoe one passage or way now used as a Garden, lyinge on ye West parte of the afforesayd house called the Laundrie house, and leadeth from the house to ye said garden, called ye Laundrie Garden, conteyninge in length 11 perches and a halfe, and in breadth on ye East parte two perches and a halfe of land, and on the West parte one pearch; and there is planted on ye North side of the walls, fower Vine trees, one Almond tree, 3 Plumme trees, 5 Barberrie trees, and on ye other side Rose trees, and ye middle dugg up for Inions, lettice, and ye like. And at ye west end of ye same one Doore way goeinge into another garden called by ye name of the Laundrie garden, conteyninge 3 roodes and fower pole, compassed aboute with a high brick wall, consistinge of one streight gravelled walke, betweene the Bricke wall and the hedge or Rainge of Gooseberrie trees and Rose trees, with two stepps discendinge into ye middle of ye garden; and round ye garden are several wall trees planted, (vizt) 5 aprecock trees, 11 peach trees, 28 vines, 55 cherrie trees, bearinge choyce and rare cherries; also 12 bay trees, with divers other trees; as also a summer or shaddow house standinge in ye middle of the affores^d garden, seated round, and built turratt fashion, and covered with slatt, with a nurcerie in ye middle of ye garden, and some apple and peare trees, wth divers other small stockes and younge plantes, moted round.

The Privie Garden.—One other Garden called the Privie Garden, alias Kitchen Garden, conteyninge 17 pole, lyinge betweene ye affors^d Garden on the east and Theobaldes Parke on ye west, wth a pleasant gravelie walke lyinge betweene the wall and a hansome quicksett hedge cutt into formes, planted in the middle of the hedges with 28 cherrie trees, goeinge East, West, and North of ye s^d garden, lyinge 8 stepps high in ascent from yᵉ middle of the garden; and yᵉ next walke 8 stepps discendinge into a levell greene grasse walke, betweene yᵉ afforesᵈ hedge, standinge 8 stepps high, and another quick sett hedge wᶜʰ goeth round yᵉ Garden, with a square knott in yᵉ middle of yᵉ Garden turned into a compleate fashion and shape, with 3 ascents, boorded and planted with Tulipps, Lillies, Piannies, and divers other sorts of flowers. Also yᵉ knott is compassed aboute with a Quadrangle or square squadron Quicksett hedge of white thorne and privett of nine foote in height, cutt into a compleate fashion, wᵗʰ fower round arbors with seates in yᵉᵐ in each corner, wᵗʰ two Doorewayes betweene each arbor, in all the fower sides, and betweene yᵉ two Doore wayes in each side runs out a Roman T: made of yᵉ same sort of hedginge, and of the same height. The head of everie T poyntes to yᵉ 3 paire of staires, discendinge downe in to yᵉ Levell from yᵉ gravell walke. At the angle corners of the outside low squadron hedge, is planted at each corner in yᵉ hedge 7 faire Cherrie trees. Also A Dooreway in yᵉ sayd garden, leading into yᵉ Mulberrie walke. Wᶜʰ sayd garden is walled round, and there is growinge to the walls, 25 Apricock trees, 3 figg trees, fower plumme trees, one peach tree, and one cherrie tree. All wᶜʰ sayd house and gardens (except yᵉ 5 Roomes under yᵉ afforesayd Divided gallerie) are in yᵉ occupacion of Mr. John Southworth, and is worth per annum 20ˡ.

The Greate Garden.—One Garden called the Greate Garden, adioyninge North on yᵉ afforesᵈ Cloyster lyinge under yᵉ Kinges Presence Chamber, and others, incompassed East, South, and West with a good brick wall, and North with the Capitall house; wᶜʰ Garden conteynes by admeasurem^t 7 acres, 2 roodes, 37 pole, and is worth per annum £14. Memorandum, in the sᵈ Garden there are nine large compleate squares or knotts lyinge upon a Levell in yᵉ middle of yᵉ sᵈ Garden, whereof one is sett forth with box borders in yᵉ likenesse of yᵉ Kinges armes, verrie artificiallie and exquisitely made; one other plott is planted with choice flowers; the other 7 knotts are all grasse knotts, handsomely turfed in the intervalls or little walkes. All the afforesᵈ knotts are compassed aboute with a Quicksett hedge of White thorne, and privett, cutt into a handsome fashion; and at everie angle or corner standes a faire cherrie tree of a greate groth, with a Ciprus in the middle of most of the knotts, and at some of the corners; wch knotts, Quicksett hedge, and ye Flower Rootes we value to be worth £5.

The Marble Fountaine.—In the middlemost knott of the affores^d nine knotts as alsoe in the middle of ye Garden standeth a large handsome Fountaine of white Marble standinge upon 3 stone stepps, (&c.)

In the middle of two of the affores^d greene knotts, . . . standeth two figures of wainscott well carved . . .

In the South side of wch sayd Garden, and in the middle of a gravelled walke . . . a faire banquetinge house built upon stone pillars, in ye fashion of a halfe round (&c.).

There are growinge to the walles of the Capitall house side in the garden 5 Apricocke trees and 14 Muscadine Vines well ordered and planted, 4^l. 15^s.

51 Ciprus trees, 12^l. 15^s.

25 Cherrie trees, 7^l. 10^s.

240 Lyme, Elm, and Sycamore trees, worth £70.

12 Black Cherry trees, 3^l.

There are also descriptions of eight gravelled walks, two Garden houses, two small rooms or seats, Cherrie trees, the Thorne hedge, other thorn hedges, Black Cherrie trees, Fruite trees, Bay trees, two seats, two stone crosses, 6 stepps, Vine trees and Barberrie trees, the White thorne hedge, the Maze Garden, the Tripesa (of 8 triangles made of white thorne), Fruite trees, a doore leadinge into ye longe greene Mulberrie walke, the Fountaine Courte, the Middle Courte, the Diall Courte, the Base Courte, the Dovehouse Courte, the Stonie close, 111 trees each side of a walk, the 14 Elms, the Greene Walke, Mulberrie trees (72 worth 8^l. 12^s.) the Orchard, the Dovehouse.

This survey is signed by Raphe Baldwyn, Ric. Heiwood, Rowland Brasbridge, and John Brudenall, and examined by William Webb, Surveyor General.

BIBLIOGRAPHY

OF WORKS ON

ENGLISH GARDENING.

PRINTED BOOKS,

Arranged Chronologically under the names of Authors or Translators and under the date of the first edition of their earliest work;—under the title of the Book and date of the first edition, when the writers' names are unknown.

An asterisk affixed to any article implies that the book has not been seen by or for me.

1516 THE GRETE HERBALL. Imprented at London in Southwark by me Peter Treveris . . MDXVI, the xx day of June. Folio *
> First Edition. It was described by Ames in his Typographical Antiquities; but there is no record of its having been seen by anyone since.

The grete herball . . which is translated out ye Frensshe into Englysshe . . With the mark of Peter Treveris. Folio. Undated *
> Described by Hazlitt as printed in " 1525-6 ;" being certainly not earlier than 1525, and apparently anterior to the dated issue of 1526.

The grete herball . . Imprentyd at London in Southwarke by me Peter Treueris . . MDXXVI the xxvii day of July. Folio
> Several copies are extant.

The grete herball. MDXXVII. 18 April *
> Such an edition is described by Ames, as printed by Treveris for Laurence Andrew.

The grete herball . . Peter Treueris . . MDXXIX, the xvii day of Marce. Folio

The grete herball newly corrected. Londini, in edibus Thome Gybson. Anno MDXXXIX. Folio

The greate Herball . . London, Jhon Kynge, MDLXI. Folio

1523 [FITZHERBERT'S HUSBANDRY] A newe tracte or treatyse moost profytable For All Husbandmen . . Imprinted . . by Rycharde Pynson [in or before 1523] Small 4to.   *

> There is a copy of this first edition in the Bodleian. In Pinson's edition of Sir Anthony Fitzherbert's Boke of Surveying, printed in 1523, this book is mentioned as having been already published. Its date cannot therefore be later than that year. The authorship by Sir Anthony Fitzherbert is considered doubtful, and in the B. M. Catalogue it is suggested that John Fitzherbert may have been the writer; but it is clear that Pinson and Berthelet both regarded the judge as the author. For a full discussion of the subject, it will be well to consult the reprint of the Treatise, which was edited by W. W. Skeat for the English Dialect Society in 1882.

—— another edition. Small 4to. in the B. M. supposed to have been printed about 1525.

—— another edition. Thomas Berthelet, 1534. Small 8vo.

—— another edition, by the same printer in 1548. Small 8vo.

Various other editions exist.

1525 W. C. [WALTER CARY?] Here begynnyth a newe mater, the whiche sheweth and treateth of ye vertues & proprytes of herbes, the whiche is called an Herball. London, R. Banckes, 1525. Small 4to.

—— another edition. Robert Redman [1530?]. Small 8vo.

—— A boke of the propreties of Herbes called an herball . . Also a generall rule of all maner of Herbes drawen out of an auncyent booke of Phisyck by W. C. W. Copland for J. Wyght [1552?]. Small 8vo.

> The letters W. C. are supposed by the B. M. cataloguers and by the older writers to mean Walter Cary, whose name appears on the Farewell to Physicke, printed by Denham in 1583; but others, including the writer in the Dict. of Nat. Biogr., take them to mean William Copland the printer; and it is to be remembered that they occur for the first time in Copland's edition.

—— other editions by Skot and Kytson: both undated

1527 JEROME OF BRUNSWICK-ANDREW. The vertuose boke of Distyllacyon of the waters of all maner of Herbes . . compyled by . . Master Iherom bruynswyke. And now newly Translate out of Duyche into Englysshe [by Laurence Andrew]. Imprinted at London . . by me Laurens Andrewe . . Mccccxxvii. Folio

<small>The translator (who was also the printer) gives his name in the Prologue.

(An edition of 1525 is mentioned in Herbert's Ames; but it had probably no existence.)

There were two issues of the book in 1527, but only the leaves at beginning and end were reprinted. The body of the book is identical in each. The first issue is dated the 17 April; the second is dated the 18 April. They are both in the British Museum.

See also under date 1561.</small>

1530?] MACER-LINACRE. Macer's Herbal practysid by Doctor Linacro. Translated out of laten into Englysshe . . R. Wyer. London . (About 1530). 8vo.

—— A newe Herball of Macer . . R. Wyer. (About 1535.) 8vo.

1538 WILLIAM TURNER. Libellus de Re Herbaria novus, in quo herbarum aliquot nomina greca, Latina & Anglica habes . . Apud J. Byddellum, Londini, 1538. 4to. eight leaves

<small>A reprint in facsimile was edited by Mr. B. D. Jackson in 1877, with a life of Turner. Privately printed.</small>

—— The names of herbes in Greke, Latin, Englishe, Duche & Frenche wyth the commune names that Herbaries and Apotecaries use. London, John Day and William Seres [1548]

—— A new Herball, wherein are conteyned the names of Herbes in Greke, Latin, Englysh, Duch, Frenche, and in the Potecaries and Herbaries Latin. London, S. Mierdman, 1551. Folio

—— The seconde parte of W. Turners herball . . Here unto is joyned also a booke of the bath of Baeth in Englande . . Collen, A. Birckman, 1562. Folio

—— The first and seconde partes . . also a Booke of the bath of Baeth. Collen, A. Birckman, 1568. Folio

1540 [ANDREW BORDE] The boke for to Lerne a man to be wyse in buylding of his howse . . The boke for a good husbande to lerne. Robert Wyer, no date [about 1540]. Small 8vo.

    Republished a year or two later by the same printer as part of the Compēdyous Regyment bearing Borde's name.

1557 THOMAS TUSSER. A hundreth good pointes of husbandrie. R. Tottell, London, 1557. In verse

    (The copy in the British Museum is presumably unique.) Several varying and augmented editions of the " Hundreth good Pointes " were issued between 1561 and 1571. The first edition of the " Five Hundreth Points " (Tusser's enlargement of the former) came out in 1573.

—— - Five hundreth points of good husbandry united to as many of good huswiferie . . R. Tottel, London, 1573 . . 2 parts, small 4to.

    Reissued at least twice (in 1576 and 1577) before the appearance of the final and complete edition in 1580.

—— Five hundred pointes of good Husbandrie . . Henrie Denham, London, 1580. Small 4to.

    The parent edition of all which followed, and the last which was published by Tusser himself.

    The reprints were numerous from that time onwards.

—— Fiue hundred pointes of good husbandrie. The edition of 1580 collated with those of 1573 and 1577 . . with a reprint of "A Hundreth good pointes . . 1557." Edited by W. Payne and Sidney J. Herrtage. Early English Dialect Society, London, 1878. 8vo.

1561 (BRAUNSCHWEIG-HOLLYBUSH) A most excellent and Perfecte Homish Apothecarye, or homely physick booke . . Translated out of the Almaine speche . . by Ihon Hollybush . A. Birckman, Collen, 1561. Folio

    It has been alleged that the name of John Hollybush in this book is merely pseudonymous for Miles Coverdale (as it was in the Latin-English New Testament of 1538).

1563 THOMAS HILL. A most briefe and pleasaunt treatyse, teachynge howe to dress, sowe, and set a garden . . T. Marshe, London, 1563. Small 8vo.

    Probably the earliest appearance of "the Arte of Gardening." A second edition is unknown.

—— The proffitable Arte of Gardening, now the third tyme set fourth . . To this annexed, two . . treatises . . T. Marshe, London, 1568. Small 8vo.

—— The proffitable Arte . . Whereunto is newly added a treatise of the Arte of graffing and planting of trees. H. Bynneman, London, 1574. Small 4to.

> This edition bears upon its title the statement " now the third time set forth," just as in the edition of 1568.
>
> Later editions were printed in 1579, 1586, 1593, and 1608, and in the last two of these a " Treatise on Bees " is included.
>
> The author was probably the same person as Didymus Mountain who wrote the Gardener's Labyrinth.

DIDYMUS MOUNTAIN [Thomas Hill] The Gardener's Labyrinth: containing a discourse of the Gardener's life . . . [completed by Henry Dethick]. H. Bynneman, London, 1577. Small 4to.

> Later editions appeared in 1594 and 1608.

1570 MATTHIAS DE L'OBEL Stirpium Adversaria Nova . . Londini, 1570-71. Folio

> Written in collaboration with Petrus Pena, under whose name it is usually placed.

—— Accessit altera pars . . T. Purfoot, Londini, 1605 . . 3 parts in 1, folio

—— Plantarum seu Stirpium Historia cui adnexum est Adversariorum volumen. Antwerp, Plantin, 1576. Folio

—— Plantarum seu Stirpium Icones. Antwerp. Plantin, 1581. Oblong 4to. (Reprinted in 1591.)

1572 LEONARD MASCALL. A Booke of the Arte and maner, howe to plant and graffe all sortes of trees . . . by one of the Abbey of Saint Vincent in Fraunce, with an addition of certaine Dutch practices, set forth and Englished by L. Mascall. Henrie Denham for John Wight [1572]. 4to.

> Other editions, in 1575, 1580?, 1582, and 1592.

1574 REGINALD SCOT. A perfite platforme of a Hoppe Garden . . London. Henrie Denham, 1574. Small 4to.

> Reprinted in 1576 and 1578, " newly corrected and augmented."

1577 HERESBACH-GOOGE. Foure Bookes of Husbandry, collected by M. Conradus Heresbachius . . Newely Englished, and increased, by Barnabe Googe. London, Richard Watkins, 1577. Small 4to.

<small>Other editions in 1578, 1586, 1614, 1631, and 1658.</small>

MONARDES-FRAMPTON. Joyfull Newes out of the newe founde worlde, wherein is declared the rare and singuler vertues of diuerse . . Hearbes . . Englished by John Frampton . . London. W. Norton, 1577. Small 4to.

<small>It comprises a description and a woodcut of the Tobacco-plant.

The Spanish author was Nicolas Monardes, frequently styled Dr. Monardus in the editions of the translation.

Reprinted in 1580 and 1596, with some additions.</small>

1578 (DODOENS-LYTE) A Niewe Herball or Historie of Plantes . . set fowrth in the Doutche or Almaigne tongue by that learned D. Rembert Dodoens . . Nowe first translated out of French into English by Henry Lyte . . Antwerpe, 1578. Folio

— another edition. London, 1586. 4to.

— corrected and amended. London, 1595. 4to.

—— another edition. London, 1619. Folio

<small>The English version was made from the French translation executed by Charles de l'Ecluse (C. A. Clusius).

See also 1606 : Dodoens-Lyte-Ram.</small>

1579 WILLIAM LANGHAM. The Garden of Health, conteyning the sundry rare and hidden vertues and properties of all kindes of Simples and Plants gathered by long experience and industrie. London, 1579 . . Small 4to.

<small>Reprinted in 1633.</small>

1583-7 RALPH HOLINSHED. Chronicles of England, Scotland, and Ireland . . London, 1586-87. 3 vols. folio

<small>This second edition (the first was published in 1577) contains in the description of England (by William Harrison), prefixed to the Chronicle, some references to gardens and orchards in England.</small>

1594 SIR HUGH PLATT. The Jewell House of Art and Nature. Conteining diuers rare and profitable Inuentions, together with sundry new experiments in the Art of Husbandry, Distillation, and Moulding . . London, Peter Short, 1594. 4to.

— Floraes Paradise, beautified and adorned with sundry sorts of delicate fruites and flowers, by the industrious labour of H. P. knight . . London, H. L. for William Leake . . 1608. Small 8vo.

—— the same work, reprinted under the title of The Garden of Eden. 1653. Small 4to.

<small>This edition was brought out, after the author's death, by his kinsman Charles Bellingham, who signs the dedication.</small>

1596 JOHN GERARD or GERARDE. Catalogus Arborum, fruticum ac plantarum tam indigenarum quam exoticarum in horto Gerardi nascentium. Londini, ex off. Roberti Robinson. 1596. Small 8vo.

<small>The copy in the British Museum is probably unique.</small>

— reprinted in 1876 under the title of "A Catalogue of Plants " . . . with notes. B. D. Jackson . .

— — — The Herball, or Generall Historie of Plantes. London, J. Norton, 1597. Folio

— — reprinted with considerable augmentation and improvement by Thomas Johnson in 1633.

<small>This immense work of standard character contains 19 preliminary leaves (including the engraved title), 1631 pp. of text, and 46 pp. of tables, etc. It has several hundreds of woodcut illustrations. Most of them appeared in the original edition of 1597, but a large number was added by Johnson in 1633.</small>

1599 DUBRAVIUS— . . A New Booke of good Husbandry . . Written in Latin by Janus Dubravius and translated into English at the speciall request of George Churchey . . London, William White. 1599. Small 4to.

<small>The translator's name is not known.</small>

GARDNER'S KITCHEN GARDEN. Profitable Instructions for the Manuring, Sowing, and Planting of Kitchen Gardens . . Edw. Allde for Edward White, 1599. Small 4to. *

<small>This first edition is so described by Lowndes, and is mentioned in Platt's Garden of Eden ; but no copy has been traced.</small>

1599 —— Profitable Instructions for the Manvring, Sowing, and Planting of Kitchin Gardens . . Imprinted at London by Edward Allde for Edward White . . 1603. Small 4to. *

> This second edition is so described *de visu* in Hazlitt's third Collection of Bibliographical Notes. He calls the author Richard Gardiner; under which name he seeks to indicate Richard Gardner of Shrewsbury, a dyer and draper, who is described as a public benefactor of that town, in Owen and Blakeway's Shrewsbury.

1600 ESTIENNE AND LIEBAULT-SURFLET. Maison Rustique or the Covntrie Farme, compiled in the French tongue by Charles Steuens and John Liebault, and translated by Ric. Surflet. E. Bollifant for B. Norton, 1600. Small folio

—— another edition, augmented by Gervase Markham. 1616. Folio

1601 JOHN TAVERNER. Certaine Experiments concerning Fish and Fruit . . Printed for William Ponsonby . . 1601. Small 4to.

1804 N. F. THE FRVITERER'S SECRETS . . London, printed by R. B., solde by Roger Iackson, 1604. Small 4to.

—— reprinted with a different title in 1608 and 1609. *See* below

1606 DODOENS-LYTE-RAM. Rams little Dodeon. A briefe Epitome of the new Herbal or History of Plants . . lately translated into English by Henry Lyte . . abridged by William Ram. London, Simon Stafford, 1606. Small 4to.

See 1578: Dodoens-Lyte.

1607 DE SERRES-GOFFE. The Perfect Vse of Silk-Wormes and their benefit . . done out of the French originall of Olivier de Serres, Lord of Pradel, into English by Nicholas Goffe . . London, Felix Kyngston, 1607. Small 4to.

1608-9 N. F. THE HUSBANDMAN'S FRUITFULL ORCHARD, shewing diuers rare new secrets for the true Ordering of all sortes of fruite . . London, Roger Iackson, 1609. Small 4to.

<small>This work had already appeared in 1608. It is the same book as the "Fruiterer's Secrets" of 1604.</small>

1609 W. S. INSTRUCTIONS for the increasing of Mulberie Trees, and the breeding of Silke-wormes for the making of Silke in this Kingdome. Whereunto is annexed his Majesties Letters to the Lords Liefetenants . . tending to that purpose. London, E. A. for E. Edgar, 1609. 4to.

1612 R. C. AN OLDE THRIFT newly revived . . the manner of planting, preserving, and husbanding yong Trees . . London, W. S. for Richard Moore. 1612. 4to. *

<small>Described by Hazlitt *de visu* in the Collections and Notes.</small>

1613 ARTHUR STANDISH. New Directions of Experience to the Commons Complaint . . for the planting of Timber and Fire-wood . . 1613. 4to.

    GERVASE MARKHAM. The English Hvsbandman. The First Part: contayning the Knowledge of the true Nature of euery Soyle . . . Together with the Art of Planting, Grafting, and Gardening . . By G. M. . . London, T. S. for John Browne. 1613. 4to.

—— The Second Booke of the English Husbandman. Conteyning the ordering of the Kitchen-Garden . . London, T. S. 1614. 4to.

<small>The work was reprinted "enlarged by the author" in 3 parts 4to. in 1635.</small>

—— Maison Rustique. 1616. *See* 1600: Estienne and Liebault-Surflet.

—— The Country Housewifes Garden . . Together with the Husbandry of Bees . . with diuers new knots for gardens, by G. M. 1617. 4to.

<small>The same book was reissued with Lawson's New Orchard in 1618 and alone in 1620 and 1623.</small>

1613 —— Markhams Farewell to Husbandry; or, the inriching of all sorts of barren and sterill grounds . . I. B. for Roger Jackson, 1620. 4to.

—— The Inrichment of the Weald of Kent . . G. P. for R. Jackson, 1625. 4to.

—— The Whole Art of Husbandrie, by C. Heresbach, translated by B. Googe, enlarged by Gervase Markham, 1631. 4to.

_{See 1577, Heresbach-Googe.}

—— A Way to get Wealthe . . 1638-31-38. 4to.

_{A compilation of some of Markham's agricultural works, already issued separately.}

—— The Countrymans Recreation, or the Art of Planting, Grafting, and Gardening . . London, 1640. 4to.

_{Reprinted in 1653 and 1654.}

1614 THE MASKE OF FLOWERS . . upon Twelfth Night, 1613 . . at the marriage of the Earle of Somerset . . N. O. for R. Wilson. 1614. Small 4to.

_{This poetical piece finds a place here as containing, in the *Stage-directions*, an admirable description of an Elizabethan garden.}

1615 PASSE-WOOD. A garden of Flowers, wherein . . is contained a . . discription of al the flowers contained in these foure followinge bookes . . translated out of the Netherlandish [by E. W. or rather T. Wood?]. S. de Roy for Crispin de Passe, Utrecht. 1615. 2 parts, oblong 4to. 163 plates

1618 WILLIAM LAWSON. A new Orchard and Garden. Or the best way for planting . . With the Country House-wifes Garden . . B. Alsop for R. Jackson, 1618-17. 2 parts in 1 vol. 4to.

— second edition. I. H. for Roger Iackson, 1623. 4to.

—— Now the third time corrected and much enlarged . . Whereunto is newly added the art of propagating plants (by Simon Harward). I. H. for F. Williams, 1626. 4 parts in 1 vol. small 4to.

_{Reissued in 1638 in Markham's Way to get Wealth.}

1624 SIR HENRY WOTTON. The Elements of Architecture . . London, John Bill, 1624. 4to.

1625 FRANCIS BACON (VISCOUNT ST. ALBANS) The Essayes or Covnsels, Civill and Morall, of Francis Lo. Verulam Viscount St. Albans. Newly enlarged. 4to.
This is apparently the first edition in which the Essay on Gardens was printed.

—— Sylva sylvarum : or a Naturall Historie . . Published after the Author's death. By William Rawley . . London, 1627. 4to.
The edition of 1676 is called tenth edition.
Rawley was Bacon's chaplain and secretary.

1626 ADAM SPEED. Adam out of Eden . . London, 1626. 8vo. *
Watt and Allibone give this date to the first edition, and treat the volume of 1659 as a reprint. Johnson calls him Adolphus Speed.

—— Adam out of Eden, or an abstract of divers excellent Experiments touching the advancement of Husbandry. London, printed for Henry Brome, 1659. 8vo.

—— The Reformed Husbandman. 1651
This is set down elsewhere as a work by Hartlib.

SIMON HARWARD. The Art of Propagating plants—printed with the third edition of William Lawson's New Orchard and Garden. *See* 1618

1629 JOHN PARKINSON. Paradisi in sole. Paradisus terrestris, or a Garden of all sorts of pleasant flowers . . with a kitchen garden . . and an Orchard. H. Lownes and R. Young, 1629. Folio
Reissued in 1635, with an additional letterpress title bearing that date.

—— the second impression, corrected and enlarged. 1656. Folio

—— Theatrum Botanicum. The Theater of Plants. Or an Vniversall and compleate Herball . . Tho. Cotes, 1640. Folio

THOMAS JOHNSON, M.D. Iter Plantarum investigationis ergo susceptum . . in Agrum Cantianum . . 1629 Julii 13. Ericetum Hamstedianum sive plantarum ibi crescentium observatio . . [Londini, 1629.] 4to.

—— Descriptio Itineris Plantarum investigationis ergo suscepti in agrum Cantianum anno Dom. 1632, et enumeratio plantarum in Ericeto Hampstediano . . . T. Cotes, 1632. 8vo.

1629 —— Mercurius Botanicus: sive Plantarum gratiâ suscepti Itineris anno MDCXXXIV Descriptio . . T. Cotes, 1634-41. 3 parts 8vo.

—— The Herball . . 1633. Folio. *See* 1597, JOHN GERARD

1639 GABRIEL PLATTES. A Discovery of Infinite Treasure, hidden since the Worlds beginning, whereunto all men . . are . . invited to be sharers with the discoverer. G. P. . . London. Printed for J. E. . 1639. 4to.

—— A Discovery of Subterraneall Treasure, viz. of all manner of Mines and Mineralls . . I. Okes for J. Emery. 1639. 4to.

—— [another edition] Whereunto is added a real experiment whereby every ignorant man . . may try what colour any berry, leaf or wood will give. . . 1679. 4to.

—— other editions

—— The Profitable Intelligencer . . containing many secrets and experiments [with a view to improvement of Agriculture . .]. 1644. 4to.

1640 C. DE SERCY. The Expert Gardener, or a Treatise concerning Gardening and Grafting. London, 1640. 4to. *
Mentioned by Johnson and Watt, but not recorded elsewhere.

(1645?) ISAAC DE CAUS. Wilton Garden. [Etchings of the Flower-beds, Fountains, Arbours, etc. of the Earl of Pembroke's Gardens at Wilton.] About 1645. 4to.

—— New and Rare Invention of Water-Works shewing the easiest waies to raise Water higher than the Spring . . Translated into English by Iohn Leak. London, 1659. Folio

1645 SAMUEL HARTLIB. Discourse of Husbandrie used in Brabant and Flanders. London, 1645. 4to. *
Mentioned in Watt's Bibliotheca.

—— (second edition). 1650. 4to. *
Mentioned by Weston and Watt.

—— The Third Edition, corrected and enlarged. London, Printed by William Dugard, 1654. 4to. *
: Described by Hazlitt *de visu*.

- Samuel Hartlib his Legacie; or an Enlargement of the Discourse of Husbandry used in Brabant and Flanders . . with Appendix. 1651. 4to.
: Other editions in 1652 and 1655.

—— An Essay for advancement of Husbandry-Learning: or Propositions for the errecting Colledge of Husbandry . . London, Printed by Henry Hills, 1651. 4to.

—— The Reformed Husbandman, or a brief treatise of the errors, defects . . of English husbandrie . . 1651. 4to. *
: This treatise is elsewhere attributed to Adam Speed. *See* 1626. It appears in Watt under both names.

Cornucopia, a miscellaneum of lucriferous and most fructiferous Experiments, observations and discoveries. 1652

—— A Designe for Plentie, by an Vniversall planting of Frvit-Trees. 1652. 4to.
: Also issued without a date, and in 1654.

—— A Discoverie for Division or setting out of Land, as to the best forms. Richard Wodenothe, 1653. 4to.

1648 JACOB BOBART. Hortus Medicus Oxoniensis . . Catalogus Plantarum Horti Medici Oxoniensis Latino-Anglicus et Anglico-Latinus. Oxon. 1648. 8vo. 2 parts in 1
: The title to the English part is "An English Catalogue of the Trees and Plants in the Physicke Garden of the University of Oxford, with the Latin names added thereunto. Oxford, H. Hall, 1648." The book was reproduced by Simon Paulli at Copenhagen in 1653 in his " Viridaria varia regia et academica."

1649 WALTER BLITH. The English Improver, or a New Survey of Husbandry . . London, J. Wright, 1649. 4to.

—— The English Improver Improved, or the Svrvey of Hvsbandry Svrveyed . . The third Impression . . John Wright, 1652. 4to.

(1650 ?) PETER STENT. Book of Flowers, Fruits, Beasts, Birds, and Flies . . 4to. A set of engravings   *

Ascribed by Hazlitt conjecturally to 1660, but it must have preceded the following. Stent and Simpson were two engravers in London about or before the middle of the seventeenth century.

1650 WILLIAM SIMPSON. The Second Booke of flowers, fruicts, beastes, birds, and flies exactly drawne, etc. London, 1650. 4to.

—— another issue. 1661. 4to.

WILLIAM HOW. Phytologia Britannica, natales exhibens Indigenarum Stirpium sponte emergentium Londini. 1650. 8vo.

1652 NICHOLAS CULPEPPER. The English Physitian, or an Astrologico-Physical Discourse of the vulgar Herbs of this Nation. London, Peter Cole, 1652. Folio   *

—— another issue. 1652, without publisher's name. 12mo.*

This is the work popularly known as Culpepper's Herbal. An edition said to be enlarged was printed in 1654, and was the parent of all succeeding issues which have appeared frequently down to the present century.

1653 A BOOK OF FRUITS AND FLOWERS shewing the nature and use of them . . London, 1653.

RALPH AUSTEN. A Treatise of Fruit Trees . . Oxford, 1653. 4to.

—— —— second edition, with the addition of many new experiments. Oxford, 1657. 4to.

—— —— (another edition, to which are added) Observations upon Sir Francis Bacon's Nat. Hist., also directions for planting wood. Oxford, 1665. 4to.

—— Observations upon some part of Sir F. Bacon's Naturall History as it concerns fruit trees, fruits, and flowers. Oxford, 1658. 4to.

This was the first edition of the "Observations," which were afterwards annexed as a second part to the 1665 edition of the "Treatise."

1653 JOHN BEALE. A Treatise on Fruit Trees shewing their manner of Grafting, Pruning, and Ordering, of Cyder and Perry, of Vineyards in England, etc. Oxford, 1653. 4to.

— The Hereford Orchards: a pattern for the whole of England. London, 1657. 12mo.

—— General Advertisements concerning Cyder, etc. London, 1677. 4to.

—— Nurseries, Orchards, Profitable Gardens, and Vineyards encouraged. (by Anthony Lawrence and John Beal). 1677. 4to.

1656 WILLIAM COLES. The Art of Simpling. An Introduction to the Knowledge and gathering of Plants . . London, J. G. for Nath. Brook, 1656. 12mo.

— Adam in Eden, or Nature's Paradise: the History of Plants, Herbs, and Flowers. 1657. Folio

1656 JOHN TRADESCANT. Museum Tradescantianum, or a collection of Rarities preserved at South-Lambeth near London. London, John Grismond, 1656. Small 8vo.

1658 SIR THOMAS BROWNE, M.D. Hydriotaphia, or a Discourse of Sepulchral Urns lately found in Norfolk; together with the Garden of Cyrus, &c. London, 1658.

JOHN EVELYN. The French Gardiner . . . Transplanted into English by Philocepos (i.e. J. E.). London, 1658. Small 8vo.

—— Sylva, or a Discourse of Forest-Trees . . To which is annexed Pomona, or an Appendix concerning Fruit-Trees. &c. London, 1664. Folio

— The English Vineyard Vindicated—*see* 1666, JOHN ROSE.

—— Kalendarium Hortense; or the Gardener's Almanac, directing what he is to do monthly throughout the year. The second Edition . . London, 1666. 8vo.

_{The first edition had been issued as portion of the Sylva in 1664.}

22

1658 —— A Philosophical Discourse of Earth, relating to the culture and improvement of it for Vegetation . . London. 1676. 8vo.

—— Of Gardens. 4 books. First written in Latin verse by Renatus Rapinus. now made English by J. E. London. 1673. 8vo.

—— The Compleat Gard'ner, &c. . . . by J. de la Quintinye . . made English by John Evelyn . . London, 1693. Folio. 2 vols.

Evelyn's " Directions concerning Melons " forms part of Vol. II. See also 1699 ; London and Wise.

—— Acetaria. A Discourse of Sallets. London, 1699 . . 8vo.

1659 ROBERT LOVELL. Παμβοτανολογία, or a Compleat Herbal. Oxford, 1659. 8vo.

—— The second edition, with many additions. Oxford, 1665. 8vo.

Copies of both editions are in the Bodleian.

THOMAS DUCKET. Proceedings concerning the improvement of all manner of land, &c. 1659.

1660 ROBERT SHARROCK. The History of the Propagation and Improvement of Vegetables. by the Concurrence of Art and Nature. Oxford, 1660. 8vo.

—— second edition, much enlarged. Oxford, 1672. 8vo.

—— third edition. London, 1694. 8vo.

JOHN RAY. Catalogus Plantarum circa Cantabrigiam nascentium. Cantab. 1660. 8vo.

—— Appendix ad Catalogum . . Cantab. 1663. 8vo.

These were anonymously published. A second Appendix was printed in 1685.

—— Catalogus Plantarum Angliæ et insularum adjacentium. Lond. 1670. 8vo.

—— Synopsis methodica Stirpium Britannicarum. 1690. 8vo.   *

—— another issue, enlarged. 1696. 8vo.

The Synopsis is an improved edition of the Catalogus of 1670.

—— Catalogus Stirpium in exteris regionibus . . 1673. 8vo.

—— Historia Plantarum generalis. 1686-1688-1704. 3 vols. folio

—— Stirpium Europæarum extra Britanniam nascentium Sylloge. 1694. 8vo.

—— Philosophical Letters . . 1718. 8vo.
> Various other works of this excellent botanist are recorded in Watt's Bibliotheca. His name was originally spelled Wray, but he seems to have dropped the W himself.

1664 JOHN FORSTER. England's Happiness Increased, or a sure and easie remedy against all succeeding Dear Years. By a plantation of the roots called Potatoes. London, 1664. 4to.

LE GENDRE-FORSTER. The Manner of ordering Fruit Trees . . by the Sieur Le Gendre. London, 1664. *
> "Sieur Le Gendre" is a pseudonym for Robert Arnault d'Andilly. The translator was John Forster.

STEPHEN BLAKE. The Complete Gardener's Practice. London, 1664. 4to.

JONATHAN GODDARD, M.D. Observations concerning the texture and similar parts of a Tree.

—— The Fruit Tree's Secrets.
> These treatises were papers read to the Royal Society, and were only printed in Evelyn's Sylva in 1664.

WILLIAM HUGHES. The Complete Vineyard, &c. London, 1665. 4to.

—— The Flower Garden, &c. London, 1671 and 1672. 12mo.

—— The American Physician; or a Treatise of the Roots, Plants, Trees, Shrubs, Fruit, Herbs, etc. growing in the English Plantations . . London, 1672. 12mo.

1665 JOHN REA. Flora, seu de Florum cultura; or a complete Florilege. London, 1665. Folio *

—— Flora, Ceres, et Pomona. 1676. Folio
> An enlarged edition of the preceding. John Rea, gentleman, who was resident at Kinlet, near Bewdley, in Worcestershire, in 1676, is sometimes mistaken for John Ray the learned Divine and Naturalist, but the latter was nearly thirty years younger.

Philosophical Transactions, published by the Royal Society, were begun in 1665
> The following are among the contributors to the early volumes (before 1700) of papers on Botanical or Horticultural Subjects— John Beaumont, James Cunningham, John Evelyn, Hon. Charles Howard, Anthony van Lunwenhock, Martin Lister, Christopher Merret, James Petiver, Robert Plott, John Ray, Richard Richardson, Sir Hans Sloane, John Temple, Ezekiel Yonge.

1666 JOHN ROSE. The English Vineyard Vindicated .. London, 1666. 8vo.
  With a preface by Philocepos, *i.e.* John Evelyn. It was reissued in a third edition in 1675, as an appendix to Evelyn's French Gardiner.

1667 ABRAHAM COWLEY. The Garden (a Poem).
  Printed at the end of Poems by Jeremiah Wells, which were published in 1667, in 8vo.

1669 RICHARD RICHARDSON. De cultu Hortorum, Carmen. London, 1669. 4to. *
  This title is taken from Johnson's list.† Watt gives the date as 1699.

FRANCIS DUDLEY (fourth Lord North). Observations and Advices œconomical. 1669. 8vo.

S. B. [SAMUEL BLAGRAVE, or as some say Billingsly.] "The Epitome of Husbandry (a complete plagiary, the first 181 pages being copied from Fitz-Herbert, and the rest from Mascall, &c.)." 1669 *
  This intitulation and note are taken from Johnson.

JOHN WORLIDGE. Systema Agriculturæ, The Mystery of Husbandry discovered . To which is added Kalendarium Rusticum . . London, 1669. Folio
  Other editions appeared in 1677, 1681, and 1687.

—— Vinetum Britannicum, or a Treatise of Cider . . London, 1676. 8vo.
  Other editions in 1678 and 1691.

—— Systema Horticulturæ, or the Art of Gardening . . London, 1677. 8vo.

1670 ILIFFE. The Compleat Vineyard. 1670 *
  This title is given by Johnson. It may be no more than an issue of William Hughes' book. *See* 1665.

LEONARD MEAGER. The English Gardener, &c. London, 1670. 4to.

—— The Mystery of Husbandry . . to which is added The Countryman's Almanack. 1697. 12mo.

—— The New Art of Gardening; with the Gardener's Almanack. 1697. 12mo.

1672 ROBERT MORISON. Plantarum Umbelliferarum Distributio. Oxon. 1672. Folio

† See under 1820.

—— Plantarum Historia Universalis Oxoniensis. Pars II. Oxon. 1680

—— Ejusdem Pars III, explevit Jac. Bobartius. Oxon. 1699. Folio

This general work on plants incorporates the earlier " Plantarum Umbelliferarum Distributio," but was itself never completed. Only Parts II and III were written.

—— Icones et descriptiones rariorum Plantarum Siciliæ, Melitæ, Galliæ et Italiæ. Oxon. 1674. 4to.

This is a translation of Paolo Boccone's Manifestum Botanicum.

FRANCIS DROPE. A Short and sure Guid in the practice of raising and ordering Fruit Trees. Oxford, 1672. 8vo.

1673 RAPINUS-EVELYN. *See* 1658: Evelyn.

1675 CHARLES COTTON. The Planter's Manual, being instructions for the raising, planting, and cultivating all sorts of fruit trees . . London, 1675. 8vo.

1676 MOSES COOK. The manner of raising, ordering, and improving Forrest- [and Fruit-] trees. . . London, 1676. 4to.

1677 ELSHOLT-SHERLEY. Curious Distillatory, or the Art of Distilling Coloured Spirits, Liquors, Oyls, &c. from Vegetables, by J. S. Elsholt, and Englished by Thomas Sherley. London, 1677. 8vo.

1681 T. LANGFORD. Plain and Full Instructions to raise all sorts of Fruit Trees that prosper in England. London, 1681. 8vo.

—— The Practical Planter of Fruit Trees. London, 1681. 8vo.   *

—— second edition, with two chapters of Greens and Greenhouses. 1696. 8vo.   *

The Practical Planter is mentioned by Watt as a distinct work from the Plain and Full Instructions.

1682 SAMUEL GILBERT. The Florists Vade-Mecum . . London, 1682. 12mo.

A second edition appeared in 1683, and a third enlarged in 1702.

1682 JOHN HOUGHTON. A Collection of Letters for the improvement of Husbandry and Trade. London, 1682. 4to.
<blockquote>Issued in numbers in 1681-82.</blockquote>

1683 COMMELIN-G. V. N. The Belgick or Netherlandish Hesperides, that is, the management, ordering and use of the Lemon and Orange Trees, made English by G. V. N. (from the Dutch of Commelin). 1683. 8vo.

1684 RICHARD HAINES. Aphorisms upon the new way of improving Cyder, or making Cyder-Royal . . raising and planting of Apple-trees, &c. London, 1684. Folio

1685 [WILLIAM ELLIS.] The Complete Planter & Ciderist, or choice Collections and Observations for the propagating all manner of Fruit-Trees . . By a Lover of Planting. London, 1685. 8vo.
<blockquote>The author's name does not appear in the book.</blockquote>

SIR WILLIAM TEMPLE. Upon the Garden of Epicurus, or of Gardening in the year 1685
<blockquote>A treatise on Gardening, especially relating to the Gardens at Sheen, written in 1685; published in the Miscellaneous Works of Temple, in 1705 and 1720; perhaps also in the edition which had appeared in 1689.</blockquote>

1693 DE LA QUINTINYE-EVELYN. See 1658: Evelyn.

1694 SIR DUDLEY CULLUM. A new invented Stove, for preserving Plants in the Green House in Winter. 4to.
<blockquote>Printed in the Philosophical Transactions of 1694.</blockquote>

1699 LONDON AND WISE. The Compleat Gard'ner [of J. de la Quintinye, translated by John Evelyn] . . now compendiously abridged . . with very considerable improvements. By George London and Henry Wise. London, 1699. 8vo.
<blockquote>For the unabridged translation, see 1658: John Evelyn.
Other editions in 1701, 1704 and 1710.</blockquote>

—— The Retir'd Gardener, from the French of Louis Liger. London, 1706. 2 vols. 8vo.

—— The Solitary or Carthusian Gardener, from the French of Francois Le Gentil. London, 1706. 2 vols. 8vo.
<blockquote>The same work, with a different title. Vol. I is from the French of Louis Liger. Vol. II from Le Gentil.</blockquote>

Fruit Walls improved by inclining them to the Horizon. 1699. 4to.

1700 Timothy Nourse. Campania Felix, or Discourses of the benefits and improvements of Husbandry. London. 1700. 8vo.

1702 T. Snow. Apopiroscopy, or experiments and observations on several Arts (Building, Agriculture, Gardening, &c.). London, 1702. 8vo.

1703 Le Blond—James. The Theory and Practice of Gardening, translated from the French of A. Le Blond, by John James. London. 1703. 8vo.

— - other editions. London, 1712 and 1728. 4to.

1704 Smith. The Husbandman's Magazine. 1704   *
    Only mentioned by Johnson.

Dictionarium Rusticum et Urbanicum. A Dictionary of all sorts of County Affairs, trading, &c. London, 1704. 8vo. Anonymous.

1706 Richard Bradley. Paintings of his succulent plants, with written accounts of them. 1706

— - A treatise on Succulent Plants. London, 1710

— - Historia Plantarum succulentarum. London, 1716-27. 4to.

—— New Improvements of planting and gardening, philosophical and practical. London, 1717. 8vo.
    Several later editions—the 6th, 1731.

- The Gentleman and Gardener's Kalendar. London. 1718. 8vo.

- - A Philosophical Account of the Works of Nature. London, 1721. 4to.

——— A Survey of Ancient Husbandry and Gardening. London, 1725. 8vo.

- - A general Treatise of Husbandry and Gardening. Formerly published monthly, now methodized and digested. London. 1726. 2 vols. 8vo.

A complete Body of Husbandry. London, 1727. 8vo.

— - Dictionarium Botanicum, or a Botanical Dictionary for the use of the Curious in Husbandry and Gardening. London, 1728. 2 vols. 8vo.
    " A work never before attempted."

- The Riches of a Hop Garden explained. London. 1729. 8vo.

1706 — — A Dictionary of Plants, their description and use. London, 1747. 2 vols. 8vo.
   Bradley was also the author of several less important treatises on Gardening and Agriculture.

1707 JOHN MORTIMER. Whole Art of Husbandry, and Countryman's Kalendar. London, 1707. 8vo.
   — Part II. containing additions proper for the Husbandman and Gardener. London, 1712. 8vo.
   —— Later editions, ed. by his grandson Thomas, 1716-1721 and 1761

WILLIAM FLEETWOOD, Bishop of St. Asaph and Ely. Curiosities of Nature and Art in Husbandry and Gardening. London, 1707. 8vo.   *
   Mentioned by Johnson.

CHARLES EVELYN. Ladies' Recreation: or the Pleasure and Profit of Gardening improved. London, 1707. 8vo.
   Several later editions, with slightly varying titles. That of 1719 is called Lady's Recreation or the Art of Gardening farther Improved.

1710 WILLIAM SALMON, M.D. The English Herbal, or History of Plants. London, 1710. Folio

1711 VAN OOSTEN. Dutch Gardener, or the Complete Florist. London, 1711. 8vo.   *
   Said to be from the French of Le Blond.

1712 JOSEPH ADDISON. An Essay on the Pleasures of the Garden. (*The Spectator*, No. 477)

1713 ALEXANDER POPE. Essay on Verdant Sculpture. (*The Guardian*, No. 173)

JAMES PETIVER. A Catalogue of Mr. Ray's English Herbal. 1713-15. Folio   *
   —— Historia naturalem, with 152 copper plates. London, 1764. 2 vols.
   Several papers in the Phil. Trans. relating to gardens in London, &c.

1714 JOHN LAWRENCE. The Clergyman's Recreation, shewing the pleasure and profit of the Art of Gardening. London, 1714. 8vo.
   Later editions, 1715, 1716; and the 5th, 1717.

—— The Gentleman's Recreation, &c. London, 1716. 8vo.

The Lady's Recreation; or the Art of Gardening improved . . . To which are added Observations concerning variegated greens by J. L. 1718. 8vo.

— — Gardening Improved (containing the three previous works). London. 1718. 8vo.

The Fruit Garden Kalendar. London, 1718. 8vo.

A new System of Agriculture, being a complete book of Husbandry and Gardening, &c. London, 1726. Folio

1715 G. CLARKE. The Landed Man's Assistant. 1715. 12mo.*
Mentioned by Johnson.

STEPHEN SWITZER. The Nobleman, Gentleman and Gardener's Recreation, &c. London, 1715. 8vo.

—— Ichnographia Rustica. London, 1718. 3 vols. 8vo.
This is an enlargement of the preceding work, second edition, 1741.

—— The Practical Fruit Gardener, &c. London, 1724. 8vo.

—— A Compendious Method for raising of Italian Brocoli . . and other Vegetables, &c. London. 1729. 8vo.

—— A Dissertation on the true Cythisus of the Ancients, &c. London, 1731. 8vo.   *

C. J. WOLFE and JAMES GANDON. Vitruvius Britannicus, or British Architect, containing plans, &c. of buildings and Gardens, public and private, in Great Britain; 200 copper-plates. London, 1715

1716 REV. HENRY STEVENSON. The Young Gardener's Director. London, 1716. 12mo.   *

—— The Gentleman Gardener instructed. London, 1716. 12mo.   *
Mentioned by Johnson, who says the sixth edition is dated 1769.

1717 JOSEPH CARPENTER. The Retir'd Gardener. London, 1717. 8vo.

SAMUEL COLLINS. Paradise Retrieved, or the Method of managing and improving Fruit Trees, with a Treatise on Melons and Cucumbers: 12 plates. London, 1717. 8vo.

GEORGE ANDREW AGRICOLA. The Artificial Gardener. London. 1717

1717 — Philosophical Treatise of Husbandry and Gardening. Translated by Bradley. London, 1721.

—— On Planting. Edin. 1777.
All three works translated from the German described by Watt.

GILES JACOB. The Country Gentleman's Vade-Mecum. 1717. 12mo.

1718 REV. JAMES GARDINER. Rapin of Gardens: a Latin Poem in 4 books. Englished by J. G. London, 1718.
See also Evelyn, 1658.

1720 PATRICK BLAIR, M.D. Botanick Essays. London, 1720. 8vo.

—— Pharmaco-Botanologia, or an alphabetical and classical Dissertation on all the British Indigenous and Garden Plants of the New London Dispensatory. London. 1723-28. 4to.

1722 THOMAS FAIRCHILD. The City Gardener, &c. London. 1722. 8vo.

—— The different and sometimes contrary motion of the sap in plants. Phil. Trans. 1724

—— Catalogus Plantarum. *See* Society of Gardeners, 1730

JOSEPH MILLAR or MILLER. Botanicum Officinale, or A Compendious Herbal. London, 1722. 8vo. *

1724 PHILIP MILLER. The Gardener's and Florist's Dictionary. London, 1724

—— Catalogus Plantarum Officinalium quæ in Horto Botanico Chelseiano atiextur. London, 1730

—— The Gardener's Dictionary. London, 1731-39. Folio
This was many times republished, abridged, translated, and enlarged. The second edition, 1733. In the seventh, 1759, Miller adopted the Linnean system of classification.

—— The Gardener's Kalendar. London. 1732. 8vo.
Several later editions, the thirteenth dated 1782.

—— The Method of Cultivating Madder, &c. London. 1758 *

—— The Elements of Agriculture, translated from Duhamel du Monceau. 1764. 2 vols. 8vo.

1726 BATTY LANGLEY. Practical Geometry applied to the Arts of Building, Surveying, Gardening, &c. London, 1726. Folio

—— A Sure Method of improving Estates by plantations of Oak, Elm, Ash, Beech, &c. London, 1728. 4to. *

—— - New Principles of Gardening, or the laying-out and planting Parterres. London, 1728. 4to.

- Pomona, or the Fruit Garden illustrated, &c. London, 1729. Folio

—— The Landed Gentleman's Useful Companion (reprint of "A Sure Method," &c.). London, 1741

B. TOWNSEND. The Complete Seedsman. Shewing the best and easiest method for raising and cultivating every Sort of Seed, &c. 1726. 8vo.

The Gentleman Farmer, or certain observations on the Husbandry of Flanders, compared with that of England. 1726. 12mo. Anonymous *
    Mentioned by Johnson.

1727 S. J. The Vineyard: a Treatise shewing the nature and method of planting, manuring, cultivating, and dressing Vines in foreign parts, &c. 1727 *
    Mentioned by Johnson.

ROBERT FURBER. Catalogue of English and Foreign trees. London, 1727. 8vo. *
    Mentioned by Watt.

—— Fruits for every month in the year. 12 plates. 1732. 8vo. *

—— - An Introduction to Gardening, &c. London, 1733. 8vo.

1728 ROBERT CASTEL. The Villas of the Ancients, "illustrated with remarks and cuts." London, 1728. Folio *

1729 JOHN COWELL. A true Account of the Aloe Americana, or Africana now in blossom . . . also two other exotic plants call'd the Cereus or Torch-thistle. London, 1729. 8vo.

- The Curious and Profitable Gardener. London, 1730. 8vo. *

The Curious Fruit and Flower Gardener. Second Edition. London, 1732. 8vo.
    Same work as above with different title-page.

1730 A SOCIETY OF GARDENERS. Catalogus Plantarum. A Catalogue of trees, shrubs . . in the gardens near London. Part I (rest never published). London, 1730. Folio
Thomas Fairchild's name appears first on the list of the twenty gardeners who signed the Preface, so this work is sometimes catalogued under his name, or under that of Philip Miller, whose name is also on the list.

1731 MARK CATESBY. Natural History of California, Florida, &c. &c. London, 1731. Imp. folio

— Hortus Europæ Americanus, or a Collection of 85 Curious Trees and Shrubs, &c. London, 1767. 4to.

1732 An Essay concerning the best methods of pruning Fruit Trees, &c. London, 1732. 8vo. Anonymous *

The nature and method of planting, manuring, and dieting a Vineyard. London, 1732. 8vo. Anonymous *

The great Improvement of Commons that are enclosed for the advantage of Lords of the Manor, the Poor, and the Public, with methods of enriching all soils, and raising timber. To ripen fruit at all times of the year ; an improvement in raising Mushrooms, Cucumbers, &c. 1732. Anonymous *
These three anonymous works are mentioned by Johnson.

WILLIAM HARPER. A Sermon on Gardening, preached at Malpas, Co. of Chester, at a Meeting of Gardeners and Florists, April 18th, 1732. London, 1732. 4to.

The Flower Garden Displayed. London, 1732. 4to.

—— A Second Edition, to which is added "A Flower Garden for Gentlemen and Ladies, being the Art of raising Flowers . . . also salleting, cucumbers, &c. as it is now practised by Sir Thomas More. Above 400 curious representations of the most beautiful flowers, &c. from the designs of Mr. Furber and others, coloured to the Life. London, 1734. 4to.

WILLIAM ELLIS. Complete Modern Husbandry, containing the Practice of Farming, etc. Second Edition. London, 1732. 8vo.

—— The Practical Farmer or Hertfordshire Husbandman. London, 1732

———   The Timber Tree improved, or the best practical methods of improving different lands with proper timber. London, 1738
Ellis was the author of several other Agricultural Treatises.
1736 PLUCHÉ-HUMPHRYS. Nature Displayed, translated from the French of N. A. Pluché, by George Humphrys. London, 1736. 8vo.
1737 BLAKEWELL's Herbal. 1737
1738 PUBLIC GARDENS. Collection of notes about Ranelagh, Cuper's Garden, &c. (Guildhall Library.) 1738-46
1739 SAMUEL TROWELL. A New Treatise of Husbandry, Gardening and other curious matters relating to country affairs. London, 1739. 8vo.
— — The Farmer's Instructor or Husbandman, and Gardener's useful and necessary Companion. Ed. by William Ellis. 1747   *
Mentioned by Johnson.

An Essay upon Harmony, as it relates chiefly to situation and building. London, 1739. 8vo. Anonymous   *
1740 CHRISTOPHER GRAY. A Catalogue of Trees and Shrubs . . . for sale. 1740   *
1744 Adam's Luxury and Eve's Cookery, or the Kitchen Garden displayed. London, 1744. 8vo.   *
Curious Experiments in Gardening, &c. 1744. 12mo.   *
These four works are mentioned by Johnson.

JOHN WILSON. Synopsis of British Plants, in Ray's Method, with a Botanical Dictionary. Newcastle, 1744
1745 A Plan of Mr. Pope's Garden and Grotto, &c. 1745
1746 DAVID STEPHENSON, M.A. The Gentleman's Gardener's Director of Plants, Flowers, and Trees, with a Garden Kalendar. London, 1746. 8vo.   *
The Beauties of Stowe. London, 1746
A description of the Gardens of Lord Viscount Cobham at Stowe. Northampton, 1747
A dialogue upon the Gardens of Lord Viscount Cobham at Stowe. London, 1748. 8vo.
The Gardens of Lord Viscount Cobham at Stowe. London, 1751

1746 GEORGE BICKHAM. The Beauties of Stowe. 1753. 8vo.
A description of the House and Gardens of the Marquis of Buckingham at Stowe. Buckingham, 1797

1747 The Complete Florist: 100 engravings. London, 1747. 8vo. Anonymous

1748 SIR WILLIAM WATSON. Papers published in the Philosophical Transactions. Accounts of the remains of John Tradescant's Botanic Garden at Lambeth. 1750. Account of the Bishop of London's Garden at Fulham. 1751. And several others on similar subjects.

—— A Letter to Andrew Ducarel. London, 1774. 4to.

1752 ATTIRET-BEAUMONT. An Account of the Emperor of China's Gardens at Pekin, by J. D. Attiret. Translated by Sir H. Beaumont, i.e. Joseph Spence. London, 1752

JAMES NEWTON. Compleat Herbal. London, 1752. 8vo.

1753 W. WEBB. A Catalogue of Seeds and Roots under their proper heads. 1753 *

FRANCIS COVENTRY. Essay in "The World" of April 12th, 1753 (No. XV.) entitled, "Strictures on the absurd novelties introduced in Gardening," and a humorous description of Squire Mushroom's Villa. 1753 *

BARTHOLOMEW ROCQUE. A Treatise on the Hyacinth, &c. London. 1753 *

—— A Practical Treatise on cultivating Lucerne-Grass, &c. London, 1775

1754 JAMES JUSTICE. The Scot's Gardener's Director. Edinburgh. 1754

—— The British Gardener's Director. Edinburgh, 1764

—— The Useful Herbal. Anonymous. London, 1754. 8vo.

1755 WILLIAM JOHN HALFPENNY. Rural Architecture in the Chinese taste. 1755. 8vo.

JOHN DALTON, D.D. Some thoughts on Building and planting, addressed to Sir James Lowther, Bart. London, 1755. 4to.

1756 On the Heat and Cold of Hot-houses. Anonymous. London, 1756. 8vo. *

Timothy Sheldrake (the elder). Gardener's Best Companion in a Greenhouse. London [c. 1756]. Folio.
- An Herbal of Medicinal Plants, etc. London [c. 1759].
John Hill, M.D. (Sir J. H.). The British Herbal, an History of plants . . cultivated for use or raised for Beauty. London, 1756. Folio

1757 Hale. Eden, or a Complete Body of Gardening . . . compiled by Sir John Hill from the papers left by Hale. London, 1757. Folio.
Thomas Hitt. A Treatise of Fruit Trees. London, 1757. 8vo. (2nd ed.) *
- - A Treatise of Husbandry on the improvement of dry and barren lands. London, 1760. 8vo.
Edward Lisle. Observations on Husbandry. 1757. 4to.
<small>Edited by his son, T. Lisle.</small>
Robert Maxwell. The practical Husbandman. 1757
William Mason (the Poet). An heroic Epistle to Sir W. Chambers. London, 1757. 4to.
—— An heroic Postscript. 1758
—— The English Garden, a Poem, in 4 books 8vo. 1772.
<small>Edition with Commentary and Notes, by W. Burgh, 1785.</small>
James Thompson. The distinguishing properties of a fine Auricula. Newcastle, 1757. 8vo.
—— The Dutch Florist. Newcastle, 1758. 12mo.

1758 Thomas Barnes. New method of Propagating Fruit Trees and Shrubs, confirmed by repeated and successful experience. London, 1758. 8vo. *
<small>Later editions, 1759 and 1762.</small>
Rev. William Hanbury. An Essay on Planting, and a scheme for making it conducive to the glory of God. Oxford, 1758. "An 8vo. pamphlet" *
—— A complete body of planting and gardening. London, 1770-1. Folio

1759 Richard North. A Treatise on Grasses and the Norfolk Willow. London, 1759. 8vo.
—— The Gardener's Catalogue of Hardy Trees, Shrubs, Flowers, Seeds, &c. 1759. 8vo. *
<small>Mentioned by Johnson.</small>
John Mills. Practical Treatise on Husbandry, translated from the French of Duhamel de Monceau. 1759. 4to. With plates

1759 — — A new and complete System of Practical Husbandry. 1762. 5 vols. 8vo.
—— The Natural and Chemical Elements of Agriculture, from the German of Gyllenborg. 1770. 12mo.
—— Essays on Agriculture. 1772. 8vo.

1760 SAMUEL PULLEIN. Observations towards a method of preserving the seeds of Plants in a state of Vegetation during long voyages. London, 1760. 8vo. *

JAMES LEE. An Introduction to Botany, containing an explication of the Theory of that Science, &c. London, 1760. 8vo.
—— Catalogue of Plants and Seeds sold by Kennedy and Lee, at the Vineyard, Hammersmith. 1774. 8vo.

The London Gardener. Anonymous. London, 1760. 8vo. *
    Mentioned by Johnson.

Adam Armed: or an Essay endeavouring to prove the advantages .. the kingdom may receive ... by means of a well ordered and duly rectified Charter for incorporating and regulating the Professor of the Art of Gardening: humbly offered and presented by the Master and Company of the same. London. Folio
    Johnson mentions this work and says it has no author's name or date, but was published about this year.

1762 T. LIGHTOLER. The Gentleman and Farmer's Architecture: being Plans for Parsonage and Farm Houses, with Pineries, Greenhouses, &c. With Plates. London. 1762. Folio
    Johnson mentions an edition of 1766.

1763 GEORGE RITSO. Kew Gardens: a Poem. London, 1763 *

JAMES WHEELER. The Botanists' and Gardeners' New Dictionary, containing names, classes, &c. ... according to the System of Linnæus. London, 1763. 8vo.
—— An Essay on the Theory of Agriculture, &c. London, 1763. 12mo. *

THOMAS MARTYN, F.R.S. Plantae Cantabrigienses, or a Catalogue of Plants which grow wild in the County of Cambridge. London, 1763. 8vo.
— A Short Account of the Donation to the Botanic Garden, by Dr. Walker. London, 1763 4to.

—— Catalogus Horti Botanici Cantabrigiensis, a Catalogue of the Botanic Garden at Cambridge. Cambridge, 1771. 8vo.

—— Rousseau's Letters: or the Elements of Botany, addressed to a Young Lady: with Notes, and twenty-four additional Letters, explaining the system of Linnæus. Translated from the French. London, 1785. 8vo.

— Thirty-eight Plates, with Explanations intended to illustrate Linnæus' System of Vegetables, and particularly adapted to the Letters on the Elements of Botany. London. 1788. 8vo.

—— Flora Rustica, exhibiting figures of such Plants as are either useful or hurtful in Husbandry; with Scientific Characters, &c. London, 1792-4. 8vo.

— The Language of Botany: being a Dictionary of the Terms made use of in that Science, principally by Linnæus, &c. London. 1793. 8vo.

—— The Gardeners' and Botanists' Dictionary of the late Philip Miller, corrected and newly arranged, with additions. London, 1803-7. 4 vols. folio

—— Various papers contributed to the Transactions of the Linnæan Society. 4to.

1764 REV. WALTER HARTE. Essays on Husbandry, and a Treatise on Lucerne, by W. H., Canon of Windsor. With plates. 1764 and 1770

The Dutch Florist, from the Dutch of Van Campen. 1764. 4to. *

Museum Rusticum et Commerciale, &c. London, 1764. 6 vols. 8vo. *

De Re Rustica. A similar work to the above, begun 1768, completed 1770. 2 vols. 8vo. *

<small>These three works are thus described by Johnson.</small>

WILLIAM SHENSTONE. Unconnected Thoughts on Gardening, in Essays on Men and Manners. 1764. 3 vols. 8vo.

The Complete Farmer; or, Dictionary of Husbandry. Published by David Henry. 1764.

<small>Johnson mentions a second edition, 1768.</small>

23

1766 JOHN LOCKE. Observations upon the growth and culture of Vines and Olives, &c., from his original MS. in the possession of the Earl of Shaftesbury. London, 1766. 8vo.

1767 The rise and progress of the present Taste in planting Parks, Pleasure Grounds, Gardens, &c., from the time of Henry VIII. to George III. In a poetic epistle to the Right Hon. Charles, Lord Viscount Irwin. 1767

JAMES RUTTER and DANIEL CARTER. Modern Eden, or the Gardener's universal Guide, &c. London, 1767. 8vo. *

JOHN GILES. Ananas: or a Treatise on the Pine Apple, &c. To which is added the true method of raising the finest Melons with the greatest success, &c. London, 1767. 8vo.

GEO. DIONYSIUS EHRET, F.R.S. Of a new Peruvian Plant lately introduced into the English Gardens (the nolana prostrata of Linnæus). Phil. Trans. 1767

JOHN ABERCROMBIE. Every Man his own Gardener. 1767. 12mo.

<small>This work has the name of Thomas Mawe also on the title, and went through several later editions.</small>

—— The Universal Gardener and Botanist, &c. London, 1778. 4to.

—— The Garden Mushroom, its Nature and Cultivation, &c. London, 1779. 8vo.

—— The British Fruit Garden, and Art of Pruning, &c. London, 1779. 8vo.

—— The Complete Forcing Gardener, &c. London, 1781. 12mo.

—— The Complete Wall-tree Pruner, &c. London, 1783. 12mo.

—— The Propagation and Botanical Arrangement of Plants and Trees, useful and ornamental. London, 1785. 2 vols. 12mo.

—— The Gardener's Pocket Dictionary, &c. London, 1786. 3 vols. 12mo.

—— Daily Assistant in the Modern Practice of English Gardening, &c. London, 1789. 12mo.

—— The Universal Gardener's Kalendar, &c. London. 1789. 12mo.

—— The Complete Kitchen Gardener, and Hot-bed Forcer, &c. London, 1789. 12mo.

—— The Gardener's Vade-mecum, &c. London, 1789. 8vo.

—— The Hot-house Gardener, &c. London, 1789. 8vo.

—— The Gardener's Pocket Journal, &c. London, 1791. 12mo.

<small>Most of these works went through several editions.</small>

1768 GEORGE MASON. An Essay on Design Gardening. London, 1768. 8vo.

—— Revised edition. 1795

JOHN GIBSON, M.D. The Fruit Gardener, containing the method of raising Stocks for multiplying Fruit Trees, &c. London, 1768. 8vo.

<small>Presumably by J. Gibson, although his name does not appear on the title-page.</small>

THOMAS WILDMAN. A Treatise on the culture of Peach Trees, to which is added a Treatise on the management of Bees. 1768

1769 JAMES GARTON. The Practical Gardener and Gentleman's Directory for every month in the year, &c. London, 1769. 12mo.

ANTHONY POWELL. The Royal Gardener, or complete Calendar of Gardening for every month in the year, &c. London, 1769. 12mo. *

ADAM TAYLOR. A Treatise on the Anana, or Pine Apple, &c. Devizes, 1769. 8vo.

THE HON. DAINES BARRINGTON. On the Trees which are supposed to be Indigenous in Great Britain. 1769

—— Chestnut Trees not Indigenous in Great Britain. 1771

—— Mr. Pegge's Observations on the Growth of the Vine in England, considered and answered. 1777

—— On the Progress of Gardening, in a letter to Mr. Norris, 1782.

<small>These treatises are all published in the Archæologia.</small>

1769 RICHARD WESTON. Tracts on practical Agriculture and Gardening . . . to which is added, a Complete Chronological Catalogue of English Authors on Agriculture, Gardening. &c. London, 1769. 8vo.

—— Second edition, greatly enlarged. 1773
The author's name appears in this but not in the first edition.

—— Botanicus universalis et hortulanus. &c. 4 vols. London, 1770-1777. 8vo.

—— The Gardener's and Planter's Calendar. London, 1773. 8vo.

—— The Gardener's Pocket Calendar. (2nd ed.) London, 1780

—— Flora Anglicana. London, 1775-89. 8vo.

—— A new cheap Manure . . . Alabaster or gypsum. Leicester, 1791. 8vo.

1770 HORACE WALPOLE. Essay on Modern Gardening, written in 1770, printed with a French translation on opposite pages by the Duc de Nivernois. Strawberry Hill, 1785. 4to.

JOHN DOVE. Strictures on Agriculture, wherein a discovery of the Physical course of Vegetation, of the Food of Plants, and the Rudiments of Tillage, is attempted. London, 1770. 12mo.

LINNÆUS – MILNE. Institutes of Botany: containing accurate, compleat, and easy Descriptions of all the known Genera of Plants; from the Latin of Charles Van Linné, by Rev. Colin Milne, LL.D. London, 1770. 4to.

A Botanical Dictionary: or, Elements of Systematic and Philosophical Botany, etc. London, 1770. 8vo. 2nd edition. 1777. Supplement, 1778. 3rd and enlarged edition. 1805. 8vo.

The Gardener's Alphabetical Calendar. 1770. 12mo. *

The Pocket Kitchen Gardener. 1770. 12mo. *

The Pocket Flower Gardener. 1770. 12mo. *

These three works, without author's names, are mentioned by Johnson.

THOMAS WHEATLEY, or WHATELY. Observations on Modern Gardening, illustrated by descriptions. London, 1770. 8vo.
Several later editions—the 5th, 1793.

JOHN ELLIS. Directions for bringing over Seeds and Plants from the East Indies and other distant Countries in a state of Vegetation. London, 1770. 4to.

— Description of the Mangostan and Bread Fruit Tree. London, 1775. 4to.

— An Historical Account of Coffee. With engravings and botanical descriptions of the tree . . . its culture and use. London, 1774. 4to.

1771 MATTHEW PETERS. The Rational Farmer, or a Treatise on Agriculture and Tillage. London, 1771. 8vo.

JAMES MEADER. The Modern Gardener or Universal Kalendar . . . from the Diary and MSS. of the late Mr. Flitt, corrected and improved by J. M. London, 1771. 12mo. *

— The Planter's Guide or Pleasure Gardener's Companion. London, 1779. Oblong 8vo.

JOHN DICKS. The New Gardener's Dictionary, or the whole Art of Gardening fully and accurately displayed. London, 1771. Folio.
An edition of 1769 is mentioned by Watt.

JOHN REINHOLD FORSTER, LL.D. Florae Americae Septentrionalis, or A Catalogue of the Plants of North America. London, 1771. 8vo.

— Characteres generum Plantarum, quas in itinere ad insulas Maris Australis, illegerunt, descripserunt, et delineaverunt annis 1772-3. London, 1776. 4to.

1772 LOUIS DE ST. PIERRE. The Art of planting and cultivating the Vine, &c. according to the most approved methods in France. London, 1772. 12mo.

SIR WILLIAM CHAMBERS. A Dissertation on Oriental Gardening. London, 1772. 4to.
The second, being the only other edition, 1773.

1773 ANDREW COLTEE DUCAREL, LL.B. and LL.D. A Letter to Wm. Watson, M.D., upon the early Cultivation of Botany in England; and some particulars about John Tradescant, Gardener to King Charles I. London, 1773. 4to.

1774 JOHN COAKLEY LETTSOM. Hortus Uptoneniis. A Catalogue of Dr. Fothergill's garden at Upton at the time of his decease. (No date) c. 1774. 8vo.

—— Grovehill, a rural and horticultural sketch. 1784. 4to.

—— A translation of Abbé de Commerell's account of the culture of the Mangel Wurzel or Root of Scarcity. London, 1788.

1775 LINNÆUS—JENKINSON. A Generic and Specific Description of British Plants: translated from the Genera et Species Plantarum, of Linnæus; with Notes and Observations, by James Jenkinson. Kendal, 1775. 8vo.

REV. SAMUEL WARD. A Modern System of Natural History, containing accurate descriptions and faithful histories of animals, vegetables, and minerals. London, 1775-77. 12 vols. 12mo.   *

—— An Essay on the different natural Situations of Gardens. London, 1775. 4to.   *

1776 HENRY HOME, LORD KAMES. The Gentleman Farmer. Edinburgh, 1776. 8vo.

Several later editions—the sixth in 1815.

WILLIAM WITHERING, M.D., F.R.S., F.L.S. A Botanical Arrangement of all the Vegetables naturally growing in Great Britain, with an easy Introduction to the Study of Botany. (Plates.) Birm. 1776. 2 vols. 8vo.

—— 2nd edition. London, 1778-90. 3 vols.   *

—— 3rd edition. 1796. 4 vols.   *

1777 JAMES ANDERSON, LL.D. Thoughts on Planting (first appeared in the Edinburgh Weekly Magazine), by Agricola. Edin. 1777. 8vo.   *

—— A Description of a Patent Hot-house, &c. London, 1804. 12mo.

WILLIAM WILSON. A Treatise on the forcing of Early Fruits, and the management of Hot Walls. London, 1777. 12mo.

CONRAD LODDIGE'S Catalogue of Plants and Seeds sold by C. L . . at Hackney near London. 1777. 8vo.
The names of plants are in German as well as in English.

—— The Botanical Cabinet. London, 1818-33. 20 vols. 8vo.

JOSEPH HEELEY. Letter on the Beauties of Hagley, Envil and the Leasowes, etc. London, 1777. 12mo.

—— Description of Hagley Park. London, 1777. 8vo.

1778 The Practical Gardener, &c. London, 1778. No author's name. *

N. SWINDEN. The Beauties of Flora displayed, &c. London, 1778. 8vo.

1779 ADAM NEALE. A Catalogue of Plants in the garden of John Blackburne, Esq., at Orford, Lancashire, alphabetically arranged, according to the Linnean system. Warrington, 1779. 8vo.

WILLIAM SPEECHLEY. A Treatise on the cultivation of the Pine Apple. York. 1779. 8vo.

—— A Treatise on the culture of the Vine in England. York, 1790. 4to.

THOMAS ELLIS (gardener to the Bishop of London). The Gardener's Pocket Calendar. London, 1779. 12mo.
An earlier edition is said to have been published anonymously in 1770.

1780 ALEXANDER WILSON. M.D. Some Observations relating to the Influence of Climate on Vegetables and Animal Bodies. London, 1780. 8vo.

JOHN TRUSLER. Practical Husbandry. 1780. 8vo.

—— Elements of Modern Gardening. 8vo.
The title-page has neither name nor date. The B.M. Catalogue ascribes it to Trusler, and assigns it to the year 1800. Johnson does not give any author, and dates it 1784.

1781 WILLIAM HOUSTOUN, D.D. Reliquiæ Houstounianæ, a Catalogue and Description of Plants. 1781. 4to.
Published by Sir Joseph Banks.

1781 SAMUEL FULLMER. The Young Gardener's best Companion for the Kitchen and Fruit Garden. London, 1781. 12mo.
1782 WILLIAM RALEY. A Treatise on the Management of Potatoes. London, 1782. 8vo.
1783 ERMENONVILLE—MALTHUS. An Essay on Landscape. From the French of Ermenonville. 1783. 12mo. *
  No author's name, but said by Johnson to be by Mr. Malthus.

CHARLES BRYANT. Flora Diætetica, or History of Esculent Plants, &c. London, 1783. 8vo.

—— A Dictionary of the Ornamental Trees, &c. Norwich, 1790. 8vo.

DE LILLE.—On Gardening. Translated from the French of l'Abbé de Lille. 1783. 4to. *

WILLIAM FALCONER, M.D., F.R.S. An Historical View of the Taste for Gardening and Laying out Grounds among the Nations of Antiquity. 1783. 8vo. *

—— An Essay on the Preservation of the Health of Persons employed in Agriculture, &c. Bath, 1789. 8vo.

—— Miscellaneous Tracts and Collections relating to Natural History, &c. Cambridge, 1793. 4to.

THOMAS KYLE. A Treatise on the Management of Peach and Nectarine Trees, either in forcing-houses, or on hot and common walls. Edinburgh, 1783. 8vo.

—— Catalogue of Plants, with their English and Latin Linnæan names, sold by Lucker and Smith, Dalston. 1783. 8vo.

WILLIAM GILPIN. Observations . . . relative chiefly to Picturesque Beauty. 1783 to 1809. 11 separate vols. 8vo.
  Several distinct works, descriptive of Tours in different parts of England, containing accounts of gardens, &c.

1785 SAMUEL FELTON. Miscellanies on Ancient and Modern Gardening, and on the Scenery of Nature. London, 1785. 8vo. *
  Without author's name.

—— On the Portraits of English Authors on Gardening. London, 1828. 8vo.

—— Gleanings on Gardens, chiefly respecting the ancient style in England. London, 1829. 8vo.

WILLIAM MARSHALL. Planting and Ornamental Gardening: a Practical Treatise. London, 1785. 8vo.
    Without author's name.

—— 2nd edition, with the title Planting and Rural Ornament. London, 1796. 8vo.

JAMES BOLTON. Filices Britanniæ, an History of the British proper Ferns, with plain and accurate Descriptions, &c. Leeds, 1785

JOHN, EARL OF BUTE. Botanical Tables, containing the different families of British Plants, &c. London, 1785. 9 vols. 8vo. *

JAMES DICKSON, F.L.S. Fasciculus Plantarum Cryptogamicarum Britanniæ. London, 1785
    Dickson was the author of many papers in the Horticultural Society's Transactions.

1786 FRANCIS XAVIER VISPRE. A Dissertation on the growth of Wine in England. Bath, 1786. 8vo.

ROBERT BROWNE. A Method to preserve Peach and Nectarine Trees from the effect of Mildew, &c. London, 1786. 12mo.

REV. PHILIP LE BROCQ, M.A. A Description of certain methods of Planting, Training, and Managing all kinds of Fruit Trees, Vines, &c. London, 1786. 8vo.

—— A Sketch of a Plan for making the Tract of Land called the New Forest a real Forest, and for various other purposes of the first national importance. 1793. 8vo.

The Compleat Herbal, or Family Physician, giving an account of all such Plants as are now used in the Practice of Physic, with their Descriptions and Virtues. 2 vols. No author's name. London, 1787. 8vo.

1787 WILLIAM CURTIS. The Botanical Magazine (begun by). London, 1787
    First series, 1787 to 1826, 53 vols.; Index, 1828. Second series, edited by S. Curtis and W. J. Hooker, 1827 to 1844, 17 vols. Third series, edited by Sir W. J. Hooker, 1845 to 1858, 13 vols.

1787 —— Flora Londinensis. 1777-1828. 5 vols. folio

GEORGE WINTER. A new and compendious System of Husbandry, containing the mechanical, chemical, and philosophical Elements of Agriculture. Bristol, 1787. 8vo.

1788 SIR JAMES EDWARD SMITH. Some observations on the irritability of Vegetables. London, 1788. 4to.

—— Plantarum Icones hactenus ineditæ, plerumque ad Plantas in Herbario Linnæno conservatus delineatæ. London, 1789-91. Fol.

—— Icones Pictae Plantarum rariorum descriptionibus . . illustratae. London, 1790-93. Fol.

—— Spicilegium Botanicum. Gleanings in Botany. London, 1791-2. Fol.

—— A Specimen of the Botany of New Holland. The figures by J. Sowerby. London, 1793. 4to. (only 1 vol. published).

—— Syllabus of a Course of Lectures on Botany. London, 1795. 8vo.

—— Remarks on the Generic Character of the Decandrons Papalionaceous Plants of New Holland. London, 1804. 8vo.

—— Exotic Botany . . . coloured figures . . of such new, beautiful or rare plants as are worthy of cultivation in the Gardens of Britain. The figures by J. Sowerby. London, 1804-5. 2 vols. 4to.

—— An Introduction to Physiological and Systematic Botany. London, 1807. 8vo.

—— Review of the Modern State of Botany, etc. Edinburgh, 1817. 4to.

—— A Grammer of Botany. London, 1821. 8vo.

1789 JOHN GRAEFER. A Descriptive Catalogue of upwards of 1100 species and varieties of Herbaceous Plants, &c., with a List of Hardy Ferns, &c., &c. London, 1789. 8vo.

JAMES ADAM. Practical Essays on Agriculture, containing an Account of Soils, and the manner of correcting them; an Account of the culture of all Field Plants; also on the Culture and management of Grass Lands, &c. London, 1789. 2 vols. 8vo. *

WILLIAM AITON. Hortus Kewensis. London, 1789. 3 vols. 8vo.

—— Second edition enlarged by his son, William Townsend Aiton, 1810-13. 5 vols. 8vo.

1790 RICHARD PULTENEY. Historical and biographical sketches of the progress of Botany. London, 1790. 2 vols. 8vo.

—— General View of the Writings of Linnæus. London. 1805. 4to.

—— Decorations for Plants and Garden. Published by Taylor. Cir. 1790

E. O. DONOVAN, F.R. and L.S. The Botanical Review, or the Beauties of Flora. Nos. 1-7. London, 1790

BRULLES. Hints for the management of Hot-beds, &c. Bath, 1790. 8vo. *

1791 RICHARD ANTHONY SALISBURY. Icones Stirpium variorum. Descriptionibus illustratae. London, 1791. 8vo.

—— Paradisus Londinensis. London, 1805-8. 4to.
   Salisbury contributed many valuable papers to the Trans. Horticultural Society.

ERASMUS DARWIN, M.D., F.R.S. The Botanic Garden, a poem, in two parts: part I, The Economy of Vegetation; part II, The Loves of the Plants. London. 1791. 4to.

- Phytologia, or the Philosophy of Agriculture and Gardening, &c. London, 1800. 4to.

JAMES SOWERBY, F.L.S. The Florist's Delight, &c. London. 1791. Folio

-- - Figures of English Fungi, or Mushrooms. London. 1792-1803. 3 vols. folio

English Botany, with Sir J. E. Smith. 1790-1820. 36 vols. 8vo.

1791 WILLIAM FORSYTH, F.A.S. Observations on the Diseases, Defects, and Injuries in all kinds of Fruit and Forest Trees, &c. London, 1791. 8vo.
—— A Treatise on the culture and management of Fruit Trees, &c. London, 1802. 4to.
1792 The Linnæan Society's Transactions, first published. 1791. 8vo.

Among the contributors of papers on Gardening subjects in the early numbers are the following: C. C. Babington, H. T. Colebrooke, Peter Collinson, L. W. Dillwyn, R. K. Greville, W. J. Hooker, J. Lindley, T. Martyn, R. Rudge, J. E. Smith, J. Sowerby, C. Stevens, J. Woods.

JAMES MADDOCK. Florist's Directory and Treatise on the Culture of Flowers, &c. London, 1792. 8vo.
—— - An improved edition by J. Curtis. London, 1810. 8vo.
1793 RICHARD STEELE. An Essay on Gardening, containing a Catalogue of Exotic Plants, &c. York, 1793. 4to. *
1794 WILLIAM AMOS. The Theory and Practice of Drill Husbandry. London, 1794. 4to. *
—— Minutes of Agriculture and Planting. London, 1804. 4to.
ADRIAN HARDY HAWORTH. Observations on the genus Mesembryanthemum. London, 1794. 8vo.
—— Synopsis Plantarum Succulentarum, &c. London, 1812-19. 8vo.
SAMUEL HAYES, M.R.I.A. A practical Treatise on Planting. Dublin, 1794. 8vo.
JAMES SHAW. Plans, elevations, sections, observations, and explanations of Forcing houses in Gardening. Whitby, 1794. Folio
JAMES MCPHAIL. A Treatise on the Culture of the Cucumber, &c. London, 1794. 8vo.
—— The Gardener's Remembrancer throughout the year, &c. London, 1803. 8vo.
WILLIAM MAUNSELL, LL.D. Letter on the Culture of Potatoes from the Shoots. London, 1794. 8vo.
SIR UVEDALE PRICE. An Essay on the Picturesque, as compared with the sublime and beautiful, &c. London, 1794-98. 8vo.

—— A Letter to H. Repton, Esq., on the application of the practice, as well as the principles of Landscape Painting, to Landscape Gardening, &c. London, 1795. 8vo.

—— A Dialogue on the distinct characters of the Picturesque and the Beautiful, in answer to the objections of Mr. Knight. London, 1801

—— On the Picturesque: including A Letter to H. Repton, Esq., and On Decorations near the House. Edited by Sir Thomas Dick Lauder. London, 1842

HUMPHRY REPTON. A Letter to Uvedale Price, Esq. on Landscape Gardening. London, 1794. 4to.

—— Sketches and Hints on Landscape Gardening, &c. London, 1795. Folio

—— Observations on the Theory and Practice of Landscape Gardening, &c. London, 1803. 4to.

—— An Enquiry into the changes in Landscape Gardening. London, 1806. 8vo.

—— On the introduction of Indian Architecture and Gardening. London, 1808. Folio   *

—— On the supposed effects of Ivy on Trees. Trans. Linn. Soc. London, 1810   *

—— Fragments on the Theory and Practice of Landscape Gardening. London, 1817. Folio

RICHARD PAYNE KNIGHT. The Landscape: a Didactic Poem, in 3 books, addressed to Uvedale Price, Esq. 1794. 4to.

—— Review of the Landscape: also of an Essay on the Picturesque; with Practical Remarks on Rural Ornament. 1795. 8vo.

1795 WILLIAM ROXBURGH, M.D., F.R.S. Plants of the Coast of Coromandel, &c. London. 1795

1796 FRANCIS BAUER. Delineation of Exotick Plants cultivated in the Royal Garden at Kew, drawn and coloured, and the botanical characters displayed, according to the Linnæan System. London, 1796. Folio

—— Illustrationes Floræ Novæ Hollandiæ, &c. Part I. 1813   *

1796 Rev. Charles Marshall. Introduction to the Knowledge and Practice of Gardening; with Hints on Fish Ponds. London, 1796. 12mo.
Several later editions—the third, 1800.

George Lindley. The plan of an Orchard, exhibiting at one view a select quantity of Trees, &c. 1796 *
Described by Johnson as a folio sheet.

—— An Account of the Culture of Potatoes in Ireland. 1796. 8vo. *
These two works are mentioned by Johnson.

James Don. Hortus Cantabrigensis, a Catalogue of Plants Indigenous and Exotic. 1796. 8vo.

1797 Strickland Freeman. Select Specimens of British Plants. Five plates. London, 1797. Folio
Part II. contains descriptions of the plants by G. Shaw.

Francis Duckenfield Astley. A few minutes' advice to Gentlemen of landed Property, and the admirers of Forest Scenery, &c., &c. Chester, 1797. 12mo. *

—— Hints to Planters, from various Authors of esteemed Authority. Manchester, 1807. 8vo. *

Thomas Skip Dyot Bucknal. The Orchardist, &c., &c. London, 1797. 8vo.

Thomas Andrew Knight, F.R.S. A Treatise on the Culture of the Apple and Pear, and on the manufacture of Cyder and Perry. London, 1797. 12mo.

—— Some Doubts relative to the efficacy of Mr. Forsyth's Plaister, in renovating Trees. London, 1802. 4to.

—— Report of a Committee of the Horticultural Society of London. London, 1805. 4to.

—— A Letter on the origin of Blight, &c. London, 1806

—— Pomona Herefordiensis, or a descriptive account of the old Cyder and Perry Fruits of Herefordshire. London, 1811. 4to.
Knight was the author of numerous papers in Trans. Hort. Soc. and other periodicals.

William Salisbury. Hortus Paddingtonensis, &c. London, 1797. 8vo.

—— The Botanist's Companion, &c. London, 1816. 12mo.

– Hints to the Proprietors of Orchards. 1817. 12mo.
The Cottager's Agricultural Companion, &c. 1822. 12mo.
– – Also Essay on packing Plants for exportation *
Nicholson's Journal, XXX., p. 339.

WALTER NICOL. The Scotch Forcing Gardener, &c. Edinburgh, 1797. 8vo.
—— Practical Planter, &c. Edinburgh, 1799. 8vo.
—— Villa Garden Directory, &c. Edinburgh, 1809. 8vo.
—— Gardener's Kalendar, &c. Edinburgh, 1810. 8vo. *
—— Planter's Kalendar, &c. Edinburgh, 1812. 8vo.

HENRY ANDREWS. The Botanist's Repository, &c. Edinburgh, 1797-99. 4to.
—— A Review of Plants hitherto figured in the Botanist's Repository. Edinburgh, 1801. 4to. *
—— Engravings of Ericas or Heaths, with Botanical descriptions. London, 1802. Folio.
—— The Heathery, or Monograph of the genus Erica, monthly numbers, in 6 vols. London, 1804 and 1814
—— Geraniums. London, 1805. 4to.

1798 CLEMENT ARCHER, M.R.I.A. Miscellaneous Observations on the effect of Oxygen on the Animal and Vegetable Systems, &c. Bath, 1798. 8vo.

ROBINSON. Forms of Stoves for Forcing Houses. London, 1798. 8vo. *

1799 ROBERT JOHN THORNTON. A new Illustration of the Sexual System of Linnæus. 1799.
—— The Temple of Flora. 1805. Imperial folio
Beautifully engraved titles and coloured plates.
—— Family Herbal. (Engravings by Bewick.) 1810.
—— Botanical Extracts, or Philosophy of Botany. (3 vols.) 1810.

LADY CHARLOTTE MURRAY. British Garden. Bath, 1799. 8vo.

1800 WILLIAM PONTEY. The profitable Planter, &c. Huddersfield, 1800. 12mo.
—— The Forest Pruner, &c. London, 1805. 12mo.
—— The Rural Improver. London, 1822

1800 Mrs. Montolieu. The Enchanted Plants, Fables in Verse. London, 1808. 8vo.

The Gardens, a Poem. From the French of L'Abbé J. de Lille. 1805. 8vo.

Rev. Thomas Owen, M.A. The Three Books of M. Terentius Varro, concerning Agriculture, translated into English. London. 1800. 8vo.

Agricultural Pursuits, translated from the Greek. London, 1805. 8vo.

—— Translation of the 14 Books of Palladius on Agriculture. London, 1807. 8vo.

1802 William Turton, M.D., F.L.S. A General System of Nature through the three grand kingdoms of Animals, Vegetables, and Minerals. Translated from Gmelin's last edition of the Systema Naturæ of Linnæus. 1802-6. 7 vols. 8vo.

Rural Recreation, or the Gardener's Instructor, &c. By a Society of Practical Gardeners. London, 1802. 8vo. *

1803 John Claudius Loudon. Observations on laying out the Public Squares of London in the Literary Journal. 1803 *

—— Observations on the formation and management of useful and ornamental Plantations, &c. Edin., 1804. 8vo.

— A short Treatise on some improvements lately made in Hot-houses. Edin., 1805. 8vo.

A Treatise on forming, improving, and managing Country Residences, &c. London, 1806. 2 vols. 4to.

—— Hints on the formation of Gardens and Pleasure Grounds, &c. 1812. 4to.

—— Remarks on the Construction of Hot-houses, &c. 1817. 4to.

—— Sketches of Curvilinear Hot-houses, &c. 1818

—— A comparative view of the Curvilinear, and common mode of Roofing Hot-houses. London, 1818. Folio *

— The Encyclopædia of Gardening. London, 1822. 8vo.
  This work went through several later editions.
— The different modes of cultivating the Pine Apple, &c. London, 1822. 8vo.
— The Encyclopædia of Plants. London, 1838. 8vo.
  Loudon was the author of several other works and numerous treatises, and editor of the Gardener's Magazine, &c. MRS. LOUDON also was the author of many works on Gardening, among which are the following:—
— — Flower Garden of Ornamental Annuals. London, 1840. 4to.
 — Ladies' Companion to Flower Garden. London, 1841. 12mo.
 — Ladies' Country Companion. London, 1846. 12mo.
 — British Wild Flowers. London, 1846. 4to.
— The Amateur Gardener's Calendar. London, 1857.

1804 R. W. DICKSON, M.D. Practical Agriculture, or a Complete System of Modern Husbandry, &c. London, 1804 and 1805. 4to.
  Editor of the Agricultural Magazine from July, 1807, to December, 1809.

1805 EDWARD RUDGE, F.L.S. Plantarum Guianæ rariarum, Icones et Descriptiones. London, 1805-7. 4 vols. folio.

1806 WILLIAM GRIFFIN. Treatise on the cultivation of the Pine Apple. Newark, 1806. 8vo.
  Another edition. 1810.

JOHN SIBTHORPE. Flora Græca, 1806-1840. 10 vols. folio

MRS. HENRIETTA M. MORIARTY. Viridarium; or Green House Plants. London, 1806. 8vo.

W. WALLIS MASON. Experiments on the Culture of Carrots.*
  Nicholson's Journal, XV., p. 57. 1806.

1807 WILLIAM SHAW. The Practical Gardener. London, 1807. 8vo. *

ALEXANDER MACDONALD. A complete Dictionary of practical Gardening. 1807. 2 vols. 4to.
  Johnson says the author of this was R. W. Dickson (see 1804), and that Macdonald was an assumed name.

WILLIAM WATSON. On the Culture of Turnips. *
  Nicholson's Journal, XVI., p. 14.

1809 J. ACTON. On the Germination of Seeds, in a letter. 1809.*
  Nicholson's Journal, XXIII., p. 214.

1809 JAMES DEDE. The English Botanist's Pocket Companion, containing the essential Generic characters of every British Plant, arranged agreeably to the Linnæan system. London, 1809. 12mo.

JOSEPH KNIGHT. An Essay on the cultivation of the Plants belonging to the Order of Proteæ. London, 1809. 4to.

MRS. AGNES IBBETSON. Many Contributions to Nicholson's Journal, on Plants and Seeds, &c. 1809. *

SYDENHAM EDWARDS. Sixty-one Plates, representing about 150 rare plants. London, 1809. 4to. *

—— The Botanical Register, or Ornamental Flower Garden and Shrubbery. 1815-1827. 8vo. 33 vols. (including index.)

<small>The continuation to 1847 was edited by J. Lindley.</small>

1810 THOMAS HAYNES. Improved System of Nursery Gardening. London, 1810. 8vo.

—— Interesting Discoveries in Horticulture, being an easy system of Propagating American and Bog Soil Plants, &c. London, 1810. 8vo.

—— A Treatise on the improved culture of the Strawberry, Raspberry, and Gooseberry. London, 1812. 8vo.

—— On collecting Soils, and composts. London, 1812. 12mo.

1811 PETER LINDEGAARD. On the mode of forcing the Vine in Denmark. London, 1811. 8vo. *

<small>Mentioned by Johnson.</small>

1812 TRANSACTIONS of the Horticultural Society begun in 1812. First Series, 4to. 7 vols. 1812 to 1830. Second Series. 1835 to 1848. With general index.

<small>The following are among the Authors of the articles in the early volumes:—Sir J. Banks; J. Braddick; Sir B. Boothby; A. Carlisle; J. Dickson; J. Dunbar; J. Fairweather; Fuller; Sir H. Goodriche; C. Harrison; A. K. Haworth; A. Hawkins; J. Hayward; D. Hill; S. Jeeves; D. Judd; M. Keens; W. Kent; T. A. Knight; J. Lindley; G. Loddiges; Lutterel; J. Maher; H. S. Matthews; J. Mean; W. Morgan; C. H. Noehden; J. Sabine; R. A. Salisbury; A. Seton; A. Sherbrook; J. Simpson; W. Spence; J. Turner; J. Venables; J. Warre; J. Wedgewood; R. Wilbraham; T. Wilkinson; J. Williams; J. Wilmot.</small>

GEORGE BROOKSHAW. Pomona Britannica, or Correct Delineations of British Fruits, with Descriptions. Atlas folio. London, 1812.
— another edition. Elephant 4to. 1817. 2 vols.
A Treatise on Flower Painting. (Part I.) 1816. 4to.   *

The Horticultural Repository, &c. 1817. 8vo.

JOSEPH TAYLOR. Arbores Mirabiles, or a Description of the most remarkable Trees, Plants, and Shrubs in all parts of the World. Illustrated. London, 1812. 12mo.
— The Bible Garden. A brief Description of all the trees and plants mentioned in Holy Scripture. London, 1836. 16mo.

THOMAS HOGG. A concise and practical Treatise on the growth and culture of the Carnation, Pink, Auricula, Polyanthus, Ranunculus, Tulip, &c. London, 1812. 12mo.
    Second edition. London, 1822. 12mo.

GEORGE TODD. Plans, Elevations, and Sections of Hot-houses, &c. London, 1812. Folio   *

1813 PETER LYON. Observations on the barrenness of Fruit Trees; the means of prevention and cure. Edinburgh, 1813. 8vo.
— A Treatise on the Physiology and Pathology of Trees, &c. Edinburgh, 1816. 8vo.
—— Comely Garden, Edinburgh.   *
    Mentioned by Watt.

1814 JOHN CUSHING. The Exotic Gardener. London, 1814. 8vo.
    Third edition. 1826. 8vo.

JOHN LUNAN (of the Island of Jamaica). Hortus Jamaicensis, or a Botanical Description of Indigenous Plants and Exotics growing in the Island of Jamaica. London, 1814. 2 vols. 4to.

LEONARD PHILLIPS, JUN. A Catalogue of Fruit Trees for sale. London, 1814. Folio   *

24 *

1814 —— Transactions in the Fruit Tree Nursery, Vauxhall.
London, 1815. Folio   *
E. WEEKS. The Forcer's Assistant, &c., &c. Chipping
Norton, 1814. 8vo.   *
SIR JOHN SINCLAIR. General Report of the Agricultural
state, and Political circumstances of Scotland. Edinburgh, 1814. 8vo.
—— Account of some experiments to promote the improvement of Fruit Trees, by peeling the bark. London, 1820. 8vo.
FREDERICK PURSH. Flora Americana Septentrionalis, or a Systematic Arrangement and Description of the Plants of North America. London, 1814. 8vo.
1816 MARIA E. JACKSON. The Florist's Manual. ("By a Lady.") London, 1816. 12mo.
J. SALTER. A Treatise upon Bulbous Roots, &c. Bath, 1816. 12mo.   *
GEORGE SINCLAIR. Hortus Gramineus Woburnensis, &c.
London, 1816. Folio
—— Hortus Ericaeus Woburnensis. London, 1825. 4to.*
—— An Essay on the Weeds of Agriculture. London, 1826. 8vo.
ISAAC EMMERTON. A plain and practical Treatise on the culture and management of the Auricula, &c. London, 1816   *
—— —— second edition. London, 1819. 8vo.
1817 JAMES MEAN. The Practical Gardener. London, 1817. 12mo.
—— The Gardener's Companion. London, 1818. 12mo.
Both works by John Abercrombie, edited and enlarged by James Mean.
W. B. PAGE. Page's Prodromus, or a general nomenclature of all the plants, indigenous and exotic, cultivated in the Southampton Botanic Gardens, &c. London, 1817. 8vo.
The Shrubbery Almanack (a single sheet). 1818.   *
1818 WILLIAM HOOKER (Botanic painter). Pomona Londinensis. London, 1818. Folio

JOSEPH HAYWARD. The Science of Horticulture. London, 1818. 8vo.
—— The Science of Agriculture. London, 1825. 8vo.
ROBERT SWEET, F.L.S. Hortus Suburbanus Londinensis, &c. London, 1818. 8vo.
—— The Hot-house and Green-house Manual, &c. London, 1820. 8vo. *
—— —— second edition. London, 1825.
—— Geraniaceæ. 5 vols. 8vo. 1820-30.
—— The British Flower Garden. 8vo. 1823-9. Second series. 1831-8.
—— Hortus Britannicus, &c. London, 1826. 8vo.
—— Cistineæ. London, 1830. 8vo.
JOHN BUONAROTI PAPWORTH. Rural Residences . . . with Observations on Landscape Gardening. London, 1818. 4to.
—— Hints on Ornamental Gardening, &c. London, 1823. 4to.

1820 HENRY FIELD. Memoirs . . of the Botanic Garden, Chelsea, belonging to the Society of Apothecaries. London, 1820. 8vo.

JOHN LINDLEY. Rosarum monographia; or a botanical History of Roses. London, 1820. 8vo.
— — — Instructions for collecting and planting seeds and plants in foreign countries, &c. London, 1823. 8vo.
—— Introductory lecture on Botany. London, 1829. 8vo.
—— A Synopsis of the British Flora. London, 1829. 12mo.
—— The Genera and Species of Orchidaceous Plants. London, 1830-40. 8vo.
—— An Introduction to the . . . Natural System of Botany. London, 1830. 8vo.
—— An Introduction to Botany. London, 1832. 8vo.
—— An Outline of the first principles of Horticulture. London, 1832. 8vo.
—— Ladies' Botany, &c. London, 1834. 2 vols. 8vo.
—— A Key to structural, physiological, and systematic Botany. London, 1835. 8vo.

1820 —— Flora Medica; or Botanical Account of the more important plants used in Medicine, &c. 1838. 8vo.
—— Sertum Orchidaceum. Coloured plates. 1838. Folio
—— School Botany; an explanation of the characters of the principal . . . Flora of Europe. 1839. 8vo.
—— The Theory of Horticulture; or an attempt to explain the . . . operations of gardening upon physiological principles. London, 1840. 8vo.
—— Pomologia Britannica. London, 1841. 3 vols. 8vo.
    Assisted in Vol. III. by R. Thompson.
—— Orchidaceæ Lindenianæ; or notes upon collection of orchids . . . by Mr. J. Linden. London, 1846. 8vo.
—— The Vegetable Kingdom. London, 1846. 8vo.
—— A Glossary of the technical terms used in Botany. 1848. 8vo.
—— Folia Orchidacea. London, 9 parts, 1852-59. 8vo.
—— The Symmetry of Vegetation. London, 1854. 8vo.
—— Descriptive Botany. 1858. 8vo.

RICHARD PIGOTT. A Short, plain Treatise on Carnations and Pinks. 1820. 8vo.   *

CUTHBERT WILLIAM JOHNSON. An Essay on the uses of Salt for agricultural purposes. London, 1820. 8vo.
—— Observations on the employment of Salt in Agriculture and Horticulture. 1825.
    Several later editions - the 11th in 1835.

1821 HON. AND REV. WM. HERBERT. Appendix to the Botanical Magazine and Botanical Register. London, 1821. 8vo.   *

HENRY PHILLIPS. Pomarium Britannicum, an historical and botanical account of fruits known in Britain. London, 1820. 8vo.
—— New Edition, under the title Companion to the Orchard. London, 1827. 8vo.
—— History of Cultivated Vegetables. London, 1822. 2 vols. 8vo.
    This is said to be the second edition, but the date of the first edition is not known.

—— New edition under the title, Companion to the Kitchen Garden. 1831. 2 vols. 8vo.
—— Sylva Florifera, or the Shrubbery Historically . . treated. London, 1823. 2 vols. 8vo.
—— Flora Domestica. London, 1823. 8vo.
—— Flora Historica; or the three Seasons of the British Parterre, etc. London, 1824. 2 vols. 8vo.
—— Floral Emblems. London, 1825. 8vo.
—— Sylvan Sketches. London, 1825. 8vo.

WILLIAM COBBET. The American Gardener; or a treatise on the situation and laying out of gardens, &c. London, 1821. 12mo.
—— The Woodlands; or a Treatise on Planting. London. 1825. 8vo.
—— The English Gardener . on the situation . . . and laying out of Kitchen Gardens, &c. London, 1829. 8vo.

SIR WILLIAM JACKSON HOOKER. Flora Scotica. London. 1821. 8vo.
—— Exotic Flora. Edinburgh, 1822-27. 4to.
—— A Catalogue of Plants in the Royal Botanic Garden, Glasgow. Glasgow, 1825. 8vo.
—— The British Flora. London, 1830. 12mo.
—— Botanical Miscellany. London, 1830-33. 3 vols. 8vo.
—— Icones Plantarum. London, 1836. 8vo.
—— Botanical Illustrations. London, 1837. 4to.
—— Flora Borealis Americana. London, 1840. 2 vols. 4to.
—— Genera Filicum. London, 1842. 8vo.
—— Niger Flora. London, 1849. 8vo.
—— Filices Exotica. 1857-59. 4to.
—— Garden Ferns. 1861. 8vo.

Outline of a General History of Gardening. London, 1821. 8vo. *

1822 Hortus Anglicanus; or Modern English Gardening. London, 1822. 2 vols. 12mo. *

These two works, without the authors' names, are mentioned by Johnson.

F. D. LEVINGSTON. A Practical Treatise on the Growth and Culture of the Gooseberry. London, 1822. 12mo.

1822 WILLIAM SALISBURY. The Cottagers' Agricultural Companion, &c. London, 1822. 12mo.

1823 PATRICK NEILL, M.A., F.L.S. Journal of a Horticultural Tour through some parts of Flanders, Holland, and the North of France, &c. Edinburgh, 1823. 8vo.

CHARLES HARRISON, F.H.S. A Treatise on the Culture and Management of Fruit Trees. London, 1823. 8vo.

—— Horticultural Register (with Sir Joseph Paxton). 1831-36

DONN. Catalogue of Plants. 1823.

Plan for cultivating Grapes in the Field. Liverpool, 1823. 8vo. *

<small>Mentioned by Johnson without the author's name.</small>

1824 THOMAS WATKINS. The art of promoting the growth of the Cucumber and Melon, in a series of directions for the best means to be adopted in bringing them to a complete state of perfection. London, 1824. 8vo.

1825 B. MAUND. The Botanic Garden. (Published monthly), 1825 to 1850. 4to.

RICHARD MORRIS. Essays on Landscape Gardening. London, 1825. 4to.

P. W. WATSON. Dendrologia Britannica. London, 1825. 2 vols. 8vo.

G. BLISS. The Fruit Grower's Instructor, &c. London, 1825. 8vo.

WILLIAM BILLINGTON, M.C.H.S. A series of Facts, Hints, Observations and Experiments on the different modes adopted for raising Plantations of Oak, with experimental remarks upon Fruit Trees. London, 1825. 8vo.

T. F. HUNT. Half a dozen Hints on Picturesque Domestic Architecture. London, 1825. 4to.

—— Designs for Parsonage Houses, &c. London, 1828. 4to.

1826 CHANDLER and BUCKINGHAM. Camellia Britannica. 8 plates. London, 1826. 4to. *

A Practical Essay on the culture of the Vine, and a Treatise on the Melon. By an experienced Gardener. Royston, 1826 8vo. *

A Catalogue of Fruit in the Horticultural Society at Chiswick. 1826.

WILLIAM WITHERS, JUN. A Memoir on the Planting and Rearing of Forest Trees. Holt, 1826. 8vo.

—— A Letter to Sir Walter Scott, Bart., exposing certain fundamental errors in his late Essay on the Planting of Waste Land, &c. Holt, 1828. 8vo.

1827 JAMES MITCHELL. Dendrologia; or a Treatise of Forest Trees, &c. Keighley, 1827. 8vo.

Account of the different Flower Shows in England during 1826. Ashton-under-Lyme, 1827. 12mo. *

Account of the different Gooseberry Shows in England during 1826. Manchester, 1827. 12mo. *

Catalogue of Fruits cultivated in the garden of the Horticultural Society of London, at Chiswick. London, 1827. 8vo.

W. COLLYNS. Ten minutes' advice to my neighbours on the use and abuse of Salt as a Manure. 1827. *

Described by Johnson as having passed through four editions.

1828 SIR HENRY STEUART OF ALLANTON, BART., LL.D., F.R.S. The Planter's Guide, &c. Edinburgh, 1828. 8vo.

SIR JAMES SINCLAIR, BART. On the Culture and Use of Potatoes. Edinburgh, 1828. 8vo.

CHARLES MACINTOSH. The Practical Gardener and Modern Horticulturalist. London, 1828. 2 vols. 8vo.

Practical Instructions for the formation and culture of the Tree Rose. Anonymous. London, 1828. 12mo. *

JOHN SAUNDERS. The Kitchen-Garden Directory, &c. London, 1828. 12mo. *

SIR WALTER SCOTT. On Ornamental Plantations and Landscape Gardening. (Quarterly Review.) 1828.

JAMES GRAHAM TEMPLE. The Scotch Forcing Gardener. Edinburgh, 1828.

1829 The Domestic Gardener's Manual, being an Introduction to Gardening on Philosophical Principles. By a Horticultural Chymist. 1829. 8vo. *

JOSHUA MAJOR. A Treatise on the Insects most prevalent on Fruit Trees, &c. London, 1829. 8vo.

—— Theory and Practice of Landscape Gardening. London, 1852. 4to.

GEORGE WILLIAM JOHNSON. A History of English Gardening. Chronological, biographical, literary, and critical. 1829. 8vo.

<small>This is the work to which frequent reference is made in the above list of books.</small>

—— The Gardener's Almanack. London, 1843. 12mo.

—— The Principles of Practical Gardening. London, 1845. 8vo.

—— The Potato Murrain and its Remedy. London, 1846. 8vo.

—— A Dictionary of Modern Gardening. 1846. 8vo.

—— The Cottage Gardener. Conducted by Johnson. 1849, etc. 4to.

—— The Cottage Gardener's Dictionary. 1852. 12mo.

<small>This work went through many editions and was often republished with supplements. The second edition, 1857, 8vo., has the same title as that of 1852. Later on it was called *Johnson's Gardeners' Dictionary*. The latest edition was revised by C. H. Wright, and D. Dewar. London, 1894. 8vo.</small>

—— Gardening for the Many. London [c. 1856]. 8vo.

<small>By Johnson and others.</small>

—— British Ferns Popularly described and illustrated by engravings. London, Winchester [printed], 1857. 8vo.

—— The Garden Manual. By the Editor and Contributors of the "Cottage Gardener." London, 1857. 8vo.

—— Science and Practice of Gardening. London, 1862. 8vo.

GEORGE DON. Encyclopædia of Plants. London, 1829. 8vo.

—— A General System of Gardening and Botany, founded upon Miller's Gardener's Dictionary. London, 1832-8. 4to.

1831 SIR JOSEPH PAXTON. The Horticultural Register (with Charles Harrison). 1831-36.
—— The Magazine of Botany. (Begun in 1834.) 16 vols. 8vo.
—— Practical Treatise on the culture of the Dahlia. 1838. 12mo.
—— Pocket Botanical Dictionary. 1840
JOSEPH HARRISON. Floricultural Cabinet 1833-51. 21 vols. 8vo.
1835 JAMES MAINE. The Villa and Cottage Florist's Directory. 1835.
JOHN DENNIS. The Landscape Gardener. Chelsea, 1835. 8vo.
CLEMENT HOARE. Practical Treatise on the cultivation of the Grape Vine on open walls. London, 1835. 8vo.
1836 R. MARNOCK. The Floricultural Magazine. 6 vols. 8vo. 1836-41.
LOUISA ANNE TWAMLEY. The Romance of Nature, or the Flower Seasons illustrated. London. 1836. 8vo.

ALPHABETICAL LIST
OF
AUTHORS OF WORKS ON GARDENING.

The dates refer to the first edition of each Author's earliest work.

ABERCROMBIE, JOHN	1767	Blakewell		1737
Acton, J.	1809	Bliss, G.		1825
Adam, James	1789	Blith, Walter		1649
Addison, Joseph	1712	Bobart, Jacob		1648
Agricola, George Andrew	1717	Bolton, James		1785
Aiton, William	1789	Boothby, Sir B. (Hort. Soc.)		1812
Aiton, Wm. Townsend (1810)	1789	Borde, Andrew		1540
Amos, William	1794	Bowles, Rev. W. L.		1835
Anderson, James	1777	Bradley, Richard		1706
Andrewe, Lawrens	1527	Braddick, J. (Hort. Soc.)		1812
Andrews, Henry	1797	Braunschweig (see Hollybush)		1561
Archer, Clement	1798	Brocq, Rev. Philippe le		1786
Astley, Francis Duckenfield	1797	Brookshaw, George		1812
Attiret, J. D. (see Beaumont)	1752	Browne, Robert		1786
Austen, Ralph	1653	Browne, Sir Thomas		1658
		Brulles		1790
BABINGTON, C. C. (Trans. Linn. Soc.)	1791	Brunswick, Jerome of (see Andrew)		1527
		Bryant, Charles		1783
Bacon, Sir Francis	1625	Bucknal, Thos. Skip Dyot		1797
Banks, Sir Joseph (Hort. Soc.)	1812	Bute, John, Earl of		1785
Barnes, Thomas	1758			
Barrington, Hon. Daines	1769			
Bauer, Francis	1796			
Beale, John	1613	CAREY, WALTER (see W. C.)		1525
Beaumont, John (Phil. Trans.)	1665	Carter, Daniel		1767
Beaumont, Sir H. (i.e. Jos. Spence)	1752	Castel, Robert		1728
Bickham, George	1750	Catesby, Mark		1731
Billingsby (see S. B.)	1669	Carlisle, A. (Hort. Soc.)		1812
Billington, William	1825	Carpenter, Joseph		1717
Blagrave, Samuel	1669	Caux, Isaac de		1615
Blair, Patrick	1720	Chambers, Sir William		1772
Blake, Stephen	1664	Chandler (and Buckingham)		1826

Churchy, G. (see Dubravius) 1599
Clarke, G. . . 1715
Cobbet, William . . . 1825
Colebrooke, H. T. (Trans. Linn. Soc.) . 1791
Coles, William 1656
Collins, Samuel . . . 1717
Collinson, Peter (Trans. Linn. Soc.) . 1791
Collyns, W. . . . 1827
Commelin (see G. V. N.) . 1687
Commerell, Abbé de (see Lettsom) . 1774
Cooke, Moses . . 1676
Copeland, W. (see W. C.) 1525
Cotton, Charles . 1675
Cowell, John . 1729
Cowley, Abraham 1667
Coventry, Francis 1753
Cullum, Sir Dudley 1694
Culpepper, Nicolas . 1652
Cunningham, James (Phil. Trans.) . . 1665
Curtis, S. (see J. Maddock) . 1822
Curtis, William . 1787
Cushing, John 1814

DALTON, JOHN 1755
Darwin, Erasmus 1791
Dede, James 1809
Dennis, John . . 1835
Dethick (see Thomas Hill) 1563
Dicks, John 1771
Dickson, James 1785
Dickson, R. W. . . . 1804
Dillywn, L. W. (Trans. Linn. Soc.) 1791
Dodoens, Rembrant (see Lyte) 1578
Don, George . 1829
Donn . . 1823
Donn, James 1796
Donovan, E. O. . 1790
Dove, John . 1770
Drope, Francis 1672

Dubravius, Janus (see translation) . . . 1599
Ducarel, Andrew Coltee 1773
Ducket, Thomas . . 1639
Dunbar, John (Hort. Soc.) . 1812

EDWARDS, SYDENHAM 1809
Ehret, Geo. Dionysius . 1767
Ellis, John . 1770
Ellis, Thomas 1779
Ellis, William 1685
Ellis, William . . . 1732
Elsholt (or Elsholtz, John Sigismond, see Sherley) 1677
Emmerton, Isaac . . 1816
Estienne, Charles (see Surflet) 1600
Evelyn, Charles . . 1707
Evelyn, John 1658

FAIRCHILD, THOMAS. 1722
Fairweather, J. (Hort. Soc.) . 1812
Falconer, William 1783
Farmer, J. . 1735
Felton, Samuel 1785
Field, Henry . . . 1820
Fitzherbert, Sir Anthony . 1523
Fleetwood, Wm. (Bishop of Ely) 1707
Forster, John . . . 1664
Forster, John Reinhold . 1771
Forsyth, William . 1791
Forsyth, William (the younger) 1794
Frampton, John . . 1577
Freeman, Strickland 1797
Fulmer, Samuel . 1781
Fuller (Hort. Soc.) 1812
Furber, Robert . 1727

GANDON, JAMES . 1715
Gardiner, Rev. James . 1718
Gardiner, Richard 1599
Garton, James . . . 1769
Gendre, Le Sieur Le (see Forster) 1664

Gentil, François Le (*see* London and Wise)	1699	Holinshed, Ralph	1586
		Hollybush, John	1561
Gerard or Gerarde, John	1596	Home (*see* Kames, Lord)	1776
Gibson, John	1768	Hooker, William	1818
Gilbert, Samuel	1682	Hooker, Sir Wm. Jackson	1821
Giles, John	1767	Houghton, John	1682
Gilpin, Wm.	1783	Houstoun, William	1781
Goddard, Jonathan	1664	How, William	1650
Goffe, Nicholas	1607	Howard, Hon. Charles (Phil. Trans.)	1665
Goodriche, Sir H. (Hort. Soc.)	1812		
Googe, Barnaby	1577	Humphrys, George	1736
Graefer, John	1789	Hughes, William	1665
Gray, Christopher	1740	Hunt, T. F.	1825
Greville, R. K. (Trans. Linn. Soc.)	1791		
		IBBETSON, Mrs. Agnes	1809
Griffin, William	1806	Iliffe	1670
G. V. N.	1683	Irwin, Charles (Viscount)	1767
HAINES, RICHARD	1684	JACKSON, Maria G.	1816
Hale	1757	Jacob, Giles	1717
Halfpenny, Wm. John	1755	James, John	1703
Hanbury, Rev. William	1758	Jeeves, S. (Hort. Soc.)	1812
Harpur, William	1732	Jenkinson, James	1775
Harrington, Sir John	1653	Johnson, Cuthbert, W.	1820
Harrison, Charles	1823	Johnson, George W.	1829
Harrison, Joseph	1836	Johnson, Thomas	1620
Harte, Rev. Walter	1764	Judd, Daniel (Hort. Soc.)	1812
Hartlib, Samuel	1645	Justice, James	1754
Harward, Simon	1626		
Hawes, Stephen	1554		
Hawkins, Sir C. (Hort. Soc.)	1812	KAMES, H. Home (Lord)	1776
Haworth, Adrian Hardy	1794	Keenes, M. (Hort. Soc.)	1812
Hayes, Samuel	1794	Kennedy (*see* Lee)	1760
Haynes, Thomas	1810	Kent, W. (Hort. Soc.)	1812
Hayward, Joseph	1818	Knight, Joseph	1809
Heely, Joseph	1777	Knight, Richard Payne	1794
Herbert, Hon. and Rev. William	1821	Knight, Thomas Andrew	1797
Heresbach, Conrad (*see* Googe)	1577	Kyle, Thomas	1783
Hill, Daniel (Hort. Soc.)	1812		
Hill, Sir John (*see also* Hale)	1756		
Hill, Thomas (Dydymus Mountain)	1563	LANGFORD, T.	1681
		Langham, William	1579
Hitt, Thomas	1757	Langley, Batty	1726
Hoare, Clement	1835	Lauder, Sir Thomas Dick (*see* Price)	1794
Hogg, Thomas	1812		

Lawrence, Anthony (see Beale)	1536	Marnock, R.	1836
Lawrence, John	1714	Marshall, Rev. Charles	1796
Lawson, William	1618	Marshall, William	1785
Le Blond (see James)	1703	Martyn, Thomas	1763
Lee, James	1760	Mascall, Leonard	1572
Lettsom, John Coakley	1774	Mason, George	1768
Levingston, F. D.	1822	Mason, W. Wallis	1806
Licbault, John (see Surflet)	1600	Mason, William	1757
Liger, Louis (see London and Wise)	1699	Matthews, H. S. (Hort. Soc.)	1812
		Maund, B.	1825
Lightoler, T.	1762	Maunsell, William	1794
Lille, l'Abbé J. de (see also Montolieu, 1800)	1783	Mawe, Thos. (see Abercrombie)	1767
		Maxwell, Robert	1757
Linacre, Thomas	1530	McPhail, James	1794
Lindegaard, Peter	1811	Meader, James	1771
Lindley, George	1796	Meager, Leonard	1670
Lindley, John	1820	Mean, James	1817
Linnæus (see Forsyth, Jenkinson, and Milne)		Merret, Christopher (Phil. Trans.)	1665
Lister, Martin (Phil. Trans.)	1665	Millar (or Miller), Joseph	1722
Lisle, Edward	1757	Miller, Philip	1724
Lobel, or L'Obel, Matthias de	1570	Mills, John	1759
Locke, John	1766	Milne, Rev. Colin	1770
Loddiges, Conrad	1777	Mitchell, James	1827
Loddiges, G. (Hort. Soc.)	1812	Monardes, Doctor Nicholas (see Frampton)	1577
London, George	1699		
Loudon, Mrs. (see Loudon)	1803	Montolieu, Mrs.	1805
Loudon, John Claudius	1803	Morgan, W. (Hort. Soc.)	1812
Lovell, Robert	1659	Moriarty, Mrs. Henrietta M.	1806
Lunan, John	1814	Morison, Robert	1672
Lunwenhock, Anthony van (Phil. Trans.)	1665	Morris, Richard	1825
		Mortimer, John	1707
Luttrel (Hort. Soc.)	1812	Mountain, Dydymus (see Hill)	1563
Lyon, Peter	1813	Murray, Lady Charlotte	1799
Lyte, Henry	1578		
MACDONALD, ALEX.	1807	NEALE, ADAM	1779
Macer (see Linacre)	1530	Neill, Patrick	1823
MacIntosh, Charles	1828	Newton, James	1752
Maddock, James	1792	N. F. 1604 and	1608
Maher, J. (Hort. Soc.)	1812	Nichol, Walter	1797
Maine, James	1835	Noehden, G. H. (Hort. Soc.)	1812
Major, Joshua	1829	North, Francis Dudley (Lord)	1669
Malthus	1783	North, Richard	1759
Markham, Gervaise	1613	Nourse, Timothy	1700

Owen, Rev. Thomas. 1820

Page, W. B. . 1817
Papworth, John B 1818
Parkinson, John . 1629
Passe, Crispin de (*see* Wood) 1614
Paxton, Sir Joseph . . 1831
Peters, Matthew . 1771
Petiver, James . 1713
Philips, Leonard, jun. . 1814
Philips, Henry . 1820
Pigott, Richard . 1820
Platt, Sir Hugh 1594
Plattes, Gabriel . 1639
Plott, Robert (Phil. Trans.) . 1665
Pluché (*see* Humphrys) 1736
Pontey, William . . 1800
Pope, Alexander . 1713
Powell, Anthony 1769
Price, Sir Uvedale 1794
Pullein, Samuel . 1760
Pultney, Richard . 1790
Pursh, Frederick . 1814

Quintinye, John de la (*see* Evelyn) . 1693

Raley, William 1782
Ram, William . 1606
Rapinus (*see* Evelyn) . 1673
Rawley, William (*see* Bacon) 1625
Ray, John . . . 1660
R. C. . . 1612
Rea, John . . 1665
Repton, Humphrey 1794
Richardson, Richard 1669
Ritso, George . 1763
Robinson . . 1798
Rocque, Bartholomew . 1753
Rose, John . . 1666
Roxburgh, William . 1795
Rudge, Edward . 1805
Rutter, James 1767

Sabine, Joseph (Hort. Soc.) 1812
St. Pierre, Louis de . 1772
Salisbury, Richard Anthony . 1791
Salisbury, William . 1797
Salisbury, William 1822
Salmon, William . 1710
Salter, J. . 1816
Saunders, John 1828
S. B. . . 1669
Scot, Reginald 1574
Scott, Sir Walter . 1828
Serey, C. de. . . 1640
Serres, Olivier de (*see* Goffe) 1607
Sharrock, Robert . . . 1660
Shaw, G. (*see* Freeman) . 1797
Shaw, James . 1794
Shaw, William . 1807
Sheldrake, Timothy 1756
Shenstone, William 1764
Sherley, Thomas 1667
Sibthorpe, John . . 1806
Simpson, Rev. John (Hort. Soc.) 1812
Simpson, William . 1850
Sinclair, George . 1816
Sinclair, Sir James 1828
Sinclair, Sir John 1814
S. J. 1727
Sloane, Sir Hans (Phil. Trans.) 1665
Smith 1704
Smith, Sir James Edward . 1788
Snow, T. . 1702
Sowerby, James . 1791
Speechley, William 1779
Speed, Adam . . 1626
Spence, Joseph (*i.e.* Sir H. Beaumont) . . . 1752
Spence, W. (Hort. Soc.) 1812
Standish, Arthur . 1613
Steele, Richard 1793
Stent, Peter . 1630
Stephenson, David . 1746
Steuart, Sir Henry, of Allanton 1828
Stevens, C. (Trans. Linn. Soc.) 1791
Stevenson, Rev. Henry . 1716
Surflet, Richard . . 1600

25

Sweet, Robert	1818	Watkins, Thomas	1824
Swinden, N.	1778	Watson, P. W.	1825
Switzer, Stephen	1715	Watson, Sir William	1748
		Watson, William	1807
		Webb, W.	1753
TAVENER, JOHN	1601	Wedgewood, John (Hort. Soc.)	1812
Taylor, Adam	1769	Weeks, E.	1814
Taylor, Joseph	1812	Weston, Richard	1769
Temple, James Graham	1828	Wheatley (or Whately) Thos.	1770
Temple, John (Phil. Trans.)	1665	Wheeler, James	1763
Temple, Sir William	1685	Whitmill, Benjamin	1763
Thompson, James	1757	Williams, J. (Hort. Soc.)	1812
Thornton, Robert John	1799	Wilbraham, R. (Hort. Soc.)	1812
Todd, George	1812	Wildman, Thomas	1768
Townsend, A.	1726	Wilkinson, T. (Hort. Soc.)	1812
Tradescant, John	1656	Wilmot, John Eardley	1815
Trowell, Samuel	1739	Wilson, Alexander	1780
Trusler, John	1780	Wilson, John	1744
Turner, J. (Hort. Soc.)	1812	Wilson, William	1777
Turner, William	1538	Winter, George	1787
Turton, William	1802	Wise, Henry	1699
Tusser, Thomas	1557	Withering, William	1776
Twamley, Louisa Anne	1836	Withers, William, jun.	1828
		Wolfe, C. J.	1715
		Woods, T. (Trans. Linn. Soc.)	1791
VAN OOSTEN	1711	Wooton, Sir Henry	1624
Venables, Rev. J. (Hort. Soc.)	1812	Worlidge, John	1677
Vispre, Francis Xavier	1786	W. S.	1609
		Wyer, Robert (see Macer)	1530
WALPOLE, HORACE	1770		
Ward, Rev. Samuel	1775	YONGE, EZEKIEL (Phil.	
Warre, J. (Hort. Soc.)	1812	Trans.)	1665

Letter from JOHN EVELYN *to* SAMUEL PEPYS, *dated from Says Court, 1st September,* 1686, *addressed*

For MR. SECRETARY PEPYS, &c.,
At the Admiralty in
Yorke buildings.

SR,
When I had last the honor to see & to dine wth you, there was a Captaine (multorum mores hominum qui vidit et urbes) who going to comand some forces in New-England, was so generous, as to offer me his assistance, in procuring for me, anything which I thought curious, & rare among the plants of those Countries. The Ingenuity, & extraordinary Industie of the Gent: by what I both learn'd from the Character you gave of him, & what I myselfe could observe in so short a time; together with your interest in him; makes me not willing to omitt so favourable an opportunitie, of putting this Note into his hands, thrö yours: and that if it may comply with his diversion, when he is in the Countrie, to collect any of these (or other) natural productions of the Vegetable Kingdome: You (who first were pleas'd, to recomend me to him) will give him leave they may be sent, & consigned under your auspicious name, to
      S^r
       Y^{rs}
      most humble, &
      continualy Obligd Servant
S: Court   J. EVELYN.
2^d: Sep^r:—86.

Plants of New England & Virginia known by these names:—

N: Engl:
1. The White Cedar ...	The Seedes onely
2. Cedar of N: England	Seedes
3. Larch-tree ...	Seedes & plants
4. Lime-tree ...	Seedes & plants
5. Hemlock-tree ...	Seedes & plants
6. Poplar or Tulip tree	Seedes & plants

7. Filbert Tree	Nutts & plants
8. Firrs of all kinds	Seedes
9. Pines of all kinds	Nutts
10. Wall Nutts of all kinds	... Nutts
11. Plums of all sorts	Stones & plant
12. Sarsaparilla :	... ... Plants
13. The Scarlet Nut : wch some call Wallnut ...	... ... Nutts & plants

Virginia :

14. Benjamin Tree	Seeds & plants
15. Gumme Tree	Seede & plants
16. Sugargras Tree	Seedes & plants
17. Date plum : ...	Stones & plants
18. Pappaw-tree	Seedes & plants
19. Chinquapine ...	Nutts & plants
20. Piekhickeries	Nutts & plants
21. Sumac trees, 3 kinds	Plants
22. Cedar of Virginia	Seedes
23. Maple tree bearing keys of crimson col^r	Seedes & plants
24. Peacock taile tree	Seedes & plants
25. Oakes, six kinds :	Acorns & plants

The seedes are best preserv'd in papers : their names written on them and put in a box. The Nutts in Barills of dry sand : each kind wrap^d in papers written on.

The trees in Barills their rootes wraped about mosse : The smaller the plants and trees are, the better ; or they will do well packed up in matts ; but the Barill is best, & a small vessell will containe enough of all kinds labells of paper tyed to euery sort with ye name :

I suppose most of those sen'd to be plants of Virginia May grow also in N— England :

S^r, y^r greate civilitie, & generous offer, makes me presume (thrõ Mr. Secretary Pepys's cover & recõmendation) to burden you with this catalogue from

      S^r

    Y^r most humble

      Sev^t J. EVELYN.

Saye Court neer Deptford,
 1 Sep^r. —86.

   Memd^m a Copie hereof was putt into ye hand of Capt Nicholson.

      Sep^{tr}. 3^d --86.

INDEX.

ABERCROMBIE, John, 281
Abingdon Abbey, 8, 15, 18, 26
Accounts, Churchwardens, 17
  ,,   Durham Monastery, 100
  ,,   Ely Place, Holborn, 25
  ,,   Fruit supplied to Edward I., 37
  ,,   Fruit bought by John Tradescant, 170
  ,,   Hampton Court, 75, 78, 89, 90, 93, 101, 102, 148, 217
  ,,   Henry Best's Farm, 201
  ,,   Henry Oxenden, 182
  ,,   Household, 84, 101
  ,,   Le Strange, 100, 102, 104, 182
  ,,   Lincoln's, Earl of, Garden, Holborn, 35
  ,,   Norwich Priory, 9
  ,,   Royal Household, 75, 78, 81, 88, 90, 96, 99, 114
Addison, Joseph, 238
Aiton, William, 293
Albury, 186
Alleys, 110
Almoner, 15, 18
Almond, 3
Aloe (Agave), 253
Alpine Flowers, 299
Andrewe Laurens, 103
Anemones, 188-9
Apothecaries' Garden, Chelsea, 223
Apothecary, 165, 167, 223
Apples, 37, 47, 142, 144, 179, 284

Appletun, 6
Applezerd, 6
Apuleius, 62
Apricot, 94, 147
Arbour, 33, 52, 78, 116
Architecture, 106
Artichoke, 96, 137
Artichoke Garden, 311, 316
Asia Minor, Plants from, 250
Ashley, Sir Anthony, 47
Ashmole, Sir Elias, 169
Ashridge, 26, 28
Asparagus, 96
Austen, Ralph, 177
Avenues, 197, 217

BACON, his description of a garden, 107
Badminton, 244
Banana, 167
Banqueting houses, 112, 114, 311, 313
Barberries, 179
Barmstone, 208
Basynge, Roberto (gardener at Winchester), 14
Bathurst, Lord, 244, 259
Batsford, 299
Batty Langley, 257, 260
Baynarde's Castle, 91
Beale, John, 222
Beans, 96
Beans, kidney, 139
Baculieu, 157
Beaulieu or Newhall, 91

Beaumont, 197
Bedford, 199, 214
Bedding out, 295
Beddington, 127, 146, 215
Bees, 203
"Beestes" at Hampton Court, 88
Belvoir, 250, 303
Berkeley House, 187
Bettisfield, 183, 189
Bicester Abbey, 13, 14
Bilton, 180, 241
Bisham Abbey, 87
Black Death, 43
Blenheim, 218
Blessed Thistle, 135
Blith, Walter, 177
Bobart, Jacob, 202
Bog, artificial, 271
Bog plants, 202
Bollard, Nicholas, 66
Borders, 122, 212
Boscobel, 116
Botanic Gardens, 160, 202
Botanical Society of London, 292
Botany, Progress of, 203, 245
Boughton, 207, 244
Bower, 32
Bowling green or alley, 197
Box, 122
Bradby, 206
Bradley, Richard, 244
Bramham, 199, 230
Bridgeman, 227, 241
Brithnodus, Abbot of Ely, 7
Britons, 1
Brompton Nursery, 216
Brookshaw, on "Fruit," 283
Brown, "Capability," 263
Brown, Agreement with Lord Scarbrough, 268
Buckingham, Edward Stafford, Duke of, 81
Buckingham Palace, 274
Bullaces, 49

Bulleyn, William, 146, 162
Bulwick, 180
Burghley House, 124, 154, 266
Bury St. Edmunds, 26

CABBAGE, 4, 46, 47
Camomile, 110
Canal, 194, 232, 235
Canons, Ashby, 212
Cannon Hall grape, 283
Canterbury, 26
Capel, Lord (Sir Henry), 184
Carew, Lord (Sir Francis, and Mr.), 124, 146, 154
Carlisle, 31
Carlton House, 261
Carnations, 152, 192
Carolina, Introduction of plants from, 249
Cascade, 207
Cassiobury, 184, 229
Castle Ashby, 264
Castles, gardens around, 31
Castle Howard, 218
Catalpa, 249
Catesby, Mark, 249
Caterpillars, 202
Cawarden, Sir Thomas, 112
Cecil, Sir Thomas, 125
Cedars of Lebanon, 225
Chambers, Sir Wm., 259
Chaplets of flowers, 58
Charing, 32
Chatsworth, 206
Chaucer, 43
Chelsea Physic Garden, 194, 223
Cherries, 2-4, 18, 48, 146
Cherryzerd, 6, 19
Chestnuts, 38
Chilham Castle, 180, 199
Christina, Abbess, 7
Christmas decorations, 17
Chrysanthemum National Soc., 292

Cider, 144
Cirencester, 65, 87
Citrons, 48
Clarendon, 32
Clarmont, 261
Cloysterjerd, 6
Clusius (Charles de L'Ecluse), 162
Cobham, 118
Coles, 17
Colewort, 135, 139
Collectors and importers of the 19th century, 285
Collins, Samuel, 251
Coloured earth, 122
Compton, Bishop, 216
Conservatories, 187, 193, 196
Conserves of fruit, 103
Cook, or Cooke, gardener to the Earl of Essex, 184, 217
Cook's garden, 131
Cookery recipes, 56, 179
Cooking of vegetables, 45, 46, 179
Corbie, monastery garden at, 5
Cornel tree, 149
Cornish gardens, 301
Coronæ sacerdotales, 16
Cranbourn, 218
Crocus, 100
Costard apples, 38, 96
Cucumbers, 251
Currants, 95, 149
Custos operum, 4
Cusworth, 199
Cut trees, 76, 119, 185, 209, 238
Cypress, 111

DAFFODILS, 170, 188
Dahlia, Introduction of, 288
Daisy, 59
Damsons, 49, 86
Dates, 48
De Caux, 109, 127
Dennis, J., 274

Distilling, 103
Dodoens Rembert, 162
Domesday Book, 20
Doody, 225
Double flowers, 250
Douglas, David, 286
Down Hall, 229
Draper, William, 9
Drayton, 110, 112
Drope, Francis, 178
Durham Monastery, 100
Dutch Influence, 197-207
Dwarfs, 208

EADGYTH, or Matilda, 7
Edward I., fruit supplied to, 37
Elizabethan Garden, description of, 128
Ely, 7, 22
Ely Place, 24
Enclosure, 107, 108, 209
English Gardens abroad, 275
Erbistock, 120
Esher, 261
Espalier, 142, 208
Essex, Earl of (Lord Capel), 184
Essex, Geoffrey Earl of, 26
Essex House, 194
Eton College, 16
Euston, 194, 200, 235
Evelyn, John, 185, 387
Exton, 200

FAIRCHILD, Thomas, 244, 245
Feckenham, Abbot of Westminster, 85
Fennel, 44
Fertilization, 244
Field gardener to the Earl of Bedford, 217
Fiennes Celia, 206, 208
Finedon, 273
Figs, 48-180

Fishponds, 86
Fitzherbert, 102, 103
Flowers for Arbours and Bowers, 117
„ at Church Ceremonies, 15-18
„ for decoration in houses, 159
„ in an Elizabethan garden, 123
„ introduction of, 123, 124
„ introduction of, in the 14th century, 54-59
„ introduction of, in the 18th century, 282
„ in a Tudor garden, 85
„ Wearing of, 157
Florist's Varieties, 291
Forcing, 193
Forcing Vegetables, 250
Foreign grafts, 98
Formal garden, modern examples of, 305
Formal garden, 106
Formal style changing from, 237
Forsyth, Wm., 284
Forthrights, 106, 110
Fortune, Robert, 286
Fountains and Waterworks, 206
Fountains, 54, 125-6, 312, 322
Foxglove, 219
French Gardeners, 196
French Influence, 196, 207
Frittilarias, 188
Fruit on Tusser's List, 94
Fruit trees trained on walls, 142
Fulham, 216
Fuchsia, introduction of, 288

GALLERY, 80, 81
Garlands, 58
Garlick, 44
Garden, derivation of the word, 6, 108
„ Formal, 106
„ House, 311, 313
„ Italian style of, 294
„ of Manor houses, 180

Garden, Necham's description of, 64
„ Old fashioned, 106
„ Paths, 52
„ Garden pests, 201
„ Roman, 2
„ Seats, 52
„ the size of, 187
„ tools, size of, 101
„ Vegetable, near London, 252
„ Walls, 50
Gardener's Company, 132
Gardener, John or Ion, 45, 49, 68
Gardeners, Society of, 246
„ Wages of, 101
Gardening, the Feate of, list of plantes in, 69
Gardinæ Sacristæ, 15, 16
Gardinarius, 8
Garret, 163, 166
Gates, 108, 181, 209
Geranium, 59, 295
Gerard or Gerarde, John, 107, 139, 144-146, 157, 162, 164
Gilbert Samuel, 189
Gilliflower, or clove pink, 58
Goodricke, Sir Henry, 226
Goodyer, 167
Gooseberry, 93
Grafting, 49, 66, 147, 148
Grafting, tools for, 102
Grapes, 20, 24, 149
Grapes, Introduction of new varieties of, 222, 283
Grass banks, 83
Graszerd, 6
Greenhouses, 90, 111, 152
Grindal, Bishop, 216
Grosseteste, Bishop, 63
Gubbins " Hertfordshire," 241

HACKNESS, 87
Hackney, Lord Brooke's Garden at, 193

INDEX.

Hackney, Sir Thomas Cooke's garden at, 215
Hackwood, 265
Hadham, the Palace, 83
Haddon, 208
Haghmon, Abbey of, 14
Hagley, 260, 262
Ha-ha's, 234, 237
Hall Barn, 223
Hammersmith, Greenhouse at, 215
Hampstead, 111
Hampton Court, 28, 75, 78, 80, 84, 88, 112, 119, 126, 197, 217
Hampton Court, Vine at, 283
Hanbury, Wm., 282
Hanmer, Sir Thomas, 183, 210
Hardy plants, 298
Harewood, 297
Harris, Richard, 98
Hartlib, Samuel, 175
Hatfield, 127, 130, 155
Hatton Grange, 87
Hawkers, cries of, 253
Hawthorne, 56
Hedges, 108, 111
Henricus Arboraii, 31
Herbal, grete, 160
Herbal Literature, 62, 161, 162
Herbarium, 32
Herber (or arbour), 32
Herbs in cookery, 44
Herbs in John Gardener's poem, 69
Herbs for Physic, 103
Herbs in Sloane MS. 1201, 71
Herbs, sweet-smelling, 110
Herbularis, 5
Hereford, 22
Herriott, Thomas, 136
Heslington, 120
Hewell Grange, 272
Hill, Thomas (Didymus Mountain), 108, 173, 178
Hips, 67
Holborn (Holbourne), 24, 25, 35

Holme Lacy, 176, 199
Hooker, Sir Joseph, 287
Hops, 146
Horse radish, 96
Horticultural Society, formation of, 284
Hortulanus, 8, 14, 16
Hothouses, 193
Housewife duties, 97, 103
How, William, 203
Hoxton, Fairchild's garden at, 151, 246
Huguenots, 124, 146
Hunstanton, 100
Hurley-on-Thames, 87
Hybridising, 245, 250

INFIRMARIAN'S garden, 15
Ingestre, 228
Ion or John Gardener, 45, 49
Iris, 59

JAMES I. and church decoration, 17
James I. and mulberry culture, 155
James I. of Scotland, 60
Jasmine, 192
Jerusalem artichokes, 137
Jet d'eau, 126
Johnson, George Wm., 284
Johnson, Thomas, 163, 167

KALE or Kole plants, 46
Kalm, 252
Kenilworth, 109
Kennington, 33
Kensington, 216
Kent, 256, 257, 261
Kew, Royal gardens at, 259, 292
Kip, 227
Kitchen garden, 92, 131

Kirby, 109
Knight, Richard Payne, 271
Knight, Thomas Andrew, 284
Knots or Knotted beds, 83, 122, 209
Kymer, Gilbert, 66

LABYRINTH, 32
Lambeth, 215
Landscape Gardeners, 257
Langland, 42
Lanthony Priory, 18
Larch tree, 226
Lavender cotton, 122
Lawrence, John, 244
Lawson, Wm., 173
Laws and acts, 99
Leac tun, 6
Leasowes, 258, 262
Leeds, 214
Leek, 4, 44
Legislation, 99
Le Joye, garden at Winchester, 15
Lelamour, John, 65
Lemon, 154
Le Nôtre, 196
Lete, Master Nicholas, 166
Lettuce, 4
Levens, 197
Lilies, Scarborough and Guernsey, 282
Lily, 58
Lincoln, Henry de Lacy, Earl of, 25, 35
Littlecote, 94, 127
L'Obel, or Lobel, 124, 163
Loddiges in Hackney, 289
London, gardens in and near, 34, 39, 132, 215, 253
London, petition of gardeners, 40
London, George, 217, 218
London Vegetable market, 98
Longleat, 217, 274
Loseley, 83, 126, 157
Lotus tree, 226

Lucre, gardener to the Queen Dowager, 217
Lyte, Henry, 162

MACER, 65
Mandrake, 135
Marigold, 55
Maske of Flowers, 128
Market Gardens, 40, 97, 252
Markham Gervase, 179
Marylebone, 114
Mason, George, 258
Mason, William, 261
Maze, 118, 314
Meager Leonard, 279
Medicinal Herbs, 65, 135, 139
Medical MS., Stockholm, 65
Medlars, 39, 49
Melbourne, 218
Melons, 96, 137
Melsa Abbey, 14
Miller, Philip, 225, 244, 249
Mint, 45
Mistletoe, 1
Mitford, 180
Moles and Molecatchers, 201
Monasteries, suppression of, 85
Monastery, gardens of officials in, 14
Mon amy, recipe for making, 56
Moor Park, 221, 235
More, Sir William, 127
Morgan, Hugh, 223
Morison, Robert, 203
Mounds or Mounts, 76, 77, 112, 116
Mountain Jennings, 127
Mount Surrey, 87
Mulberries, 39, 155
Mushrooms, 219
Mustard, 139
Myrtle, 154

NARFORD, 275
Nature, copying of, 257

Nauewes, 139
Neckam, A. (Abbot of Cirencester), 19, 20, 63, 64
Nectarine, 148, 180
Netherton, 221
Newcastle, 214
Newstead Abbey, 28, 87
Nonsuch, 91, 126
Norfolk's, Duke of, Garden, 199
Norwich Priory, 8, 15
Nosegays, 157
Nursery Gardens, improvement of, 175
Nuts, 37, 38, 95

O.—The O of the Gardener, 13
Oatlands, 91, 150
Old Gardens, destruction of, 263
Olive, 2, 246
Onion, 4, 44
Oranges, 48, 124, 154, 193, 208
Orangeries, 152, 193
Orange Garden, 307
Orchard or Orceard, 6, 18, 98, 140, 175
Orchids, 288, 290
„ Collectors of, 291
„ Earliest importation of, 289, 291
„ first growers of, 289
O Radix, 13
Oriental style of Gardening, 259
Ornithogalum, 219
Orto Cērsòr, 19
Ortjerd, 6
Osborne, Dorothy, 222
Oxford Botanic Garden, 194, 202
Oxford, 39

PAINSHILL, 262
Palladius, 63, 66
Paradise, 15
Parliamentary Surveys, 177, 194, 307

Parkinson, John, 167, 110, 117, 122, 125, 135, 140, 145, 149
Parsley, 4, 44
Parterre, 209
Pathenesburgh, 19
Paxton, Sir Joseph, 297
Peaches, 39, 48, 148, 180
Pears, 37, 38, 47, 96, 145, 179, 283
Pearmains, 38
Peas, 96, 139
Penny, Dr., 172
Percival, Lord, Letters to D. Dering, 232, 242
Perrault, 196
Periwinkle, 54
Pescodde, 47
Petition of Gardeners to the Lord Mayor, 40
Pheasant Garden, 316, 319
Phillips, Henry, 284
Pineapple, 251, 283
Pippins, 142
Plants sent from abroad, 218
„ collectors of, 166
„ in the 12th century, 64
„ of the 14th century, 59
„ in the 15th century, 69
„ in Tudor gardens, 85
„ in an Elizabethan garden, 123
„ mentioned by Evelyn, 187, 387
„ in the 18th century, 248, 282
„ in the 19th century, 286, 291
Platt, Gabriel, 178
Platt, Sir Hugh, 171
Pleached Alley, 111
Plums, 48, 146
Pomatum, 144
Pomegranate, 48, 154
Pomerium, 18, 19
Pond Garden, 90
Pontefract, 199
Pools, 87, 125
Pope, Alexander, 238, 241
Porkington Treatise, 66

Potato, 135
Potato, sweet, 136
Powis Castle, 199, 266
Presents of fruit or flowers, 104
Price, Sir Uvedale, 277
Priest, Dr., 162
Primroses, 56
Privet, 121
Prizes, 192, 291
Probus, Emperor, 2
Propagation of plants, 219
Public parks and gardens, 212, 216
Pumpkins, 138
Pyracantha, 111

QUINCES, or coynes, 38, 142
Quintinye, Jean de la, 196

RADISH, 4
Railed flower beds, 74, 89
Raleigh, Sir Walter, 136
Ramsey Abbey, 9, 15, 26
Ranelagh, 215
Ranunculus, 188
Raspberry, 93, 149
Ray, John, 203
Rea, John, 183, 187, 195
Reformation, 85
Regent's Park, 278
Repton, Humphrey, 271
Ribston, 18, 155, 226
Richmond, 90, 241
Robinson, Wm., 297, 303
Roche Abbey, 266
Rockingham, 77, 120
Rock garden, 273, 299
Roger le Herberur, 31
Rolls, Ely, 25
„ Norwich Priory, 9
Roman gardens, 1, 2, 3
Romsey Abbey, 7

Rosamond, Fair, 32
Rose, John, 194, 196, 222, 251
Rosemary, 121, 157
Rosery or roseria, 23
Roses, 4, 56, 85, 111, 287
Rousham, 261
Roxburgh, Dr. Wm., 289
Runcival peas, 97
Rushes, strewing of, 17, 156

SACRISTAN, gardens of the, 14
Saffron, 46, 100
Saffron Walden, 32, 46, 199
Saint Armand, 4
„ Barbe, Sir John, 209
„ Etheldreda, 7
„ Gall, 5
„ Germain des Prés, 4
„ James' Park, 196, 274
„ John, Knights of, 28
„ Martin Outwich, churchwarden accounts, 17
„ Mary Hill, churchwarden accounts, 17
„ Remy, 4
Salisbury, Earl of, 124, 150
Salle, Robert, 50, 66
Sandwich's, Lord, house near Huntingdon, 209
Says Court, Deptford, 185
Saxon names of plants, 3
Scilly Isles, narcissus culture in, 302
Scott, Sir Walter, 276, 278
Scudamore, Lord, 176
Sensitive plant, 192
Serpentine, the, in Hyde Park, 259
Shelters for delicate plants, 152, 196
Shenstone, William, 258
Sherard, William, 250
Shows and Floral Exhibitions, 192, 291
Shrublands, 297
Skirret, 138
Sloane, Sir Hans, 225

Sopwell, 28
Southampton House, Bloomsbury, 216
Southcote, Philip, 262
Spalding, 26
Speechley, William, 283
Speed, Adolphus, 177
Spring Gardens, 214
Spring gardening, 303
Statues, 260
Stinging-nettle, 2
Stoake, 208
Stoves, 193
Stowe, 241
Strawberry, 48, 92
Street cries, 253
Streams, 127
Strewing rushes and flowers, 156
Sturtivant, Simon (Hatfield), 127
Sub-tropical gardening. 299
Sundials, 200
Swiring, 18
Switzer, Stephen, 227, 237
Syon or Sion, 111, 156

TACITUS, 2
Taragon, 96
Temple, Sir Wm., 221
Templars, Knights, 28
Tenham (cherry orchards), 98
Terrace, 106, 109
Tradescant, John, 124, 146, 151, 169
Treasurer's garden, 14
Transplanting, 217
Theobalds, 112, 127, 155, 201, 319
Thornbury, 81
Topiary work, 75, 119, 209
Topiary art, the decay of, 238
Tower of London, 91
Town gardening, 39, 212
Tuggy, Ralph, 167
Tulips, 135, 188
Turner, William, 160
Tusser, Thomas, 94, 161

Turnips, 139
Twickenham, Pope's garden at, 239

URNS, 261

VARIEGATION, 250
Vases, 125
Vegetable, culture of, 42, 96, 131, 251
Vegetable Seeds from abroad, 132
Views and plans of gardens, 227
Vine in Hampshire, 2
Vines, 19, 149, 222, 283
Vines, forcing of, 250
Vine pruners, 20
Vineyards, 2, 20, 149, 316
Vineyards, Roman, 19
Vinitor, 20
Violets, 56
Viridi succo, verjuice, 23
Vynour, Adam, 24

WALDEN, Bishop Roger de, 16
Walks, 52, 110
Waller, Edmond, 234
Wall fruit, 140, 317
Walls of stone or brick, 107
Waltham, 91
Wanstead, 91, 218
Warwick Castle, 199
Weeding, 101
Wardens, 18, 47, 96
Waterworks, 206
Watts, John, 223
Watton, 199
Wentworth, Stainborough, 200
Westminster, Botanic garden at, 223
Westminster, 31, 34, 86, 91
Whitehall, 91, 114, 126, 196
Woad, 1
Woodford, 273
Woodstock, 32, 91

Wookey, 16
Women weeders, 101
Wort or Wortes, 6, 44, 46
Wotton in Surrey, 186
Wilderness, 314
William Rufus, 6
Wild gardening, 303
Wilton, 109, 127, 206
Wimbledon, 125, 193, 307
Winchester, 14, 15, 127
Window boxes, 159, 294
Windsor, 31, 33, 60, 83, 91
Wine, 20, 26, 131

Wise, Henry, 217, 218
Wright, 263
Wurt or wurtes or wyrt, 6, 46
Wyn moneth, 20
Wyrtun, 6
Wyrtȝerd, 6

YEWS, 78, 83, 108, 120
York Place, Whitehall, 91

ZOUCHE, Lord, 124

www.ingramcontent.com/pod-product-compliance
Lightning Source LLC
Chambersburg PA
CBHW050850300426
44111CB00010B/1201